STUDIES ON VOLTAIRE AND THE EIGHTEENTH CENTURY

SVEC

2001:03

Manuscripts should be prepared in accordance with the *SVEC* style sheet, available on request and at the Voltaire Foundation website (www.voltaire.ox.ac.uk). One paper copy should be submitted to the *SVEC* general editor at the Voltaire Foundation, 99 Banbury Road, Oxford OX2 6JX, UK; an electronic version, with a summary of about 750 words, should be sent to jonathan.mallinson@trinity.ox.ac.uk.

Dieu créant le monde, drawn by Nicholas Blakey and engraved by Etienne Fessard, in Buffon's *Histoire et théorie de la terre* in vol.i of the *Histoire naturelle* (photograph courtesy of the Lloyd Library, Cincinnati, Ohio)

Rhetoric and natural history: Buffon in polemical and literary context

JEFF LOVELAND

SVEC
2001:03

VOLTAIRE FOUNDATION
OXFORD

ISBN 978-0-7294-0747-2
ISSN 0435-2866

Voltaire Foundation
99 Banbury Road
Oxford OX2 6JX, UK

First printing 2001

A catalogue record for this book
is available from the British Library

The correct reference for this volume is
SVEC 2001:03

This series is available on annual subscription

For further information about this series
and other Voltaire Foundation publications see
www.voltaire.ox.ac.uk, or email svec@voltaire.ox.ac.uk

Printed and bound by Lightning Source UK Ltd

Contents

Acknowledgements

IN 1994 I completed a doctoral dissertation entitled 'Rhetoric and science in Buffon's natural history'. Little of that effort survives here, but my research and experiences as a graduate student were important in preparing the way for this study. Philip Stewart and Seymour Mauskopf in particular deserve my profound thanks for directing my dissertation and continuing to read my work and encourage me in subsequent years.

Since arriving at the University of Cincinnati I have benefited from the Charles Phelps Taft Memorial Fund, which allowed me to purchase needed materials for the Langsam Library, present parts of my book at conferences, and pursue my research without interruption in the summers of 1996 and 1998. I am also grateful to the librarians at Langsam Library's Interlibrary Loan for finding me a number of rare books, to Michael Flannery and his staff at the Lloyd Library for their hospitality and assistance, and to the ARTFL project at the University of Chicago for allowing me early access to the electronic *Encyclopédie*.

William Christmas, Julie Hayes, Elizabeth Liebman, John Pappas, Phillip Sloan, and Downing Thomas were all generous in letting me take advantage of their work in progress. For their comments and suggestions regarding parts of the manuscript I would like to thank Heather Arden, Alex Berman, Bruce Craddock, Julie Hayes, Ellen McNiven Hine, John McEvoy, Mary Terrall, Downing Thomas, and two anonymous readers for *SVEC*. To Frank Kafker I owe a special word of gratitude for his constant encouragement and his careful reading of my manuscript. Having not always heeded my readers' advice, I am ultimately responsible for any flaws.

Without the support of my friends and family I could never have completed this project. Above all, Caroline Benforado encouraged me through years of study and uncertainty and carefully edited everything I wrote. I dedicate the book to my mother Carol Jackson Loveland in memoriam.

Jeff Loveland

Abbreviations

BMHN	Bibliothèque centrale du Muséum d'histoire naturelle, Paris
Correspondence Haller-Bonnet	Otto Sonntag (ed.), *The Correspondence between Albrecht von Haller and Charles Bonnet* (Bern 1983)
D	Voltaire, *Correspondence and related documents*, ed. Th. Besterman, 51 vols (Geneva, etc. 1968-1977)
DPV	Denis Diderot, *Œuvres complètes*, ed. Herbert Dieckmann *et al.* (Paris 1975–)
HARS	*Histoire de l'Académie royale des sciences*
HN	*Histoire naturelle, générale et particulière, avec la description du Cabinet du roi*, 15 vols (Paris, Imprimerie royale 1749-1767)*
HNM	Buffon, *Histoire naturelle des minéraux*, 5 vols (Paris 1783-1788)
HNO	*Histoire naturelle des oiseaux*, 9 vols (Paris, Imprimerie royale 1770-1783)*
HNS	Buffon, *Supplément à l'histoire naturelle*, 7 vols (Paris, Imprimerie royale 1774-1789)
JHB	*Journal of the history of biology*
Journal de Trévoux	*Mémoires pour l'histoire des sciences et des beaux arts*
MARS	*Mémoires de mathématiques et de physique ... tirés des registres de cette Académie* (bound together with *HARS*)
M	Voltaire, *Œuvres complètes de Voltaire*, ed. Louis Moland, 52 vols (Paris, Garnier Frères 1879-1883)
RHS	*Revue d'histoire des sciences et de leurs applications*
Voltaire	*The Complete works of Voltaire*, ed. W. H. Barber *et al.* (Oxford 1968-)

* Unless noted otherwise, references to *HN* and *HNO* are to sections of the series credited to Buffon.

Introduction

This book examines the rhetorical dimension of six contested issues of eighteenth-century natural history. Its focus is on the French naturalist Georges-Louis Leclerc de Buffon and the quarrels in which he fought – quarrels over style, final causes, the chain of being, systems, mathematics, and nominalism. Buffon is not well-known today, as none of his 'discoveries' entered the canon of great scientific achievements. The same can be said of most eighteenth-century naturalists. But natural history as Buffon and his contemporaries knew it remains a fascinating object of study for its breadth, its cultural openness, and its disorderly energy.

From a modern perspective eighteenth-century natural history lacked proper boundaries between its various levels. Texts of natural history straddled modern lines between literature and science. They mixed mythological allusions and poetic bravura with matter-of-fact descriptions. They mixed salacious innuendo and sensual odes to love with tables of data and numbered experiments. They mixed detailed observations with vast theological and philosophical principles. They humanised some animals and examined others clinically.

Corresponding to this 'confusion' of what we now consider separate activities was the complex intermingling of natural history's participants. Since there were few degrees and positions in natural history, most naturalists were amateurs in some sense of the word. Natural history was the business of academic experts, but also, just as importantly, of writers, collectors, teachers, curators, patrons, explorers, and thousands of readers of books on the subject. These latter were not a homogeneous or merely passive group of spectators. Nor did 'producers' of natural history – famous observers and theorists, for example – exercise their livelihoods without regard for the *curieux*. Sometimes they did, but in a period lacking funding for the sciences in general, many sought to make a 'product', often a textual one, that would be pleasing to a public. The community of natural history may have been a hierarchy, as Carl Linnaeus imagined the community of botany, but it was full of ambiguities regarding power and status.[1]

In the absence of clear-cut boundaries regarding sub-cultures and languages, eighteenth-century natural history found a large role for rhetoric – that is, for linguistic tools for attracting and persuading readers. On the one hand there were constant debates among naturalists. The sciences have always been contentious, of course, and they remain a site for powerful if subtle forms of rhetoric. In the eighteenth century scientific rhetoric enjoyed

1. On Linnaeus's hierarchy see Sten Lindroth, 'The two faces of Linnaeus', in *Linnaeus: the man and his work*, ed. Tore Frängsmyr, revised edn (Canton, Mass. 1994), p.23.

considerably more latitude – polemical latitude as well as latitude to invoke arguments from a wide range of sources. Disputants made grandiose and implausible methodological claims to shore up their conclusions. They insisted their results were entirely factual, free of supposition, or perfectly certain, while accusing their rivals of dogmatism, loose thinking, and intellectual dishonesty. They tried to intimidate each other and impress neutral parties by associating their programmes – often hypocritically – with mathematical certainty or religious orthodoxy. They caricatured each other's work in the most unfair terms. And they couched attacks and insults in conspicuously literary language, as in this passage from Jean-Jacques Dortous de Mairan's diatribe against the scientific ideas of Gabrielle-Emilie Du Châtelet: 'Il y a certainement ici quelqu'un qui a tort, qui s'abuse par les préjugés de l'autorité, ou de l'amour-propre, et dont les raisonnements applaudis aujourd'hui par un nombre de savants, fourniront à la race future un exemple de plus de la faiblesse de l'esprit humain.'[2]

Meanwhile, the use of different rhetorics – tools for attracting readers – allowed writers to reach out to different groups of readers at a time when natural history was both a science and a very popular pastime. A writer for the *Mémoires* of the Parisian Académie royale des sciences did not write in the same way as a writer for a provincial journal or one claiming young people as his or her readers. Similar adjustments to audience can be seen in today's natural histories, which range from academic papers to books for children, but many eighteenth-century natural histories slipped in and out of different rhetorics within the same text in the interest of attracting specialists as well as various kinds of amateurs. Illustrated natural histories, at one extreme, were often calculated to appeal visually to bibliophiles and the worldly even as their textual side presumed considerable scientific knowledge. But texts without pictures were aimed at multiple audiences too. Jacques-Christophe Valmont de Bomare was typical of many naturalists in recommending his extremely popular *Dictionnaire raisonné universel d'histoire naturelle* to specialists, *l'homme du monde*, and well-born youths of either sex.[3] No doubt such prescriptions should be interpreted with caution, but by mixing different styles, images, themes, and approaches, many naturalists clearly sought to write for diverse groups of readers.

Rhetoric, then, may be an element of science in all times and places, but it was especially important in the freewheeling world of eighteenth-century natural history. Buffon's 36-volume masterwork, the *Histoire naturelle* (1749-1789), epitomises the tendency. Not only was Buffon embroiled in scientific polemic throughout his career, his *Histoire naturelle* was both a best-seller and a work

2. 'Lettre de Mr de Mairan', in Gabrielle-Emilie Le Tonnelier de Breteuil Du Châtelet, *Institutions physiques. Nouvelle édition (1742)*, ed. Sonia Carboncini (Hildesheim 1988), p.504. The whole exchange between Mairan and Du Châtelet was extraordinarily rhetorical. For a characterisation of their rhetorics see Mary Terrall, 'Emilie Du Châtelet and the gendering of science', *History of science* 33 (1995), p.296-98.

3. 5 vols (Paris, Didot *et al.* 1764), i.xvi-xvii.

consulted by specialists. But rhetoric was not just a fact of Buffon's and others' lives as writing naturalists. It also provides a critical tool for understanding natural history. First, attention to rhetoric adds complexity and subtlety to eighteenth-century debates, which are often confused with the simplified polarities of the polemics they gave rise to. In particular, rhetoric provides a means of making sense of Buffon's notorious exaggerations and contradictions. Second, by taking account of variously 'frivolous' as well as serious aspects of natural history – nearly all of which were designed to attract groups of readers – rhetorical analysis situates natural history within society. So considered in its polemical and literary context, eighteenth-century natural history becomes a richer and more comprehensible enterprise.

Natural history

Now as always natural history is hard to define. In the abstract it can be said to study visible nature, especially living beings, with an eye to description and classification, but this definition remains vague until supplemented with information about the field's social position. As a subject for museums, television, and other avenues for popular science, natural history enjoys a vast following. Books on the subject have been best-sellers almost constantly since the eighteenth century. In the United States these days 'nature writing' is selling more than ever before.[4] Within universities there remain departments, chairs, and museums of natural history, as well as courses on the subject. But to the extent that it still exists more than merely nominally, academic natural history is an unprestigious discipline within the sciences. Increasingly, proponents of reductive, laboratory-orientated approaches to the study of nature have questioned the relevance of a discipline premised on 'mere' observation and classification. Since the mid-nineteenth century, when natural history flourished in European and American universities, the decline in the field's professional fortunes has been considerable.[5]

In the eighteenth century natural history had different roles and relationships. Strictly speaking, early natural history was not part of science or biology, since neither of those subjects existed in their modern forms. Unlike the word 'biology', which only came into usage in the early nineteenth

4. Robert Finch and John Elder (eds), intro. to *The Norton book of nature writing* (New York 1990), p.24. On the literary origins of American nature writing see Peter A. Fritzell, *Nature writing in America* (Ames, Ia. 1990), p.71.

5. James A. Secord, 'The crisis of nature', in *Cultures of natural history*, ed. N. Jardine, J. A. Secord and E. C. Spary (Cambridge 1996), p.447-59. For more on the history of natural history since the eighteenth century see Lynn K. Nyhart, 'Natural history and the "new" biology', in *Cultures of natural history*, p.439-42; Paul Lawrence Farber, *Discovering birds: the emergence of ornithology as a scientific discipline, 1768-1850* (Baltimore, Md. 1997), p.xviii-xxi; James L. Larson, *Interpreting nature: the science of living form from Linnaeus to Kant* (Baltimore, Md. 1994), p.183-89; Peter F. Stevens, *The Development of biological systematics: Antoine-Laurent de Jussieu, nature, and the natural system* (New York 1994), p.199-218, 233-61; Phillip R. Sloan, 'Natural history', in *Cambridge history of eighteenth-century philosophy*, ed. Knud Haakonssen (Cambridge, forthcoming).

century, the word 'science' was common in early modern Europe, but its meaning was broader than today. Science was knowledge in general, not necessarily scientific knowledge. Used in the plural, as in the title of the Académie royale des sciences, the word 'science' was associated with disciplines now seen as science, but in this sense eighteenth-century natural history was one of many sciences, not part of science. Within the Académie des sciences, founded in 1666, the sciences were divided into two groups through the late eighteenth century, one representing the mathematical sciences (with sections on geometry, mechanics, and astronomy), the other representing the physical sciences (with sections on chemistry, botany, and anatomy). Other sciences were relegated to one or the other category in the table of contents of the Académie's annual *Histoire*, with geography, hydrography, optics, and chronology going to the mathematical side, for example, and the whole of natural history going to the physical side under the rubric 'physique générale'. But identifying *physique* with the non-mathematical grew problematic as mathematical 'experimental physics' gained a broad following. In 1785, significantly, a section on experimental physics was added to the mathematical, not the physical side of the Académie.[6] Placing natural history in *physique* was becoming an anachronism.

Another general designation for early modern sciences was 'natural philosophy', often considered the period's 'equivalent' for the modern term 'science'. Natural philosophy excluded the classic mathematical sciences, but as the study of nature in the word's broadest sense, it included natural history according to some. Usually, however, natural philosophy was restricted to theoretical sciences, those intended to uncover causes, and it was opposed to more empirical fields such as *physique particulière*, 'pure' botany, and natural history. But if eighteenth-century natural history emphasised description and classifying, its distance from causality and natural philosophy should not be overstated. Like botany, natural history had difficulty maintaining boundaries with neighbouring fields in natural philosophy.[7] Naturally there were different traditions in natural history. Some naturalists focused on taxonomy, anatomy, or external description, while others stressed physiology, temporal change, or natural law – all concerns inseparable from natural philosophy.[8] Different traditions were intermixed for the most part, and many verged sporadically on 'external' subjects. The article 'Histoire naturelle' in the *Encyclopédie* plausibly points out several such subjects – astronomy, anatomy, medicine, agriculture, and chemistry – and places natural history at their centre and origin.[9]

6. See Roger Hahn, *The Anatomy of a scientific institution: the Paris Academy of Sciences, 1666-1803* (Berkeley, Calif. 1971), p.98-100.

7. On the transgressions of the Académie's 'Baconian' *Mémoires des plantes* see Alice Stroup, *A Company of scientists* (Berkeley, Calif. 1990), p.65-79, 89.

8. See Farber, *Discovering birds*, p.124-25; Richard W. Burkhardt, Jr, *The Spirit of system: Lamarck and evolutionary biology*, revised edn (Cambridge, Mass. 1995), p.192-93.

9. *Encyclopédie, ou Dictionnaire raisonné des sciences, des arts et des métiers*, ed. Denis Diderot and Jean Le Rond d'Alembert, 17 vols of text ([Paris] 1751-1765), viii.225-28. For more on the diversity of eighteenth-century natural history see Sloan, 'Natural history', forthcoming.

Buffon's concept of natural history, which stretched it toward investigations of causality and temporal change, was especially broad. The 'interdisciplinary' character of the *Histoire naturelle* can be gauged by his many apologies for encroaching on other disciplines. 'Mais finissons', he suggests after a discussion of products made from cattle, 'car l'histoire naturelle doit finir où commence l'histoire des arts.'[10] Similar statements mark his withdrawal from – and thus his presence in – philosophy, medicine, agriculture, and physics.[11] Predictably chemistry and anatomy also figure prominently in the *Histoire naturelle*. Buffon's ambitious view of natural history dictates a similarly broad view for anyone studying the *Histoire naturelle*, but it is important to recognise that his view coexisted with narrower ones. Chrétien-Guillaume de Malesherbes, for example, criticised Buffon for not being a *naturaliste* but instead a *physicien*. All that this meant, in the end, was that Buffon lacked expertise regarding living beings and their taxonomy – a debatable point.[12]

A section devoted to natural history was not opened in the Académie des sciences until 1785,[13] but even in the Académie it had been a recognised field of study for over a century. In the world at large the activity of studying natural history can be traced through millennia. Concern with nature is common to people in all times and places. Similarly structured 'folk taxonomies' are found in all cultures and can plausibly be considered a cognitive first step toward natural history.[14] Among the ancient Greeks and Romans, *historia naturalis* was an enquiry into nature consisting of *historiae* – noteworthy anecdotes, almost always from the boundaries of the known world. Ancient natural history did not necessarily study nature chronologically, but the common equation of the distant and the primitive sometimes pushed it toward a form of developmental analysis.[15] Three natural historians of classical antiquity – Aristotle, Theophrastus, and Pliny the Elder – were critical sources for later works on nature. All three receive praise and prominent billing in the opening pages of Buffon's *Histoire naturelle*. Yet despite their importance as sources and 'precursors' for modern naturalists, ancient natural historians were involved in a project specific to their context. Roger French has argued, in particular, that ancient natural history was not scientific, since it was neither objective, secular, nor experimental, and since it was not

10. *HN*, iv.473. See also *HNM*, ii.491.

11. *HN*, ii.79, 515; *HNM*, i.425; *HNS*, i.35. In 1749, if not 1774, Buffon seems to have considered his work to be part of *physique*. See for example *HN*, i.127, 131, ii.161.

12. Chrétien-Guillaume de Lamoignon de Malesherbes, *Observations sur l'histoire naturelle générale et particulière de Buffon et Daubenton*, 2 vols (Paris, Pougens 1798), ii.20, 249, 253, 272-77, 320. On Buffon's early acquaintance with botany and even zoology see Lesley Hanks, *Buffon avant l'"Histoire naturelle"* (Paris 1966), p.79, 137, 153, 159-60, 187-88.

13. Hahn, *Anatomy of a scientific institution*, p.98-100. On an earlier attempt to create a section on natural history see E. C. Spary, 'The "Nature" of Enlightenment', in *The Sciences in Enlightened Europe*, ed. William Clark, Jan Golinski and Simon Schaffer (Chicago 1999), p.273-74.

14. Scott Atran, *Cognitive foundations of natural history: towards an anthropology of science* (Cambridge 1990).

15. Roger French, *Ancient natural history* (London 1994), p.1-5.

concerned with universal laws or with manipulating nature.[16] This is a restrictive view of science, but the point is a good one and applies all the more strongly to works written on nature in the European Middle Ages, whether bestiaries, fables, herbals, or scholastic treatises.

In the Renaissance natural history acquired newly firm contours as an erudite discipline through the cultivation of museums, catalogues and even 'naturalists' by humanistic, patrician culture, first in Italy and later in France and elsewhere. Natural historians such as Ulisse Aldrovandi, Konrad Gesner, Pierre Belon, Leonhard Fuchs, and Caspar Bauhin compiled and corrected ancient natural history, extending its results north of the Mediterranean. Though attentive to experience and observation as well as literary precedent, they retained many of the goals and values of ancient natural history.[17] 'Emblematic' natural history, as practised by Gesner and others, implied recording vast amounts of lore about natural objects with seeming disregard for its truth or falsity. Immensely popular in the sixteenth century, this form of study was largely displaced by narrower, more critical natural histories by 1650. This 'scientific revolution' was later attributed to Francis Bacon, whose posthumous *Silva sylvarum, or a natural history* (1627) illustrated his plan for a new natural history, empirical and practical. In fact 'Baconian' natural history seems to have grown up in ignorance of Bacon's proposals.[18] Still, 'scientific' naturalists of the Enlightenment saw themselves as successors to Bacon or even Locke or Newton. Regard for past natural history, especially emblematic natural history, plummeted at this time, despite continuities between old and new practices.[19] Like natural philosophers, eighteenth-century naturalists proclaimed their determination to follow facts and experience, rather than hypotheses, hearsay, or the bookish erudition they associated with the Renaissance. Disproving myths about nature inherited from the Renaissance, the Ancients, or amorphous tradition came to be a favourite pastime, conducted with zeal in scientific societies. Typical were Pierre-Louis Moreau de Maupertuis's experiments around 1727 to prove that the salamander was not immune to fire.

16. French, *Ancient natural history*, p.ix-xiii.

17. Paula Findlen, *Possessing nature: museums, collecting, and scientific culture in early modern Italy* (Berkeley, Calif. 1994), p.1, 4-5, 69, 107-108, 165, 170; Phillip R. Sloan, 'Natural history, 1670-1802', in *Companion to the history of modern science*, ed. R. C. Colby *et al.* (London 1990), p.296-97. On natural history as a discipline, see Findlen, *Possessing nature*, p.153, 291; Paula Findlen, 'Francis Bacon and the reform of natural history in the seventeenth century', in *History and the disciplines: the reclassification of knowledge in early modern Europe*, ed. Donald R. Kelley (Rochester, N.Y. 1997), p.240n.

18. William B. Ashworth, Jr, 'Natural history and the emblematic world view', in *Reappraisals of the scientific revolution*, ed. David C. Lindberg and Robert S. Westman (Cambridge 1990), p.303-32. See also Peter Harrison, *The Bible, protestantism, and the rise of natural science* (Cambridge 1998), p.1-5.

19. Harriet Ritvo, *The Platypus and the mermaid and other figments of the classifying imagination* (Cambridge, Mass. 1997), p.16-18, 20-21, 38; Findlen, *Possessing nature*, p.94, 344-45, 393-94, 400-405.

However much they were exaggerated in the Enlightenment, there were real discontinuities between older natural histories and scientific natural history as it came to be practised in the seventeenth and eighteenth centuries. But if scientific continuity is problematic in the 'lineage' of natural history from antiquity to the present, literary continuity is less so.[20] Not only did eighteenth-century naturalists borrow and repudiate the accounts of their forebears, they were also writing in longstanding though mutable literary genres. In addition many natural histories from Pliny's onward were written to compete for the attention of a public of non-specialists. The task of winning readers was one shared by naturalists as different as Pliny, Gesner, Buffon, and Jean-Henri Fabre. Natural history's readers varied considerably from epoch to epoch, but variously constructed, often moralising accounts of animals, prodigies, and distant regions were regularly popular, sometimes immensely so. In the Middle Ages, it is said, only the Bible was more widely read than Pliny's *Naturalis historia* and the *Physiologus*, a compendium of allegories regarding animals, plants, and stones.[21] In eighteenth-century England and France, natural histories were again best-sellers, led by Noël-Antoine Pluche's *Spectacle de la nature* (1732-1750), Buffon's *Histoire naturelle*, and Oliver Goldsmith's *History of the Earth and animated nature* (1774).[22]

These books' success was not just a consequence of the century's 'love affair' with natural history. Eighteenth-century Europe was certainly impassioned by the subject in both its textual and more material forms – museums, collections, gardens, and menageries.[23] Written in French more often than Latin, the sciences were becoming matter for study and conversation in fashionable French society by the late seventeenth century. With its colourful descriptions and close ties to literature, natural history was especially suitable for a broad readership. But writers of successful natural histories sought out their audiences as much as they found them. They wrote to attract readers and adapted their subjects to make them more marketable.[24] Professional esteem may have been more important than readership to the author of a paper for

20. See French, *Ancient natural history*, p.263-64; Harrison, *Bible, protestantism*, p.191-92.

21. Charles Singer, 'Biology', in *Encyclopaedia Britannica*, 14th edn (Chicago 1948), iii.611; Harrison, *Bible, protestantism*, p.23-24. On the *Physiologus* see French, *Ancient natural history*, p.276-86.

22. Daniel Mornet, 'Les enseignements des bibliothèques privées (1750-1780)', *Revue d'histoire littéraire de la France* 17 (1910), p.460; Daniel Mornet, *Les Sciences de la nature en France, au XVIIIe siècle* (New York 1911), p.247-49; G. S. Rousseau, 'Science books and their readers in the eighteenth century', in *Books and their readers in eighteenth-century England*, ed. Isabel Rivers (New York 1982), p.218-27.

23. Yves Laissus, 'Les cabinets d'histoire naturelle', in *Enseignement et diffusion des sciences en France au XVIIIe siècle*, ed. René Taton (Paris 1964), p.659-712; Findlen, *Possessing nature*, p.393-405; Mornet, *Sciences de la nature*, p.1-12; David Elliston Allen, *The Naturalist in Britain: a social history* (New York 1976), p.26-51. See also the essays in Jardine, Secord and Spary (eds), *Cultures of natural history*.

24. See Jean-Marc Drouin and Bernadette Bensaude-Vincent, 'Nature for the people', in *Cultures of natural history*, p.408-10; Roger Cooter and Stephen Pumfrey, 'Separate spheres and public places: reflections on the history of science popularization and science in popular culture', *History of science* 32 (1994), p.247-51.

the *Mémoires* of the Académie des sciences, but writers seeking fame or fortune had to engage a broader audience. Best-selling naturalists such as Buffon and Pluche were clearly concerned with attracting and pleasing readers. Natural history was often in this way a rhetorical enterprise, and it makes sense to wonder what techniques authors used to reach such large audiences.

To study the rhetoric of natural history we need to take an inclusive attitude toward the scientific community. As Roger Hahn has noted, the notion of a scientific community is inevitably anachronistic when applied to the Ancien Régime.[25] At a most restrictive level we might define it by membership in national scientific societies, among them the Parisian Académie des sciences. This allows precision in deciding who was in the scientific community and who was not, but it is an illusory precision, since the criterion excludes some of the period's retrospectively unmistakable 'scientists'. Avoiding this pitfall entails accepting imprecision and according at least partial membership to a host of non-academicians – members of provincial learned societies, lecturers, demonstrators, professors, technical assistants, travellers, instrument makers, and, of course, writers.[26] For our purposes it is particularly important to avoid hasty judgements about the boundaries between 'vulgar' and 'professional' science. For many of their critics, even today, Buffon and Pluche were vulgarisers, pleasing to the public but not to professionals. This is a misleading simplification, regardless of the 'quality' of Buffon's or Pluche's science. Works on natural history ought to be distinguished according to the audience they sought and attained, but these distinctions are never as neat as binary oppositions.[27] Moreover, comparisons among variously popular and professional natural histories are crucial for understanding scientific rhetoric, since adjustments to audience can be seen via contrasts. Toward that end this study will juxtapose less well-regarded, popular naturalists such as Pluche and Jacques-Henri Bernardin de Saint-Pierre with 'giants of science' such as René-Antoine Ferchault de Réaumur and Bernard de Jussieu.

Buffon

Georges-Louis Leclerc, later comte de Buffon, was born in Montbard in 1707 and died in Paris in 1788.[28] By the end of his life he was commonly considered one of the great literary figures of the French Enlightenment, along with Voltaire, Rousseau, and Montesquieu. His father became a magistrate in the

25. *Anatomy of a scientific institution*, p.85.

26. Stroup, *Company of scientists*, p.180-217; Spary, '"Nature" of Enlightenment', p.273-74, 288. See also Findlen, *Possessing nature*, p.9-11.

27. On this and other misconceptions regarding popular science see Cooter and Pumfrey, 'Separate spheres', p.237-67.

28. Good general treatments of Buffon and the *Histoire naturelle* can be found in Jacques Roger, *Buffon. Un philosophe au Jardin du roi* (Paris 1989); Otis E. Fellows and Stephen F. Milliken, *Buffon* (New York 1972); David Goodman, *Buffon's natural history* (Milton Keynes 1980); Pierre Gascar, *Buffon* (Paris 1983).

Dijon parlement in 1714, and the family moved to Dijon, where Buffon attended the Jesuit Collège des Godrans. Afterwards he studied law, a move consistent with expectations that he would take his place among the political and administrative elite of the city. By at least 1726, however, Buffon had decided that he was more interested in the sciences. The next year he began corresponding with the Swiss mathematician Gabriel Cramer, and he was soon in touch with other potential allies and mentors. He concentrated on mathematics, and was admitted to the Académie des sciences in 1734 on the basis of an innovative solution to the problem still known as 'Buffon's needle problem'. Said to have inaugurated the study of geometrical probability,[29] the problem was to calculate the probability that a needle dropped randomly on a regularly tiled floor would come to rest in a position where it did not cross any tiles.

Buffon's interests soon evolved in other directions. His correspondence with Cramer suggests that he saw himself as a devoted if distracted geometer for several years after 1734. In 1737 he translated Newton's *Methodus fluxionum et serierum infinitorum* into French. Mathematical probability continued to intrigue him, and he never lost his affection for small-scale calculation. But in the 1730s and 1740s his research dealt more and more with non-mathematical fields such as chemistry, agriculture, and natural history. In 1735 he translated Stephen Hales's *Vegetable statics*, an exploration of chemistry, pneumatics, and plant physiology. Between 1737 and 1744, responding to the pleas of the naval minister Jean-Frédéric Phélypeaux de Maurepas, Buffon carried out research on trees and wood strength, partly in collaboration with Henri-Louis Duhamel Du Monceau. In 1739 he changed sections within the Académie, moving from the section on mechanics, his original placement, to the one on botany.[30]

In 1739 Buffon was named intendant of the Jardin du roi thanks to his expertise in arboriculture, his favour with Maurepas, a little lobbying by his friends, and a mysterious recommendation from his moribund predecessor Charles-François de Cisternay Dufay. The intendancy had been widely coveted, notably by Duhamel Du Monceau, and the appointment scandalised many in the Académie in view of Buffon's youth and lack of seniority. This was not the first time Buffon made enemies through his ambition, but his enemies would be numerous from this date onward.[31] Within the Académie the powerful Réaumur came to dislike him around 1740. In 1747 Buffon became an international celebrity for assembling a group of 'burning mirrors' capable of starting fires through reflected sunlight at distances of up to two hundred feet,

29. For a caveat on this truism see Charles Lenay, 'Le hasard chez Buffon. Une probabilité "anthropologique"', in *Buffon 88. Actes du colloque international pour le bicentenaire de la mort de Buffon (Paris, Montbard, Dijon, 14-22 juin 1988)*, ed. Jean Gayon *et al.* (Paris 1992), p.621.

30. On the unusualness of this move see Eric Brian, 'L'Académie royale des sciences de l'absolutisme à la Révolution', in *Histoire et mémoire de l'Académie des sciences. Guide de recherches*, ed. Eric Brian and Christiane Demeulenaere-Douyère (Paris 1996), p.20.

31. Hanks, *Buffon avant l'*'*Histoire naturelle*', p.20-21, 147-50.

and jealousies presumably grew all the worse. Compounding the offence of Buffon's ambition and youthful success was his independence. Unlike other members of the Académie, for example, Buffon abandoned Paris – and thus seemingly his duties – for much of the year in order to spend extended summers on his estate in Montbard. Later he would flout protocol and publish the *Histoire naturelle* outside the Académie, probably to save time and avoid a potentially unfriendly review.[32]

Lesley Hanks has argued that Buffon's way of writing changed with his arrival at the Jardin du roi. His style, she suggests, became more polished and elegant, and he no longer hesitated to announce bold conclusions.[33] Whether new or not, Buffon's pomp and provocativeness fuelled animosity. Kant among others took offence at his overconfidence in making pronouncements, Horace-Bénédict de Saussure claimed his character aroused 'hostility',[34] and for many of his enemies Buffon's grandiloquence seemed affected. All of this made him, as William Morton Wheeler put it, a 'hyperkinetic *arriviste* with rhetorical pretensions', at least to those disposed to accent the negative.[35] By 1749, Montesquieu claimed, Buffon had many enemies among French *savants*, and this was before his feuds with d'Alembert and his allies.[36] The fact that Buffon alienated a certain number of his colleagues is unsurprising. To his credit, he inspired friendship and affection as well. But in a study of scientific rhetoric it is important to remember the polemical context in which he found himself and the provocative style in which he wrote. In the eighteenth-century sciences, polemics and rhetoric were not unique to Buffon, but his situation and character make them exceptionally visible.

Around 1740 Buffon began planning a written natural history centred on the collections at the Jardin du roi. Early on he conceived it as a fifteen-volume series covering animals, plants, and minerals – the three kingdoms of nature – but the work came to be both longer and less comprehensive.[37] Buffon's *Histoire naturelle, générale et particulière, avec la description du Cabinet du roi* was ultimately published in quarto in thirty-six volumes between 1749 and

32. Roger, *Buffon*, p.45, 117, 257. More generally see Phillip R. Sloan, 'Buffon studies today', *History of science* 32 (1994), p.471-72.

33. *Buffon avant l'"Histoire naturelle'*, p.182-87.

34. Jean Ferrari, 'Kant, lecteur de Buffon', in *Buffon 88*, p.161; Kenneth L. Taylor, 'The *Epoques de la nature* and geology during Buffon's later years', in *Buffon 88*, p.373n. See also Roger, *Buffon*, p.133-34, 324, 334; Malesherbes, *Observations sur l'histoire naturelle*, i.4-5, 75, ii.201.

35. Intro. to René-Antoine Ferchault de Réaumur, *The Natural history of ants* (New York 1926), p.20.

36. Charles-Louis de Secondat de Montesquieu to Gaspard-Joseph de Cérati, 11 November 1749, in *Correspondance de Montesquieu*, ed. François Gébelin and André Morize, 2 vols (Paris 1914), ii.235.

37. Buffon's outline of 1748 can be found in *Journal des savants* (October 1748), p.639-40. Another prospectus claimed that Buffon would cover 'mythical' animals, among other things. While it seems unlikely that he ever planned to do so, the prospectus sheds light on the fundamentally literary expectations many readers had for natural history. See *Mercure de France* (October 1748), p.161. On the tradition of 'cataloguing' natural history collections see Findlen, *Possessing nature*, p.36-44.

1789, as well as in many other editions and formats. Early title pages did not list an author, but Buffon and Louis-Jean-Marie Daubenton both signed the dedication to Louis XV. In addition to writing much of the series, Buffon directed and edited it, but the *Histoire naturelle* was otherwise highly collaborative. Hundreds of pages on animal anatomy, usually omitted from later editions, were written by Daubenton for the first fifteen volumes. Much of the *Histoire naturelle des oiseaux* was the work of two collaborators, Philibert Guéneau de Montbeillard and Gabriel-Léopold Bexon. Less tangible but still significant contributions came from Barthélemy Faujas de Saint-Fond, Charles-Nicholas-Sigisbert Sonnini de Manoncourt,[38] and probably others. On top of all this, hundreds of 'correspondents' mailed in observations and anecdotes, many of which Buffon cited in the *Histoire naturelle*. Nor did Buffon's death put an end to the series. Eight further volumes, apparently based on Buffon's notes, were written and published by Bernard-Germain-Etienne de Lacépède between 1788 and 1804.

Volumes i to iii, published together in 1749, were and remain the most controversial and carefully scrutinised. Volume i contains a methodological 'Premier discours' entitled 'De la manière d'étudier et de traiter l'histoire naturelle'. Here Buffon rejects Linnaean taxonomy, proposes an anthropocentric way of ordering nature, and places strict limits on the use of mathematics in the sciences. Volume i continues with the *Histoire et théorie de la terre*, an account of the Earth's development making erosion and sedimentation the main agents of change. The *Histoire et théorie* features Buffon's famous 'hypothesis' that the Earth came into being as a result of a near collision between the sun and a comet. Volume ii contains his theory of reproduction, which departed from the prevailing view that living beings were pre-formed in one parent's body. According to Buffon, both male and female genitals house indestructible 'organic molecules' derived from their food. Mixed in the womb, these organic molecules are organised into a complete being by Newtonian micro-forces and an 'interior mould'. In 1748 Buffon 'confirmed' the existence of female sperm and organic molecules in a series of microscopic observations with the English microscopist John Turberville Needham. These observations, possibly vitiated by over-magnification,[39] are reported in detail in the *Histoire naturelle*.

Buffon's treatment of animals begins with human beings. Volume ii contains sections on human nature and the stages of human life – childhood, puberty, maturity, and old age. Besides examining the senses and narrating the first day of a newly created 'Adam' to show that all knowledge derives from experience,[40] volume iii includes 'Variétés de l'espèce humaine', a survey of human

38. Elisabeth Anderson, 'La collaboration de Sonnini de Manoncourt à l'*Histoire naturelle* de Buffon', *SVEC* 120 (1974), p.329-58.

39. Phillip R. Sloan, 'Organic molecules revisited', in *Buffon 88*, p.415-38.

40. Imagining adults without experience was a common means of illustrating the thesis of sensationism in the mid-eighteenth century. Both Diderot and Buffon accused Condillac of plagiarising 'their' idea to create the famous learning statue of the *Traité des sensations*, but the

culture and character throughout the world. Volumes iv to xv, the *Histoire des quadrupèdes* (1753-1767), cover the world's land mammals – as Carl Linnaeus would call them – in sequential articles. Early on, Buffon doubted that even the horse and the ass were genetically related, but by the 1760s he had come to believe that 'degeneration' from just a few dozen original species had given rise to the hundreds of currently known animals. Limited though it is, this 'transformism' later earned him a questionable reputation as a precursor to Darwin.[41] Volumes xvi to xxiv, the *Histoire des oiseaux* (1770-1783), and volumes xxv to xxix, the *Histoire des minéraux* (1783-1788), give individual treatment to different birds and minerals. A seven-volume *Supplément* (1774-1789) consists mainly of addenda to previous volumes and reprints of Buffon's early papers for the Académie des sciences, but it also features new material. In 1753 Buffon was elected to the Académie française, and his inaugural discourse appears in the fourth *Supplément*. This volume also contains his 'Essai d'arithmétique morale', a conglomerate of old and new texts on probability theory. Finally, volume v of the *Supplément* contains the *Epoques de la nature* (1778), Buffon's second treatment of the Earth's evolution. Unlike the *Histoire et théorie de la terre*, the *Epoques* makes heat – specifically the Earth's dwindling internal heat – a major determinant of geological change, and it gives the Earth a definite and heterodox age – 75,000 years, a figure Buffon adjusted considerably in various manuscripts.[42]

Public reaction to the first several volumes of the *Histoire naturelle* prefigured reactions to the series in general. A first printing of between 500 and 1000 copies sold in six weeks, with more than 250 copies reserved for Louis XV. Other printings and editions followed soon after, ranging in size from folio to duodecimo, and the series was translated into German, Dutch, English, Italian, and Spanish in Buffon's lifetime.[43] The public at large was obviously taken with the *Histoire naturelle*. The series' success was often attributed to women readers, a suggestion in keeping with its known popularity in certain salons.[44] Subscribers to the series included, pre-eminently, aristocrats, provincials, and representatives of the noble professions.[45] Thanks to their enthu-

charges seem dubious. See Laurence L. Bongie, 'A new Condillac letter and the genesis of the *Traité des sensations*', *Journal of the history of ideas* 16 (1978), p.83-94.

41. See Fellows and Milliken, *Buffon*, p.112-24. The term 'transformism', an anachronism when applied to the eighteenth century and even to Lamarck, will occasionally be used here as a convenient designation for Buffon's views on organic change, but the term is usually reserved for more ambitious 'transformisms'. See Jacques Roger, *Les Sciences de la vie dans la pensée française au XVIIIe siècle*, 2nd edn (Paris 1993), p.577-78, 665-68.

42. Jacques Roger, intro. to Georges-Louis Leclerc de Buffon, *Les Epoques de la nature* (Paris 1962), p.lxiv-lxvii.

43. E. Genet-Varcin and Jacques Roger, 'Bibliographie de Buffon', in *Œuvres philosophiques de Buffon*, ed. Jean Piveteau (Paris 1954), p.525-27; Paul-Marie Grinevald, 'Les éditions de l'*Histoire naturelle*', in *Buffon 88*, p.631-37.

44. On Buffon and *salonnières* see Roger Saban, 'Le testament de Buffon', in *Buffon 88*, p.99-101; Roger, *Buffon*, p.495-98.

45. Pietro Corsi, *The Age of Lamarck*, translated by Jonathan Mandelbaum (Berkeley, Calif. 1988), p.1, 22; Spary, '"Nature" of Enlightenment', p.295-96; Alain-Marie Bassy, 'A l'heure des

siasm the *Histoire naturelle* became one of the century's greatest best-sellers.[46] For a time all of Paris was abuzz with Buffon. Adding to the furor, the Sorbonne condemned the *Histoire naturelle* in 1751, perhaps to avoid being outdone by the Jansenists. Eager to avoid problems, Buffon made a show of recanting fourteen suspect propositions in a preface to volume iv, and the Sorbonne backed down. The series went on, and the offending early volumes continued to be printed without any changes. Buffon was probably too powerful and popular to fear anything as harsh as prison or exile, but he took care in later volumes to protect himself with occasional pious allusions and statements of faith. Coming from a thinker who was certainly no Christian and who may have been an atheist for much of his life, these declarations must be considered thoroughly rhetorical.[47]

Among the scientific elite the *Histoire naturelle* was received with less favour. It was read with attention by natural historians and philosophers throughout the world. Buffon's theories formed an almost obligatory starting point for speculations about geology, reproduction, and other subjects in the second half of the century, and many of his colleagues appreciated the series as a source of information.[48] Buffon's epistemology and concept of species influenced Kant and German biology, while his vision of natural history inspired a 'school' of French naturalists in the several decades after his death.[49] But his principal theories were never much accepted. D'Alembert and Condorcet, the Académie's leaders through most of Buffon's heyday, pronounced the *Histoire naturelle* loose and hypothetical, a *galimatias physique*.[50] Many of the period's most distinguished naturalists, notably Réaumur and Albrecht von Haller, also considered and rejected Buffon's natural history. Georges Cuvier was not alone among Parisian scientists of the later eighteenth century in hoping for Buffon's demise so that French natural history could chart a fresh course.[51]

grandes synthèses. L'œuvre de Buffon à l'Imprimerie royale, 1749-1789', in *L'Art du livre à l'Imprimerie nationale* (Paris 1973), p.184.

46. Spary, '"Nature" of Enlightenment', p.295-96; Mornet, 'Enseignements des bibliothèques privées', p.460. Mornet's results remain significant though subject to the caveats expressed in Robert Darnton, *The Literary underground of the Old Regime* (Cambridge, Mass. 1982), p.167-82.

47. On Buffon's irreligion see Gurdon Wattles, 'Buffon, d'Alembert and materialist atheism', *SVEC* 266 (1989), p.285-317; Roger, intro. to Buffon, *Epoques*, p.xciv-cxiv; Jacques Roger, 'Diderot et Buffon en 1749', *Diderot studies* 4 (1963), p.231-36. For ironic commentary on Buffon's pious rhetoric see Thomas-Marie Royou, *Le Monde de verre réduit en poudre, ou analyse et réfutation des Epoques de la nature de M. le comte de Buffon* (Paris, Mérigot 1780), p.4-6, 65-66, 80-82, 92; *Nouvelles ecclésiastiques, ou mémoires pour servir à l'histoire de la constitution Unigenitus* (6 February 1750), p.21, 25.

48. See for example Shirley Roe, 'Buffon and Needham: diverging views on life and matter', in *Buffon 88*, p.439; P. B. Wood, 'Buffon's reception in Scotland: the Aberdeen connection', *Annals of science* 44 (1987), p.180; Roger, *Buffon*, p.206, 226.

49. Phillip R. Sloan, 'Buffon, German biology, and the historical interpretation of biological species', *British journal for the history of science* 12 (1979), p.109-53; Timothy Lenoir, 'The Göttingen school and the development of transcendental Naturphilosophie in the Romantic era', *Studies in history of biology* 5 (1981), p.119-35; Corsi, *Age of Lamarck*, p.1-39.

50. Roger, *Buffon*, p.263-66; Elisabeth Badinter and Robert Badinter, *Condorcet (1743-1794). Un intellectuel en politique* (Paris 1988), p.119-20, 145; Ronald Grimsley, *Jean d'Alembert, 1717-83* (Oxford 1963), p.95-96, 282.

51. Corsi, *Age of Lamarck*, p.3-7.

Even after his death Buffon remained a potent presence. Augmented and re-edited by Lacépède and others, the *Histoire naturelle* continued to be a best-seller throughout the first third of the nineteenth century.[52] Nor did Buffon's scientific influence immediately disappear. Enthusiasts of his literary, philosophical kind of natural history defended it as scientifically proper into the early nineteenth century. But Cuvier was opposed to popular natural history in its scientific aspirations, and he and his allies succeeded in relegating the *Histoire naturelle* to the margins of science.[53] Like many outmoded intellectual figures Buffon was subject to periodic revivals, especially as a 'precursor' to 'evolutionary biologists' such as Jean-Baptiste de Lamarck or Darwin.[54] Thanks to his orthodox front and his distance from the *philosophes*, he was portrayed as an anti-Enlightenment hero in 1919 by Louis Dimier, a sympathiser with the ultra-conservative Action française. But Buffon's reputation was manifestly in decline by the early twentieth century. By 1950 Diderot had supplanted him in the standard list of the four luminaries of eighteenth-century French letters. Jean-Louis de Lanessan's 1884-1885 critical edition of the *Histoire naturelle* was the only one available until 1954.[55]

Since the 1950s research on Buffon as an intellectual figure has grown considerably. Hundreds of articles and books have been written, and a better-understood Buffon has been returned to the mainstream of intellectual history. Within this recent literature the studies of Lesley Hanks, Jacques Roger, and Phillip Sloan deserve special mention. Like almost everything written on Buffon in the last several decades this study is profoundly indebted to their work. But interest in Buffon as a writer has steadily languished, despite his new vogue as a philosopher and scientist.[56] Significantly, none of the published papers from the 1988 colloquium 'Buffon 88' focuses on his style, his rhetoric, or his literary identity. Histories of literature, which devoted whole chapters to Buffon during the nineteenth century, nowadays grant him a few pages at most. To a degree this is a welcome reaction to previous efforts to trivialise Buffon's thinking and to glorify the aesthetics of the *Histoire naturelle*. But the current emphasis on Buffon's thought, as reconstructed in its most distilled form, is equally unbalanced. Buffon was a best-selling author as well as a philosopher, and natural history was a literary genre as well as a science.

52. Martyn Lyons, 'Les best-sellers', in *Histoire de l'édition française*, ed. Henri-Jean Martin, Roger Chartier and Jean-Pierre Vivet, 4 vols (Paris 1982-1986), vol. iii, *Le Temps des éditeurs. Du romantisme à la Belle Epoque*, p.373-75; Corsi, *Age of Lamarck*, p.21-22.

53. Corsi, *Age of Lamarck*, p.1-39; Drouin and Bensaude-Vincent, 'Nature for the people', p.411-14.

54. See for example Stuart M. Persell, 'The revival of Buffon in the early Third Republic', *Biography: an interdisciplinary quarterly* 14 (1991), p.12-24; Joy Harvey, 'Buffon and nineteenth-century French anthropologists', in *Buffon 88*, p.649-65. The expression 'evolutionary biologist' is an anachronism with respect to Darwin as well as Lamarck and Buffon, but it captures the spirit of the Buffon-to-Darwin genealogy.

55. Persell, 'Revival of Buffon', p.12, 14. On Buffon's historiographical image and his 'replacement' by Diderot see Roger, *Buffon*, p.10-12; Fellows and Milliken, *Buffon*, p.16-18.

56. Sloan, 'Buffon studies today', p.469-72.

Most of all, the *Histoire naturelle* is a highly rhetorical text, not just a repository of scientific ideas. Buffon wrote it to communicate knowledge about natural history, but he also wrote it to be convincing, comprehensible, and appealing to a range of readers. Touching different readers required different strategies. Humanist literati could be wooed with poetic style, grand generalities, and intelligible theories. Bibliophiles could be captivated with superbly illustrated quartos. Fashionable audiences could be titillated with descriptions of animals' 'loves'. The unconvinced could be won over by simplifying issues and caricaturing rivals. The pious could be placated with tributes to God. The uninitiated could be assisted with analogies and overviews. And serious naturalists could be drawn in with the promise of new information. As Alain-Marie Bassy has suggested in discussing the plates in the *Histoire naturelle* – notably the incongruous biblical illustration in Buffon's anti-biblical *Histoire et théorie de la terre* – there was something for almost everyone somewhere in the series.[57] In natural history, Emma Spary has writen, 'meaning functioned simultaneously on several distinct and irreducible levels'.[58] Though her assertion specifically concerns transitions from specimens to images and texts, it could just as well be applied to different levels of text within natural histories. Such pluralism created tensions and contradictions, as Buffon's critics have long observed. In Arthur O. Lovejoy's words, '[Buffon] was little careful of consistency – and extremely careful of rhetorical effect.'[59] A rhetorical approach can help explain the inconsistencies of the *Histoire naturelle* and re-establish the series' identity as a particular text.

Studying Buffon's rhetoric need not mean renouncing recent research on his ideas. Many scholars have acknowledged at least marginally that rhetoric plays a role in the *Histoire naturelle*. Sloan, for example, has argued that Buffon's nominalism in the 'Premier discours' masks a hidden realism, Roger has been diligent in pointing out Buffon's perfunctory religious rhetoric, and Hanks has suggested that Buffon's denigrations of mathematics are often

57. Bassy, 'A l'heure des grandes synthèses', p.176-80. For another example of an image contradicting the text of the *Histoire naturelle* see Alain-Marie Bassy, 'Le texte et l'image', in *Histoire de l'édition française*, vol. ii, *Le Livre triomphant, 1660-1830*, p.159-60. On others' efforts to accommodate different audiences for the sciences see Mary Terrall, 'Generation and desire in Maupertuis's science of life', *Isis* 87 (1996), p.216-29; Marie-Françoise Mortureux, *La Formation et le fonctionnement d'un discours de la vulgarisation scientifique au XVIIIe siècle à travers l'œuvre de Fontenelle* (Paris 1983), p.615-17; Erica Harth, *Ideology and culture in seventeenth-century France* (Ithaca, N.Y. 1983), p.250-78. On the necessity of recognising different kinds of 'popular' readers see Cooter and Pumfrey, 'Separate spheres', p.251.

58. Emma Spary, 'Codes of passion: natural history specimens as a polite language in late 18th-century France', in *Wissenschaft als kulturelle Praxis, 1750-1900*, ed. Hans Erich Bödeker, Peter Hanns Reill and Jürgen Schlumbohm (Göttingen 1999), p.105-107.

59. 'Buffon and the problem of species', in *Forerunners of Darwin, 1745-1859*, ed. Bentley Glass, Owsei Temkin and William L. Straus, Jr (Baltimore, Md. 1968), p.108. On Buffon's inconsistency within chapters and articles see also Sloan, 'Buffon, German biology', p.116, 119-20; Jorge Martinez-Contreras, 'Des mœurs des singes. Buffon et ses contemporains', in *Buffon 88*, p.559; Roger, *Buffon*, p.126-27, 344, 426-27; Roger, *Sciences de la vie*, p.542; Paul Mengal, 'La psychologie de Buffon à travers le traité *De l'homme*', in *Buffon 88*, p.606-608.

polemical.[60] These insights will be developed in the following pages. More generally this study is premised on recent arguments by Sloan and others that Buffon's epistemology was in many ways stable from the 1740s to his death.[61] But here it is necessary to distinguish between Buffon's stated epistemology and the vaguer assumptions that guided his research. The epistemology Buffon professed was in large part a credo of absolutes (for example, only individuals exist, mathematics has no reality, all knowledge comes from comparison). Many of these principles were staples of eighteenth-century philosophical discourse. Some had become so common by Buffon's time that experts remain divided over whether to call them Lockean or Leibnizian.[62] To compound the problem, Buffon's conscious epistemology mostly comes out in snippets in polemical contexts, not in sustained philosophical analyses. But if his 'official' epistemology is hard to reconstruct for being fragmented, it clearly diverges in certain key respects from his working assumptions about scientific knowledge, as discernible in less self-conscious parts of the *Histoire naturelle*. The relation between these two epistemologies – one implicit and intuitive, the other conscious and highlighted – will be examined below in detail, but in general it can be said that Buffon's official epistemology exaggerated and distorted beliefs manifest in his practice. With its righteous separation of good from bad science it was ideal for polemics, and this is how it functioned in the *Histoire naturelle*, though Buffon probably believed in it consciously.

Recognising that there is rhetoric in Buffon's professed epistemology can help disentangle longstanding debates regarding the evolution of his ideas and attitudes. If, for example, Roger's contention that Buffon evolved toward greater rationalism in the 1760s is now justly questioned, it is nonetheless true that Buffon evolved toward greater acceptance of the term 'system', among other things, as we will see in chapter 4.[63] Here Buffon's rhetoric changed more than his underlying assumptions. At other times bits of rhetoric remained in his repertory long after he had outgrown the conceptual and epis-

60. Phillip R. Sloan, 'The Buffon–Linnaeus controversy', *Isis* 67 (1976), p.356-75; Roger, intro. to Buffon, *Epoques*, p.c-ci, cxii; Roger, *Buffon*, p.210-22, 251-54, 345n.; Hanks, *Buffon avant l'‘Histoire naturelle’*, p.52n., 60n., 206, 213.

61. Sloan, 'Buffon–Linnaeus controversy', p.358; Phillip R. Sloan, 'From logical universals to historical individuals: Buffon's idea of biological species', in *Histoire du concept d'espèce dans les sciences de la vie* (Paris 1987), p.132-33; François Duchesneau, *La Physiologie des Lumières. Empirisme, modèles et théories* (The Hague 1982), p.264-67; M. J. S. Hodge, 'Two cosmogonies (theory of the Earth and theory of generation) and the unity of Buffon's thought', in *Buffon 88*, p.246; Hanks, *Buffon avant l'‘Histoire naturelle’*, p.224-34. A good example of the old view of Buffon as inconstant can be found in Pierre Flourens, *Histoire des travaux et des idées de Buffon* (Paris 1844), p.15-20.

62. See for example Sloan, 'Buffon studies today', p.474-75; Hodge, 'Two cosmogonies', p.245-46.

63. Roger's view can be found in Jacques Roger, 'Buffon, Georges-Louis Leclerc, comte de', in *Dictionary of scientific biography*, ed. Charles Coulston Gillispie, 16 vols (New York 1970-1980), ii.577-78; Roger, *Buffon*, p.565; Roger, *Sciences de la vie*, p.564-67. For contrary arguments see the two preceding notes and Stephen Milliken, 'The lyricism of the intellect: Buffon's *Epoques de la nature*', *Diderot studies* 6 (1964), p.299-301.

temological bases that originally supported them. As noted in chapter 3, the chain of being was one schema for ordering nature that Buffon continued to use for its literary, polemical, and pedagogical value even after he turned to more complicated models. Elsewhere both his rhetoric and his practical assumptions seem to have shifted, as in his dealings with classification (see chapter 6). Coupled with awareness that beliefs can be more or less deep and superficial, partial and inconsistent, rhetorical analysis provides a way of understanding Buffon's contradictions that considers both utterances and philosophical undertones.

Rhetorical analysis is not meant to bring back the tradition of seeing Buffon as a pretentious imposter. Thanks to recent historiography his reputation as a naturalist is secure enough now to survive recognition of a certain amount of superficialness. It would be unfair, in any case, to deal with him alone in a study of scientific rhetoric. For placed in the broad context of eighteenth-century natural history, as it is in each of the following chapters, Buffon's rhetoric appears only modestly exceptional. Other authors wrote tactically, exaggerating and deforming their deeper beliefs in the interest of routing rivals, attracting a readership, or making strong points. Though self-satisfied myths dating back to his lifetime have portrayed him as a gross anomaly, Buffon was not the only rhetorically minded philosopher of the Enlightenment. As we shall see, scientific discussions of style, final causes, the chain of being, systems, mathematics, and nominalism were regularly rhetorical in most everyone's hands, not just Buffon's. Once account is taken of the functions as well as the meanings of intellectual discourse, the famous 'issues' in eighteenth-century natural history appear less clear-cut and philosophical than in standard presentations. The issues were real ones, but it would be an error to abstract them from concerns about readers and polemical strategy.

Scientific rhetoric

As should be clear now, rhetoric, in this study, is to be understood as a set of linguistic techniques for attracting and persuading readers. This definition – not especially true to eighteenth-century notions[64] – draws on both classical and modern views of rhetoric. It is classical in conceiving rhetoric as a general art of persuasion, but modern in focusing on the written word and in considering persuasion in philosophical abstraction, apart from rhetoric's traditional canons and niceties. Much of what will be discussed here could be slighted as 'mere rhetoric', that is, as rhetoric betraying some truth. Authors misrepresenting their beliefs to be persuasive or orthodox can be said to practise such

64. Claude Buffier defined rhetoric as 'l'art de dire ce qu'il faut pour persuader' in *Traité philosophique et pratique d'éloquence*, in *Cours de sciences sur des principes nouveaux et simples* (Paris, Cavelier and Giffart 1732), p.299. But this eminently classical definition hardly represents rhetoric as the eighteenth century conceived it. On the evolution of the notion of rhetoric in the Enlightenment see Thomas M. Conley, *Rhetoric in the European tradition* (New York 1990), p.188-205; Peter France, *Rhetoric and truth in France: Descartes to Diderot* (Oxford 1972), p.3-33.

rhetoric. So too can authors who make persuasive arguments that are nonetheless not formally or empirically compelling, as judged, for example, by the standards of the contemporary scientific community. Rhetoric and truth will be set off against each other frequently in the following pages, but the dichotomy is proposed heuristically, not as an absolute. Above all, the opposition gives the term 'rhetoric' a needed specificity. In any case a great deal of the rhetoric discussed here can also be seen positively. As we will see, the rhetoric of eighteenth-century natural history could be a thing of great beauty, a means of welcoming readers, a pedagogically beneficial shorthand, and more.

To the modern reader the concept of scientific rhetoric may seem arcane, if not contradictory. As we all know, rhetoric is kept under close check in the highly evolved genre of the modern scientific paper. Literary style is constrained, blatant appeals to emotion, common sense, and aesthetic sensibilities are mostly taboo, and authors write for audiences that are to a certain extent held captive by professional identity. Truth, it might seem, is the only determinant of which texts the scientist reads and accepts. This view, however, considerably exaggerates the formalistic character of scientific thinking and writing. Even in mathematics, where the reduction of inference to mechanical symbol shuffling has been accomplished in principle for many domains, it is not clear that mathematicians' actual written arguments are mere abbreviations or summaries of perfectly rigorous, lower-level arguments.[65] Elsewhere in science it is difficult to imagine how argumentation could be stripped of its human contingency. Terminology and titles are chosen, papers, paragraphs, and sentences are organised and structured, and certain statements are made to the exclusion of others. Subtle though they are in effect and intention, these choices are in part rhetorically motivated.

Awareness of science's rhetorical dimension is not new. Although many of the canonical founders of modern science in the sixteenth and seventeenth centuries expressed a desire to purge natural philosophy of rhetoric, their interactions with rival thinkers reveal an implicit recognition that polemics had a place in written science. Many a rival's methodological claims were written off, for example, as disingenuous efforts at self-legitimation. If only marginally, historians of ideas have also long noted the tactical, audience-pleasing character of certain ostensibly scientific pronouncements. Yet only since the 1980s has there appeared a large number of scholarly works focusing on scientific rhetoric.[66] Interest in the subject has come from a wide range of

65. Philip Kitcher, 'The cognitive functions of scientific rhetoric', in *Science, reason, and rhetoric*, ed. Henry Krips, J. E. McGuire and Trevor Melia (Pittsburgh, Penn. 1995), p.52-54; Philip Kitcher, 'Persuasion', in *Persuading science: the art of scientific rhetoric*, ed. Marcello Pera and William R. Shea (Canton, Mass. 1991), p.4-8.

66. For histories and bibliographies of work on scientific rhetoric see Randy Allen Harris (ed.), intro. to *Landmark essays on rhetoric of science: case studies* (Mahwah, N.J. 1997); Peter Dear (ed.), intro. to *The Literary structure of scientific argument* (Philadelphia, Penn. 1991), p.1-9; J. V. Golinski, 'Language, discourse, and science', in *Companion to the history of modern science*, p.110-23.

academic disciplines – history, philosophy, literature, language, and communication – and the goals and presuppositions of those studying scientific rhetoric have been correspondingly diverse. The movement's motivations should not be simplified, but it has obviously been favoured by the collapse of positivistic views of methodology and epistemology, the development of academic programmes in technical writing, the growth of more culturally oriented histories of science, and the new interest in language emanating from French post-structuralist philosophy. It would be an exaggeration to say that a 'rhetorical turn' has taken place generally in the history of science, but scientific rhetoric has come to be a popular and respectable object of study.

Two normative tendencies have been common in work on scientific rhetoric. Following a long line of efforts to debunk positivistic idealisations of science, a number of studies of scientific rhetoric have aimed to undercut the myth of science's independence from human activity.[67] Working from the opposite direction, many studies of scientific rhetoric have sought to establish rhetoric as a far from irrational means of persuasion and to cleanse it of its modern image as ornament or trickery. Inspiration here comes from such champions of rhetoric as Kenneth Burke, Chaim Perelman, Stephen Toulmin, and Robert Scott.[68] Both crusades make good points. It is certainly worth stressing that faith in science as a superhuman, infallible truth is naive and dangerous. It is also worth noting that rhetoric need not be the enemy of reason that it has often been made out to be in the last several centuries. But when taken to extremes, efforts to humanise science and rehabilitate rhetoric can lead to the conclusion that little distinguishes science from rhetoric. This is counter-intuitive and counter-productive if nothing else.

In itself there is perhaps nothing wrong with making science and rhetoric equivalent forms of imperfect rationality. Words can be defined as we wish to define them. But useful conceptual distinctions repudiated in one vocabulary have a way of sneaking back in other guises. In particular, scholars insistent on the near identity of science and rhetoric seem to experience a recurrent need to establish similar dichotomies in other terms. Thus Alan Gross, having informed us that 'no line can be successfully drawn between rhetoric and scientific knowledge', goes on to oppose 'brute facts of nature' to the science he has just equated with rhetoric.[69] A similar displacement of the problem

67. On the relation of post-positivism to the study of scientific rhetoric see Lawrence J. Prelli, *A Rhetoric of science* (Columbia, S.C. 1989), p.1-8; John A. Schuster and Richard R. Yeo (eds), intro. to *The Politics and rhetoric of scientific method* (Dordrecht 1986), p.ix-xxvi; Harris, intro. to *Landmark essays*, p.xiii-xvii.

68. 'Classics' often mentioned in studies of scientific rhetoric include Kenneth Burke, *A Rhetoric of motives* (New York 1950); Chaim Perelman, *Traité de l'argumentation* (Paris 1958); Stephen Toulmin, *The Uses of argument* (Cambridge 1958); Robert L. Scott, 'On viewing rhetoric as epistemic', *Central states speech journal* 18 (1967), p.9-17. The more limited view of rhetoric rejected by these authors is described in Gérard Genette, 'La rhétorique restreinte', in *Figures III* (Paris 1972), p.21-40.

69. J. E. McGuire and Trevor Melia, 'The rhetoric of the radical rhetoric of science', *Rhetorica* 9 (1991), p.303-304. For a similar critique of the strong 'constructivism' of Barry Brummett see

can be found in Marcello Pera's attempt to make 'dialectics' rather than methodology the driving force behind science. Pera's goal is to situate dialectics, a 'canon of validation' of rhetorical forms, between logically consequent methodology and the irrational so that 'we are no longer forced to choose between an algorithm and a chat at the bar'.[70] Yet Pera's excellent examples of scientific rhetoric recreate the dichotomy in miniature, since he ends up distinguishing elements of reason and trickery in particular arguments. Significantly the word 'rhetoric', which Pera defines neutrally as 'persuasive forms of reasoning or argumentation', comes to take on its usual meaning, as in his contention regarding Darwin that 'rhetoric seems to have led him astray in this case'.[71] In essence we are still forced to choose between the rationally justified and the rhetorical, an opposition not all that different from the more hyperbolic ones Pera sets out to vanquish.

Implicitly both Gross and Pera distinguish rhetoric from at least certain parts of science – 'brute facts of nature' or logical consequentialness – and this is in some sense a necessity of their enterprise. Without such a distinction the study of scientific rhetoric lacks specificity. Why not look at epistemology, methodology, science, or rhetoric if the expression 'scientific rhetoric' is just a pleonasm?[72] More theoretical, less pragmatic reasons for resisting the equation of science and rhetoric can also be offered. Central to all is the recognition of a certain resistance, 'recalcitrance' in the words of J. E. McGuire and Trevor Melia, that constrains science in a unique way.[73] But arguments of this kind quickly lead into the vast, indeterminate debates over the validity of realism in its various forms.[74] For our purposes it is sufficient to note that everyday and even intellectual usage authorises a distinction, however murky, between science and rhetoric. Someday, conceivably, a naturalised epistemology based on better understanding of mental operation may provide a stable middle ground between these two terms.[75] In the meantime

Richard A. Cherwitz and Thomas J. Darwin, 'Beyond reductionism in rhetorical theories of meaning', *Philosophy and rhetoric* 27 (1994), p.315n.

70. *The Discourses of science*, translated by Clarissa Botsford (Chicago 1994), p.10-12, 136.

71. *Discourses of science*, p.57-102.

72. Warnings against over-extension of the term 'rhetoric' have become commonplace as the word is applied to ever larger and more abstract domains. By some judgements the present study would be more properly characterised as a study of scientific 'intertextuality' or of scientific literariness than as one of scientific rhetoric. See Brian Vickers, *In defence of rhetoric* (Oxford 1988), p.439n.; Harris, intro. to *Landmark essays*, p.xxiv-xxviii; Dilip Parameshwar Gaonkar, 'The idea of rhetoric in the rhetoric of science', *Southern communication journal* 58 (1993), p.258-95.

73. 'Some cautionary strictures on the writing of the rhetoric of science', *Rhetorica* 7 (1989), p.87-99. Rhetoric, Pera argues, is constrained in its own way, much as a judge is constrained by legal precedent (*Discourses of science*, p.57-58).

74. For arguments about realism by scholars studying scientific rhetoric see for example Cherwitz and Darwin, 'Beyond reductionism', p.323n.; Alan G. Gross, *The Rhetoric of science* (Cambridge, Mass. 1990), p.193-207; Ernan McMullin, 'Rhetoric and theory choice in science', in *Persuading science*, p.55-76.

75. Philip Kitcher's 'psychologistic' approach to the rhetoric of mathematics seems promising in this light. See for example Kitcher, 'Persuasion', p.3-27. The literature on naturalised epistemology is extensive, but few follow Kitcher in relating their results to persuasion and rhetoric.

the persistence of the contrast in spite of efforts to expunge it suggests the advisability of making peace with it, if only provisionally. This does not mean concluding that rhetoric and science constitute polar opposites or that rhetoric is irrational. Science is nothing without persuasion, and rhetoric is unpersuasive unless it has a measure of rational appeal. But the rhetoric of science is interesting precisely insofar as it brings together different aspects of one human project and elucidates their interaction in particular cases.

Given their intertwinement, identification of science vis-à-vis rhetoric is bound to be difficult. With regard to early sciences and natural philosophy, most of the evidence available to scholars is textual. In the mathematical sciences a simplistic solution would be to restrict science proper to symbolic formulae and dismiss everything else as presentation and rhetoric. But this is unsatisfactory since formulae have no meaning without textual explanation, and since mathematics was long done verbally without special symbols. In any case proposals for finding mathematics' epistemic core in its formulae have little bearing on a discipline such as natural history where specialised symbolism never played a big role. Here it is clear that, as Steven Shapin puts it, 'no type of talk, taken by itself, offers the historian direct access to historical reality'.[76] The very persuasiveness of rhetoric makes it impossible to count on being able to 'unmask' it. The orthodox rhetoric of certain early modern heretics was sufficiently convincing, for example, that it was taken at face value by contemporaries as well as by subsequent historians, only to be proven hypocritical by the discovery of other documents.

But if texts are the main artifacts of past natural history, it would be wrong not to recognise that there are distinctions among texts and even within them as to purposes, levels of seriousness, and intended modes of reading. These distinctions cannot be codified in absolute rules or with total certainty, but a few general guidelines can be sketched out. Public, highlighted portions of an author's production such as preambles, prefaces, and texts presented orally are often hotbeds of rhetoric, designed as they are to establish reputations, legitimate research, and engage a broad readership. Works of this kind should be compared with less public, less focalised works: correspondence, private notebooks, first drafts, and published materials composed with a minimum of self-reflection. Inconsistencies suggest the possibility of rhetorical distortion. But even if authors are perfectly consistent at every level, rhetorical analysis may still be relevant. More important than repeated pronouncements in defining authors' beliefs are the implicit assumptions that animate their research. Sweeping declarations about nature, the universe, or scientific method need to be read in light of implicit knowledge evident in other parts of a text, particularly in cases where rhetorical motives are not hard to imagine for the declarations in question.

Kitcher himself casts his views in a vocabulary that downplays rhetoric in 'The naturalists return', *The Philosophical review* 101 (1992), p.53-114.

76. 'Talking history: reflections on discourse analysis', *Isis* 75 (1984), p.126.

Conducted in this way, rhetorical analysis is an important complement to intellectual history in two respects. First, it situates science socially. This is particularly desirable in the case of natural history, which has long been connected with educated society via museums, collections, and literary genres. Natural historians of the Enlightenment were in many cases addressing broad segments of society, and their rhetoric sheds light on their relations with readers. Most of these readers left no trace of their modes of reading or reactions to texts, and the supposition of ideal readers can lead to over-interpretation.[77] Here extrapolation from a known best-seller such as Buffon's *Histoire naturelle* can be instructive. Certain owners of the series – a potential 'coffee table' work in its more lavish printings – may not have examined it beyond illustrations, but references to Buffon's ideas among eighteenth-century readers suggest that many read parts of it. Combined with information about actual contemporary readings of the *Histoire naturelle*, the rhetoric of the series offers clues about how and why people read natural history and 'vulgar' science in the Enlightenment.

Second, the study of scientific rhetoric provides a framework for recognising degrees of belief and disbelief. To identify an utterance as a product of rhetoric is merely a first step toward explaining its relation to other utterances and to authors' beliefs. For rhetorical white lies, often inherited half-consciously from colleagues and contemporaries, need not be recognised as such by their users.[78] Instances of consciously hypocritical, 'pure' rhetoric are probably as rare as instances of completely unrhetorical 'plain speaking'. Most writers, it seems likely, manage to believe in their rhetoric to a certain extent, though perhaps not at deeper, less conscious levels. Rhetorical analysis opens the way for distinguishing pronouncements as more or less hypocritical or reflective of their authors' beliefs and behaviour. These distinctions are important because past intellectuals are often quickly pigeon-holed on the basis of a small number of easily quoted statements, and because quarrels between past intellectuals are often read at face value as being mythically polarised. This is not to imply that intellectual history is merely a 'comedy of errors' in which 'the great minds would have realised they were not so far apart had they bothered to read each other's works a little more closely', as Steve Fuller characterises over-idealised studies of scientific rhetoric.[79] But neither are conflicts over ideas typically as simple and Manichaean as participants and later commentators often present them.

77. For critiques of the notion of the reader in studies of scientific rhetoric see Steve Fuller, '"Rhetoric of science": a doubly vexed expression', *Southern communication journal* 58 (1993), p.308-10; Steve Fuller, 'The strong program in the rhetoric of science', in *Science, reason, and rhetoric*, p.100-105.

78. Acceptance of 'unconscious' rhetoric situates this study nearer a 'discourse-centered model of "intertextuality"' than a more classical 'agent-centered model of "intentional persuasion"'. See Gaonkar, 'Idea of rhetoric', p.276-82.

79. 'Strong program', p.103.

Each of these goals of rhetorical analysis – understanding readers and literary culture, evaluating degrees of belief and of polemical posturing – animates roughly half of the following study. The first three chapters concentrate on means of attracting a readership, notably by making natural history intelligible, literary, and entertaining. The first chapter, on style, deals almost exclusively with rhetoric as pleasing an audience. The passage from this chapter to following ones on final causes and the chain of being may seem disconcerting, since these latter subjects are traditionally considered part of the conceptual substance, not the form, of natural history. Yet as we will see, evocations of final causes and the chain of being were frequently little different from stylistic flourishes in eighteenth-century natural history. Buffon, in particular, was prone to using such images for their literary and pedagogical value even when they clashed with his deeper opinions.

Still, final causes and the chain of being were more than just tools for ingratiating readers. They were also contested ideas of eighteenth-century natural history, each giving rise to arguments and partisan rhetorics. Accordingly, already in chapters 2 and 3 we will discuss rhetoric as persuasion, sketching out different groups' polemical tactics. Scientific polemic and rhetorical ploys will become an almost exclusive emphasis in the final three chapters, dealing with systems, mathematics, and nominalism, respectively. All three of these topics were located near the centre of fervid debates in eighteenth-century natural history and philosophy. Just as importantly, they were intimately linked in the *Histoire naturelle*, which combined denunciations of systems and mathematics with a seemingly nominalistic resistance to classification. These intertwined stances, none wholly true to Buffon's approach to natural history, constituted the point of departure for a great many of his general polemics against rival naturalists.

1. Style

UNDER various names and in various senses, style has always been an important element of rhetoric. Levels and types of style, *genera dicendi*, have been distinguished and regulated since Greek and Latin antiquity. More generally, *elocutio*, one of the five traditional divisions of rhetoric, was regularly equated with style in the seventeenth and eighteenth centuries and remains a near equivalent. Manuals of rhetoric from antiquity onward have devoted considerable, even disproportionate attention to stylistic concerns such as vocabulary, figures, and sentence construction. Even when rhetoric is considered apart from its disciplinary history as any means of attracting and persuading an audience, style is clearly a significant part of it. Not only is stylish writing more persuasive in many contexts, it can also function as a lure to potential readers, drawing them into a work in the same way that gilt pages, pleasing illustrations, or an author's reputation might.

Style was a particularly important rhetorical tool in the case of the *Histoire naturelle*. 'Le style est l'homme même', Buffon declared famously in his 'Discours sur le style' for the Académie française (*HNS*, iv.11). In accord with this maxim his style became an object of widespread admiration soon after he started the *Histoire naturelle*, so much so that it threatened to obscure other aspects of the series. Enthusiasm for his style – which Buffon cultivated with public readings, anecdotes about his rigorous writing habits, and his 'Discours sur le style' – was a major contributor to his popularity in his lifetime and after. As Friedrich Melchior Grimm observed in 1756, high style was one of the series' primary selling points.[1] Having what Rousseau called 'la plus belle plume de son siècle' could not fail to be advantageous in a society in which success as a writer depended on appealing to a public with broadly literary tastes and education.[2] But even as it helped him attain international renown, the acknowledged excellence of his style provided a means for Buffon's detractors to dismiss the *Histoire naturelle* as superficial. How, they inquired pointedly, could Buffon pretend to write serious yet stylish natural history? Was attention to aesthetics not inherently contrary to the severity required of the natural philosopher? The 'two cultures' of science and literature were already at odds.

Understanding Buffon's style, its attractiveness to the public, and its offensiveness to some in the scientific community requires a broad examination of

1. Roger, *Buffon*, p.443. Less plausibly Gilbert Romme claimed that style was the series' only attraction. See Alessandro Galante Garrone, *Gilbert Romme. Histoire d'un révolutionnaire (1750-1795)*, translated by Anne and Claude Manceron (Paris 1971), p.44-45.

2. Jean-Jacques Rousseau to Pierre-Alexandre Du Peyrou, 4 November 1764, in R. A. Leigh (ed.), *Correspondance complète de Jean-Jacques Rousseau*, 52 vols (Geneva, etc. 1965-1998), xxii.5.

the role of style in the Ancien Régime and in natural history and the sciences in particular. Buffon's successes and failures with style were only extreme manifestations of a general ambiguity surrounding style, the sciences, and 'vulgarisation' during the period. These successes and failures were also to some degree alien to the actual contents of the *Histoire naturelle*. The style attributed to him by awed contemporaries and later connoisseurs is more a select distillation than a realistic cross-section of the *Histoire naturelle*. In the same way, complaints about Buffon's style as inimical to natural history almost always hinged on more general, non-stylistic features of his approach. To attack Buffon's style was to attack both his readership and his philosophical habits. Style was thus critical not only to Buffon's rhetoric and overall project but also to the polemics his series gave rise to.

Style and natural history

Traditionally style was associated with the principle of decorum: different styles were prescribed for different audiences, subjects, and narrative goals. This association became particularly rigid in seventeenth-century France. The tripartite division of styles, the most simple of several inherited from antiquity, was made to correspond to social class, with high style sitting atop a hierarchy above the temperate and humble styles.[3] Comparing style – often bad style – to dress, cosmetics, or bodily appearance had been common since antiquity. When appropriate, wrote Quintilian, style was like clothing that reflected a body's inner state.[4] The comparison had special relevance in the seventeenth century, as detailed sumptuary and vestiary codes mirrored those imagined by grammarians for style. But just as regulations and presuppositions regarding clothing were weakened in the next century by the pressure of changing social and commercial circumstances, neoclassical prescriptions regarding style lost a good deal of their strength in the course of the eighteenth century. High style in particular lost much of its formal exclusiveness. In the seventeenth century it had been restricted to genres of obvious grandeur: sermons, funerary orations, royal panegyric, and the tragedy in verse. In the eighteenth century its usage was expanded, a tendency epitomised by the *Histoire naturelle*.[5]

Further jeopardising French neoclassical hierarchies in aesthetic matters was the sublime. Already in the seventeenth century the sublime acted as a tacit complement to the formal niceties of aesthetic theory, 'a rule for evading rules', as Peter Gay put it.[6] Under its aegis one could admire the irregular, the

3. France, *Rhetoric and truth in France*, p.10.

4. Quintilian, *Institutionis oratoriae libri XII*, ed. Ludwig Radermacher, 2 vols (Leipzig 1965), ii.68-69 (bk VIII: preface, no.18-22).

5. John Porter Houston, *The Traditions of French prose style* (Baton Rouge, La. 1981), p.91-92, 120.

6. Peter Gay, *The Enlightenment: an interpretation*, 2 vols, vol. ii, *The Science of freedom* (New York 1969), p.304.

terrible, the obscure, and the stark – all qualities lacking sanction in official aesthetics. Important in defining the sublime as the eighteenth century would know it was the first-century treatise *De sublimitate*, long falsely attributed to the third-century Greek philosopher Cassius Longinus. Nicolas Boileau-Despréaux's French translation of 1674 brought 'Longinus' and the sublime to the forefront of aesthetic thought. Without neglecting the technical side of sublimity, *De sublimitate* drew attention to the sublime as an emotional, inspired experience. Sublime style would continue to be regarded as equivalent to high style throughout the eighteenth century, but the sublime itself would increasingly be situated outside the three traditional styles. Defined by its impact more than its form, it could be found in even laconic, plain phrases, provided they managed to awe readers in context. 'Simple sublimity' on the order of the 'Fiat lux' of Genesis was in fact regularly contrasted with the prolix, 'Asian' pomp of high, sublime style.[7] As a subjective, emotionally defined quality, the sublime would ultimately menace the whole notion of formally defined style, shifting attention away from technique and decorum toward inner experience and personal perspective.

But these developments were only beginning in the time of Buffon. Despite the growing importance accorded to genius and feeling as determinants of style, neoclassical decorum remained a potent force in eighteenth-century letters.[8] Protests against rote style were quickly forgotten when the time came to write panegyric and dedications to protectors. Emphatic *style académique* flourished as ever in the Académie française. Jesuit colleges and provincial academies continued to encourage high style and traditional eloquence, little affected by avant-garde experiments with 'personal' style. Theorists of style continued to waffle between style as decorum and style as an expression of subjective experience. Buffon is typical in his 'Discours sur le style'. On the one hand he associates style with genius and creative vision, while on the other he asserts that it follows from subject, regardless of an author's personal perspective.[9] Theories of style have rarely reflected practice very accurately, but here they reveal a deep ambivalence in eighteenth-century letters.

In short, though attitudes toward style were considerably more liberal than in the preceding century, eighteenth-century authors still had reason for stylistic conservatism. Style was no longer as closely bound to genre, but neither was it something to be cultivated without thought of decorum. Many audiences looked to style for aesthetic pleasure, yet the 'entertainment' they sought and found in it was constrained by a strong sense of convention and

7. France, *Rhetoric and truth in France*, p.26. The professor of rhetoric Charles Rollin argued against distinguishing sublime style from sublimity or simple sublimity from high sublime style, but his definition of sublime style wavers between formal and psychological criteria. See *Traité des études. De la manière d'enseigner et d'étudier les belles-lettres par rapport à l'esprit et au cœur*, 2 vols (Paris, Veuve Estienne 1740), i.378, 389.

8. See Jean-Pierre Seguin, *La Langue française au XVIIIe siècle* (Paris 1972), p.166-70.

9. Ana María Gómez Torres, *Las Ideas de Buffon sobre retórica y poética en los inicios de la teoría literaria moderna* (Málaga 1996), p.50.

appropriateness. Though not necessarily indicative of epistolary practice, the many prescriptive guides to letter-writing written in the century suggest how important decorum remained in even personal correspondence. Book reviews, too, point to the coexistence of appropriateness and aesthetics as criteria for good style. The compliment 'beau style' paid to any number of quite plainly written works was sometimes a token one, especially when directed toward women writers, but it more often referred to a style's simple appropriateness to a subject or genre. Readers continued to prize such suitability all through the century.

Style in natural history was influenced by decorum and precedent in several ways. As a genre, natural history was little studied or regulated by eighteenth-century literary theorists. But it inherited stylistic tendencies in virtue of its contents and its relations with better-defined genres. Many writers on natural history were clerics with theological ambitions, and most natural histories dealt with the theologically charged matters of nature's abundance and providence. Both factors pushed the genre toward the higher realms of style characteristic of religious discourse. Even the impious Buffon described nature's power and plenitude in hymn-like language, to the obvious delight of his more reverent readers. At the same time the specificity and multiplicity of the natural world worked against stylistic elevation. With so much to be named, natural history could hardly aspire to the general, polite vocabulary required of high style, except perhaps fitfully or through vagueness or periphrase.

Above all, natural history was considered one of the sciences by the late seventeenth century, and its style fell prey to judgements about scientific decorum. Not all works entitled natural histories or dealing with the natural world were read and criticised as scientific, but those with interests and emphases approaching the academic sciences typically were. Buffon's *Histoire naturelle*, written by an academician and treading on obviously scientific territory, was inevitably subjected to scientific scrutiny – an outcome in keeping with Buffon's intentions. As a scientific text, however, the *Histoire naturelle* elicited criticism for its style, which many claimed was incompatible with scientific enquiry. Voltaire's famous apostrophe to Buffon summed up the incongruity: 'Dans un style ampoulé parlez-nous de physique.'[10] As Jean-François Marmontel reported, literary critics reproached Buffon for having 'fastueusement écrit dans un genre qui ne voulait qu'un style simple et naturel'. The genre in question, Marmontel implied, was scientific natural history, a genre wholly different from poetic endeavours.[11]

In view of such reactions many scholars have concluded that style became incompatible with the new sciences sometime in the Enlightenment. Judgements abound regarding the precise timing and meaning of this theorised

10. Cited in Fellows and Milliken, *Buffon*, p.157.

11. *Mémoires d'un père pour servir à l'instruction de ses enfants* (Paris 1819), in *Œuvres de Marmontel*, vol. i-ii, at i.248.

separation of scientific and polite culture.[12] Seventeenth- and eighteenth-century commentators had mixed views on the subject. The founders of the new sciences, from Bacon and Descartes to the Royal Society and the Académie des sciences, expressed considerable hostility toward traditional rhetoric, in part because of impatience with scholastic philosophy. 'Le règne des mots et des termes est passé', declared Bernard Le Bovier de Fontenelle, secretary of the Académie des sciences, in 1733.[13] But hostility toward rhetoric was commonplace, even among rhetoricians, and if early natural philosophers succeeded in being less obviously stylish and rhetorical than their predecessors and contemporaries, it is clear in retrospect that they did not abandon rhetoric as completely as their words might imply.[14] Descartes, for example, was quick to condemn rhetoric in philosophy when his rival Pierre Gassendi's style was the object under scrutiny, and his own style was in many ways plain and nondescript – a rhetoric of non-rhetoric that would prove highly persuasive in the subsequent history of the sciences. But he also used numerous similes, paid attention to rhythm, and tailored his presentation to the needs of his audience. These were all traditional rhetorical concerns.[15]

In natural history Buffon's rival Linnaeus took an especially harsh stand against rhetoric. In practice he coined poetically and sexually suggestive terminology and wrote playful, dramatic prose in prefaces, Swedish works, and in certain other contexts.[16] Still his well-known Latin writings were mostly dense and unstylish, so much so that they aroused complaints among the French for being unreadable and unimaginative.[17] By the eighteenth century, in fact,

12. See for example Hahn, *Anatomy of a scientific institution*, p.54-55; Geoffrey V. Sutton, *Science for a polite society: gender, culture, and the demonstration of enlightenment* (Boulder, Col. 1995), p.273; Jeanne Peiffer, 'L'engouement des femmes pour les sciences au XVIIIe siècle', in *Femmes et pouvoirs sous l'Ancien Régime*, ed. Danielle Haase-Dubosc and Eliane Viennot (Paris 1991), p.215; Scott L. Montgomery, *The Scientific voice* (New York 1996), p.98-99; Harth, *Ideology and culture*, p.222-309; David Locke, *Science as writing* (New Haven, Conn. 1992), p.6; Londa Schiebinger, *The Mind has no sex? Women in the origins of modern science* (Cambridge, Mass. 1989), p.233-41.

13. Cited in Hahn, *Anatomy of a scientific institution*, p.7n. For condemnations of language, rhetoric, and even books in general by early natural philosophers see Pera, *Discourses of science*, p.130; Hahn, *Anatomy of a scientific institution*, p.7-8, 30-32; Ulrich Ricken, *'Gelehrter' und 'Wissenschaft' im Französischen: Beiträge zu ihrer Bezeichnungsgeschichte vom 12.-17. Jahrhundert* (Berlin 1961), p.115-17; Harrison, *Bible, protestantism*, p.194-95.

14. France, *Rhetoric and truth in France*, p.27, 37-39; Montgomery, *Scientific voice*, p.73-99. For a rhetorical analysis of the new sciences' campaign against rhetoric see Brian Vickers, 'The Royal Society and English prose style: a reassessment', in *Rhetoric and the pursuit of truth: language change in the seventeenth and eighteenth centuries*, ed. Brian Vickers and Nancy S. Struever (Los Angeles 1985), p.3-76.

15. France, *Rhetoric and truth in France*, p.64-67; Sutton, *Science for a polite society*, p.64-82. For a more abstract Cartesian rhetoric see Thomas M. Carr, Jr, *Descartes and the resilience of rhetoric: varieties of Cartesian rhetorical theory* (Carbondale, Ill. 1990), p.1-61.

16. See Lisbet Koerner, 'Carl Linnaeus in his time and place', in *Cultures of natural history*, p.155; James L. Larson, *Reason and experience: the representation of natural order in the work of Carl von Linné* (Berkeley, Calif. 1971), p.120, 126; Wilfrid Blunt, *The Compleat naturalist: a life of Linnaeus* (New York 1971), p.34-35, 148, 152-54, 162-64, 196-97; Londa Schiebinger, 'Gender and natural history', in *Cultures of natural history*, p.165-70; Lindroth, 'Two faces of Linnaeus', p.6-11.

17. Pascal Duris, *Linné et la France (1780-1850)* (Geneva 1993), p.20, 29, 122-24; Lindroth, 'Two faces of Linnaeus', p.5; Rousseau, 'Science books and their readers', p.222.

prominent French thinkers were lamenting the sciences' tendency to exclude literary virtues. Diderot worried that the 'austere mathematical language' of post-Newtonian natural philosophy had cut it off from the public.[18] In his article 'Géomètre' for the *Encyclopédie*, d'Alembert argued that mathematics and literature, different as they were, were not incompatible.[19] Voltaire prefaced the 1741 edition of his soberly written *Eléments de la philosophie de Newton* (1738) with a similar protest. To his old teacher Pierre-Joseph Thoulier d'Olivet he wrote: 'Je ne vois pas pourquoi l'étude de la physique écraserait les fleurs de la poésie. La vérité est-elle si malheureuse qu'elle ne puisse souffrir les ornements?'[20] It is tempting to interpret these defences of 'scientific literature' as nostalgia for a disappearing ideal, but testimonials about rhetoric are notoriously unreliable, shaped as they are by their own rhetorical interests. Significantly neither Voltaire nor d'Alembert hesitated to mock the literary character of scientific texts by Fontenelle and Buffon.

Institutions offer a promising avenue for investigating relations between style and the sciences. In 1666 Colbert established the Académie royale des sciences with a narrow scientifico-technical focus, but only after carefully considering Charles Perrault's proposal for a general academy that would have included sections on belles-lettres, history, philosophy, and mathematics. Colbert ultimately opted for what he hoped would be the more useful, politically expedient arrangement,[21] thus setting up a stark opposition between the Académie des sciences and the literary Académie française, between the 'académie des choses' and the 'académie des mots', as Voltaire liked to put it. How intimate belles-lettres and the sciences might have been in Perrault's academy is unclear. French provincial academies were comprehensive for the most part, including the arts and sciences in one institution. Empirical natural history set the tone for their scholarship, bringing the study of nature near other forms of history, especially antiquities. Scientific articles published in the provincial academies tended to be more literary than those in the *Mémoires* of the national Académie des sciences. Typically, an article on the physics of hail in the *Mémoires de l'Académie de Dijon* dwelt on the mythic resonances of experiments with the lightning rod: 'Quel fruit doit-on espérer des tentatives de ces nouveaux Prométhées? Peut-on ravir impunément le feu du ciel? Il est vrai qu'on n'a pas à craindre d'être attaché sur le Caucase, et d'y être dévoré par des vautours affamés; mais on a des maux moins imaginaires à redouter.'[22]

18. Gay, *Science of freedom*, p.156-62.

19. Jean Le Rond d'Alembert, 'Géomètre', in *Encyclopédie*, vii.627-29.

20. Voltaire, *Eléments de la philosophie de Newton*, in Voltaire 15 (1992), p.192; Voltaire to Pierre-Joseph Thoulier d'Olivet, 20 October 1738, D1631. But compare Voltaire to Claude-Nicolas Le Cat, 15 April 1741, D2463.

21. David J. Sturdy, *Science and social status: the members of the Académie des sciences, 1666-1750* (Woodbridge 1995), p.63-79; Hahn, *Anatomy of a scientific institution*, p.4-14.

22. Denis Barberet, 'Dissertation sur la nature et la formation de la grêle', *Mémoires de l'Académie de Dijon* 1 (1769), p.21.

But provincial academies fostered little scientific innovation, and their scientific importance declined after 1750.[23]

In the end the presence of literary style in the sciences must be gauged through the period's texts. Scientific vocabulary offers one telltale index. Correlated with dreams of literary, humanistic sciences were worries about scientific vocabulary throughout the Enlightenment. In the seventeenth century 'purists' intent on preserving polite society from regional, technical, and vulgar language promoted the use of a general, polite vocabulary modelled primarily on conversation in the royal court. This purism created tension as the sciences sought and won the attention of *honnêtes gens*. Not only were the sciences dependent on specialised terminology, their terminology was often redolent of the 'pedantic' languages Latin and Greek. The first three editions of the dictionary of the Académie française, published in 1694, 1718, and 1740 respectively, deliberately omitted scientific terminology, leaving it for an independent, separately published supplement.[24] Only in 1762, with its fourth edition, did the Académie française capitulate and implement a more liberal lexical policy. Other dictionaries had long included scientific vocabulary,[25] and the Académie's purism was regularly criticised from the late seventeenth century onward, most famously in Antoine Furetière's *Factums* (1685-1688) and François de La Mothe-Fénelon's *Lettre à l'Académie* (1714). But neoclassical restrictions on vocabulary still had strong support in word if not deed.[26]

Esoteric vocabulary in limited quantities was perhaps never incompatible with worldly success.[27] Even in the heyday of Boileau, Claude Perrault's *Mémoires pour servir à l'histoire naturelle* (1671-1676) was apparently very popular, perhaps thanks to its illustrations and sporadic moralising, despite its frequent recourse to specific, inelegant anatomical words: 'verge', 'thorax', 'méat', 'rate', 'pancréas', and so on.[28] The popularity of dictionaries of the arts and sciences throughout the Enlightenment testifies to a widespread interest in technical vocabulary. Early naturalists nonetheless hesitated to use 'barbarous' terminology and neologisms. Réaumur used technical language in

23. James E. McClellan III, *Science reorganized: scientific societies in the eighteenth century* (New York 1985), p.89-99, 133-37; Daniel Roche, 'Natural history in the academies', in *Cultures of natural history*, p.127-44.

24. The supplement in question, Thomas Corneille's *Dictionnaire des arts et des sciences* (1694), was nonetheless given volume numbers that made it seem a continuation of the *Dictionnaire de l'Académie*. See Jean-Jacques Thomas, *La Langue volée. Histoire intellectuelle de la formation de la langue française* (Bern 1989), p.78. On the partial return to this model in the early nineteenth century see Bernard Quemada, *Les Dictionnaires du français moderne, 1539-1863. Etude sur leur histoire, leurs types et leurs méthodes* (Paris 1967), p.176-77.

25. Quemada, *Dictionnaires du français moderne*, p.48-52, 58-59, 206-207. On the eighteenth-century situation see Seguin, *Langue française au XVIIIe siècle*, p.161-66.

26. W. von Wartburg, *Evolution et structure de la langue française*, 10th edn (Bern 1971), p.192.

27. See Drouin and Bensaude-Vincent, 'Nature for the people', p.414-16.

28. Sutton, *Science for a polite society*, p.123. Along with its considerable anatomical vocabulary, inelegant for its specificity and its evocation of the body's ignoble functions, the *Mémoires* featured lavish illustrations celebrating poetic myths which the text largely belied. See Harth, *Ideology and culture*, p.261-78.

his *Mémoires pour servir à l'histoire des insectes* (1734-1742), but he also used abstract periphrase to avoid coining new words or relying on esoteric ones such as 'abdomen', 'organisme', or 'envergure'. The idea of 'abdomen' he rendered at one point as 'le bout du derrière'. Similar combinations of generic anatomical and spatial language allowed him to do without specialised terms for many insect structures.[29] Buffon's assistant Louis-Jean-Marie Daubenton, by profession an anatomist and thereby linked to a group with a specialised language, also used periphrase from time to time, referring, for example, to 'parties naturelles', 'côtés du corps', and 'extrémités inférieures' instead of to genitals, flanks, or lower legs.[30] Rousseau, an admirer of Linnaean vocabulary in less 'vulgar' works on botany, denounced it as 'pedantic' in his *Lettres élémentaires sur la botanique* (1771-1774).[31] Anxiety about upsetting the *génie de la langue* persisted in the reform of chemical nomenclature late in the century – and chemistry, unlike natural history, had relatively little to gain from literati.[32]

Beyond vocabulary, style in general offers clues about relations between the sciences and polite culture. Erica Harth and Geoffrey Sutton have examined the literary properties of a number of variously scientific and para-scientific texts of the late seventeenth and early eighteenth centuries and arrived at somewhat different results. Both compare among other things two first-hand descriptions of chameleons, one by Madeleine de Scudéry published in her *Nouvelles conversations de morale* (1688) and the other by her friend Perrault for the *Mémoires pour servir à l'histoire des animaux*. Writing from opposite sides of the 'door' separating the Académie des sciences from educated culture at large, Perrault and Scudéry come to conflicting conclusions regarding the chameleon's diet, Scudéry agreeing with ancient fabulists that it lives on air alone, Perrault insisting that it lives on flies.[33] But their texts are also distinctive from a literary perspective. Sutton correctly points out that Perrault enlivens his account with moralising, allegorical passages similar to those in Scudéry's version,[34] but these are in fact rare in a text overflowing with anatomical vocabulary and simple descriptive sentences. The science of Perrault's *Mémoires*

29. Max Fuchs, 'La langue des sciences', in *Histoire de la langue française des origines à nos jours*, ed. Ferdinand Brunot (Paris 1966), vol. vi, pt 1, fasc. 2, p.560-63, 572-81. Buffon, while susceptible to the same kinds of periphrase, did use the word 'abdomen', refused by Réaumur. See Ernst Höhne, *Der Stil Buffons: eine stilistische Untersuchung der 'Histoire naturelle, générale et particulière'* (Marburg 1914), p.75.

30. See for example *HN*, iii.203-205, ix.174. On the word 'génital' see Fuchs, 'Langue des sciences', p.563n. See also Corsi, *Age of Lamarck*, p.31.

31. Duris, *Linné et la France*, p.104-105.

32. Fuchs, 'Langue des sciences', p.641-75. See also the remarks on vocabulary in English 'polite science' in Alice N. Walters, 'Conversation pieces: science and politeness in eighteenth-century England', *History of science* 35 (1997), p.127-28.

33. Erica Harth, *Cartesian women: versions and subversions of rational discourse in the Old Regime* (Ithaca, N.Y. 1992), p.103-105. On the role of chameleons in early modern natural history see Finden, *Possessing nature*, p.299-302.

34. *Science for a polite society*, p.124-26. See also Harth, *Ideology and culture*, p.261-78.

may not be cut off from style and polite culture, but its points of attachment are limited.

Harth and Sutton pursue their analyses with regard to Fontenelle and the later publications of the Académie des sciences. From 1702 to 1797 the Académie published two jointly bound periodicals, the *Mémoires de mathématiques et de physique* containing original papers by academicians, and the *Histoire de l'Académie royale des sciences* containing summaries of academic research, book reviews, untheorised, sometimes fantastic empirical 'observations', and eulogies of dead members, all written or compiled by the secretary – that is, by Fontenelle for the issues dated 1699 to 1740, then by Mairan (1741-1743), Jean-Paul Grandjean de Fouchy (1744-1776), and Condorcet (1773-1790).[35] The *Histoire* was designed as a work of vulgarisation. As Fontenelle wrote in the opening volume, it was meant for those with only a 'médiocre teinture de mathématique et de physique'.[36] The success of Fontenelle's *Entretiens sur la pluralité des mondes* (1686), a 'vulgar' astronomy, made him an ideal choice for introducing the academic sciences to a broad readership. The *Histoire de l'Académie* was said to have become fashionable thanks to his efforts.[37]

As Harth observes, Fontenelle's *Histoire* is quite different from his *Entretiens*. In the *Entretiens*, a work that highlights its accessibility to women readers, the gallant narrator emphasises pleasure and humanistic moralising in his lessons on Cartesian astronomy to the curious marquise. In the *Histoire*, in contrast, Fontenelle stresses utility over pleasure and moralising.[38] Replete with metaphors, digressions, allusions, and personification, the *Entretiens* are in general more literary than the *Histoire*. Their 'floweriness' would become legendary, and thus exaggerated, in the course of the eighteenth century. Under Fontenelle's direction the *Histoire*, for its part, was not always serious or stylistically neutral. On occasion, Sutton demonstrates, it could rival the *Entretiens* with its humour and playfulness.[39] Nor did such literariness vanish from the *Histoire* with Fontenelle's departure after the 1740 volume. His successors Mairan and Grandjean de Fouchy were no less assiduous in speckling the *Histoire* with high-style moralising, as in the following: 'Les muses des mathématiques ne méritent jamais tant l'encens et le respect des hommes, que quand elles daignent sortir en quelque sorte de leur sanctuaire.'[40] Even Condorcet elevated his language for academic eulogies. But regardless of its author, the *Histoire* never attained the sustained literariness of Fontenelle's *Entretiens*.

35. On the *Histoire* and *Mémoires* see Anne-Sylvie Guénon, 'Le XVIIIe siècle', in *Histoire et mémoire de l'Académie*, p.113-15. For the dates of appointment of the various secretaries see Brian, 'L'Académie royale des sciences', p.19; Sturdy, *Science and social status*, p.352-53.

36. *HARS* 1699 (1702), p.à1*v* ('Préface').

37. *HARS* 1757 (1762), p.192-93.

38. Harth, *Cartesian women*, p.143-44. See also Leonard M. Marsak, *Bernard de Fontenelle: the idea of science in the French Enlightenment* (Philadelphia, Penn. 1959), p.5, 45.

39. *Science for a polite society*, p.164-68. See also *HARS* 1757 (1762), p.192. As suggested below, Sutton's claim that the *Histoire* was 'something like a serialized version' of the *Entretiens* seems overstated.

40. *HARS* 1746 (1751), p.87.

The comparison can be extended to the Académie's *Mémoires*. Here there are traces of literary language: occasional exclamations and allusive digressions, even sublime vocabulary for describing natural catastrophes such as earthquakes and epidemics.[41] Buffon's 1747 paper on burning mirrors, published in the *Mémoires*, foreshadows his later elevation with its elegant exordium.[42] But the *Mémoires*, overall, are less literary than the *Histoire*. Here at the heart of the Académie des sciences, style had little function and risked inviting ridicule or perturbing implicit strictures against stylistic extravagance.[43] Lower on the 'scale' of vulgarisation, stylistic liberties were increasingly permitted and increasingly functional as the sciences reached out toward less professional, more literary readers.

Not only did scientific style vary as a function of vulgarisation, it also varied diachronically as a function of patronage. As Christian Licoppe has shown, changing patterns of patronage had a considerable effect on seventeenth- and eighteenth-century scientific rhetoric. In the second half of the seventeenth century, natural philosophers emphasised surprises and marvels in their presentation of experiments to arouse the curiosity of their well-to-do patrons. From about 1720 to 1760 natural philosophers stressed utility over marvels, dependent as now they were on the largesse of the state. Finally, in the late eighteenth century scientists emphasised their experiments' precision for the benefit of colleagues, a sign that the scientific community was now secure in society.[44] While silent on the question of scientific style, Licoppe's periodisation suggests that styles of scientific writing should have grown dryer and less literary in the course of the Enlightenment. This is true in some contexts, certainly in the *Histoire de l'Académie* and probably in the Académie's other publications. On the other hand the 'culture of curiosity' which Licoppe associates with the patronised sciences of the late seventeenth century overlapped his later cultures of utility and precision and continued on a massive scale throughout the eighteenth century. Along with marvels and surprises, style was a way of appealing to such curiosity. It was neither appreciated by all readers nor cultivated by every successful scientific vulgariser. Réaumur's *Mémoires pour servir à l'histoire des insectes*, modelled on the *Mémoires* of the Académie des sciences, captured the public's attention with a great many marvels

41. See for example Guillaume-François Rouelle, 'Sur le sel marin (première partie)', *MARS* 1745 (1749), p.74; Jean-Jacques Dortous de Mairan, 'Discours sur la propagation du son dans les différents tons qui le modifient', *MARS* 1737 (1740), p.17; René-Antoine Ferchault de Réaumur, 'Sur la manière de conserver les œufs', *MARS* 1735 (1738), p.465; Jean-Etienne Guettard, 'Mémoire sur quelques montagnes de la France qui ont été des volcans', *MARS* 1752 (1756), p.27-28; Jean-Etienne Guettard, 'Mémoire sur les causes qui peuvent rendre les chevaux poussifs', *MARS* 1745 (1749), p.91; Charles-Marie de La Condamine, 'Mémoire sur l'inoculation de la petite vérole', *MARS* 1754 (1759), p.615.

42. 'Invention de miroirs ardents pour brûler à une grande distance', *MARS* 1747 (1752), p.82.

43. See Roger, *Sciences de la vie*, p.173-74, 179.

44. Christian Licoppe, *La Formation de la pratique scientifique. Le discours de l'expérience en France et en Angleterre (1630-1820)* (Paris 1996).

and a style little remarkable beyond its diffuseness.[45] But it is probably no accident that the eighteenth century's greatest scientific best-sellers, Buffon's *Histoire naturelle* and Pluche's *Spectacle de la nature*, were patently literary – Buffon with his grandeurs and Pluche with his dialogues.[46]

In this sense the eighteenth century's jibes against Fontenelle's *Entretiens* require explanation, especially insofar as they often emanated from vulgarisers. In the *Rêve de d'Alembert* Diderot has 'Théophile de Bordeu' counter 'Julie de Lespinasse''s inclination for philosophy *à la* Fontenelle with a typical devaluation of the *Entretiens*: 'Franchement, je ne sais si ce ton frivole convient aux sujets graves.'[47] Some authors matched repudiations of the *Entretiens* with minimally stylistic vulgarisations – for example, Voltaire with his *Eléments de la philosophie de Newton*, Du Châtelet with her *Institutions de physique*, and Joseph-Jérôme Le François de Lalande with his *Astronomie des dames* (1786).[48] Many more authors, Diderot among them, merely substituted one sort of stylishness for another. In 1773 François Rozier's *Journal*, originally entitled *Observations et mémoires sur la physique, sur l'histoire naturelle, et sur les arts et métiers*, pronounced Fontenelle's style to be out of date and noted that nowadays the secretary of the Académie had a more difficult task: '[Il s'agit] d'instruire, d'intéresser et de plaire sans emprunter des ornements étrangers ou même recherchés, sans recourir à l'antithèse, ni à l'épigramme, ni à aucune des parures trop artificielles du style.'[49] Yet the scientific articles in Rozier's *Journal* were often considerably more stylish than those in the *Mémoires* of the Académie des sciences. Six months after dismissing Fontenelle, for example, the *Journal* printed an anonymous article on mountains exalting Buffon's style and featuring a

45. On Réaumur's style see Roger, intro. to Buffon, *Epoques*, p.cxxn.; Roger, *Sciences de la vie*, p.772; Wheeler, intro. to Réaumur, *Natural history of ants*, p.xv-xvi; *Réaumur's memoirs on steel and iron*, ed. Cyril Stanley Smith, translated by Anneliese Grünhaldt Sisco (Chicago 1956), p.xv-xvi; Albrecht von Haller to Charles Bonnet, 25 May 1771, in *Correspondence Haller-Bonnet*, p.939.

46. On Pluche's style see Dennis Trinkle, 'Noël-Antoine Pluche's *Le Spectacle de la nature*: an encyclopedic best-seller', *SVEC* 358 (1997), p.130-31. On dialogue in natural history see Spary, '"Nature" of Enlightenment', p.284-85. As Schiebinger notes, Pluche abandoned dialogue halfway through the *Spectacle*, just as Formey abandoned dialogue halfway through his vulgarisation of Christian Wolff's philosophy, *La Belle Wolfienne*. But Schiebinger's argument that 'as the century wore on, even science written specifically for women was stripped of poetic ornaments' is questionable. Vulgarisers may have reacted against Fontenelle's ostentation, but few chose to forego stylistic ornament in general. See Schiebinger, *Mind has no sex*, p.238-40; Aileen Douglas, 'Popular science and the representation of women: Fontenelle and after', *Eighteenth-century life* 18 (1994), p.7.

47. Denis Diderot, *Le Rêve de d'Alembert*, in *DPV*, xvii.133. For other critiques of Fontenelle see Harth, *Cartesian women*, p.148; Fuchs, 'Langue des sciences', p.562; Mornet, *Sciences de la nature*, p.196-97.

48. On these authors see Terrall, 'Emilie Du Châtelet', p.292-94; Alain Niderst, *Fontenelle* (Paris 1991), p.343-47; Schiebinger, *Mind has no sex*, p.240.

49. *Observations et mémoires sur la physique, sur l'histoire naturelle, et sur les arts et métiers* 1 (March 1773), p.232. A similar prescription for vulgarisers can be found in *Mercure de France* (November 1744), p.2-3.

Latin epigraph, noble vocabulary, periodic sentences, anaphora, hyperbole, and other stylistic flourishes.[50]

In all fairness the style of the *Entretiens* was in some respects more 'ornamental' than those of the vulgarisers who later condemned it. Artifice and affectation were keys to Fontenelle's identity as a literary Modern, and his scepticism about classical and natural unity made his style more a pleasant camouflage than a meaningful ensemble.[51] But some of his stylistic traits, notably his comparisons, were essential to his project of vulgarisation. Major elements of his style – his explicative metaphors, his playful use of dialogue, and his personifications, for example – remained staples of scientific vulgarisation long after the *Entretiens* became an object of mockery.[52] Clearly eighteenth-century scorn for the *Entretiens* was to some degree rhetorical. Rejecting Fontenelle's precedent became a convenient way for authors to distance themselves from more 'vulgar' writing in a period in which nearly everyone lamented the overproduction of books and the superficiality of certain amateurs. Pluche and Buffon would later play the same role as scapegoats for the world of vulgarisation.

After Buffon's death Rozier's *Journal*, now owned and edited by Jean-Claude Delaméthrie, became an important weapon in the hands of Buffon's disciples. True to its origins – and despite its rejection of Fontenelle's extravagance – the *Journal* continued to promote the cause of natural history as a literary enterprise accessible to amateurs as well as professionals. Cuvier and fellow anti-Buffonians, in contrast, campaigned for a 'severe' scientific style and for the more closed scientific community this would imply. By the early nineteenth century they had taken control of the Institut and official French natural history in general, yet new editions of Buffon's works and Rozier's *Journal* continued to outsell Cuvier's writings.[53] In tune with public taste, vulgarising stylists thrived on science's margins during the nineteenth century. Even in the mainstream of the scientific community Cuvier's victory was not complete, for the same debates about style resurfaced in the late nineteenth century, with Ernst Renan cautioning French scientists to abandon style or be left behind by German scientists.[54] Even today French scientific articles are often stylish in a way Anglophone articles are not – a pale reflection of

50. 'Essais sur l'étude des montagnes', *Observations et mémoires sur la physique, sur l'histoire naturelle, et sur les arts et métiers* 2 (November 1773), p.416-32.

51. See Roger, *Sciences de la vie*, p.773-74; Niderst, *Fontenelle*, p.24-27, 286.

52. France, *Rhetoric and truth in France*, p.76-77. On the English situation see Douglas, 'Popular science', p.7.

53. Corsi, *Age of Lamarck*, p.1-39; Pietro Corsi, 'Buffon sous la Révolution et l'Empire', in *Buffon 88*, p.639-48. Not coincidentally, Rozier's *Journal* also took the lead in combating the new chemistry. See Bernadette Bensaude-Vincent, *Lavoisier. Mémoires d'une révolution* (Paris 1993), p.257-60; Corsi, *Age of Lamarck*, p.34.

54. Wolf Lepenies, *Das Ende der Naturgeschichte: Wandel kultureller Selbstverständlichkeiten in den Wissenschaften des 18. und 19. Jahrhunderts* (Munich 1976), p.136-38, 140-41.

the French Enlightenment's rumoured weakness for verbal effect. As Charles Bonnet put it in 1761, 'Le Français sera toujours prenable par des oreilles.'[55]

All in all it is difficult to assign a date to the 'separation' of style from the sciences. Style was at issue in the new sciences from their inception, just as rhetoric had been at issue in philosophy sporadically since antiquity. The new sciences, while castigating rhetoric, gave rise to subtle new rhetorical forms, one ideal of which was plain style. Compared to works popular in polite society, the official publications of the Académie des sciences were on the whole plain, and they grew slightly plainer in the course of the eighteenth century. Yet the publications of the Académie, especially the *Histoire*, continued to harbour stylish passages, a sign of ongoing contact with the public at large. Some in the scientific academies spoke of good style as an asset, but testimony on the subject was contradictory. In any case academic strictures against extravagant style were bound to be broken by authors trying to capitalise on public interest in the sciences. Here in the much ridiculed area of vulgarisation, styles varied widely, but many authors sensed the need for something distinctively literary to match the literary upbringing of the reading public.

Of all the sciences, natural history was perhaps best suited for vulgarisation. Many of its usual subjects – animals, flowers, precious stones, and so on – came with symbolic and literary prehistories which helped guide their packaging for polite audiences. In particular the natural world had long been exploited for allegory, and the theological orientation of much early natural history created a precedent for high-style presentation. Eager to describe exotica and academic novelties, eighteenth-century natural histories were lexically permissive with respect to the standards of contemporary grammarians, but many authors still hesitated before resorting to neologism or specialised language. In its more popular forms natural history learned stylistic lessons from successful vulgarisations in other areas, notably from Fontenelle's *Entretiens*, though few would avow it. When it was stylish, natural history, like other sciences, aroused ire and derision inside and outside the Académie des sciences. Such reactions were all the more vehement in that style in the sciences was emblematic of a culture of curiosity that went along with it.

Buffon's style and its critics

On 25 August 1753 Buffon delivered a speech before the Académie française to solemnise his induction. The 'Discours sur le style', as it soon came to be known, won immediate acclaim and became a canonical presence in the

55. Bonnet to Haller, 31 January 1761, in *Correspondence Haller-Bonnet*, p.231. For a similar citation see the epigraph in France, *Rhetoric and truth in France*, p.68. On Bonnet's own stylishness see *The Private memoirs of Madame Roland*, ed. Edward Gilpin Johnson (New York 1976), p.169. On French scientists' current stylishness see Montgomery, *Scientific voice*, p.299-301.

French educational system.[56] Though inconsistent on this matter and others, the 'Discours' anticipates modern views of style by presenting it as inseparable from thought, more a matter of *inventio* and *dispositio* than simple ornament.[57] Its most famous line, 'le style est l'homme même', has been interpreted as anticipating the Romantic notion of 'personal' style, but Buffon was probably thinking of style as reflecting a person's rational distinctiveness rather than biographical or psychological quirks.[58] With regard to the word 'style' in Buffon's famous aphorism, the context suggests a meaning midway between *elocutio* and more cognitive qualities. Yet even in his own time Buffon's contentions that 'le style est l'homme même' and 'les ouvrages bien écrits seront les seuls qui passeront à la postérité' (*HNS*, iv.10) were construed as asserting that style outweighs substance.[59] With the provocative formulae of the 'Discours sur le style' – consistently one of his better-known texts – Buffon paved the way for his dismissal as a mere stylist.

Buffon's interest in style did not begin with the 'Discours'. Much of his youthful correspondence was stilted and precious – qualities perhaps encouraged by his Jesuit education and his contacts with the learned, humanistic elite of 'eloquent' Dijon.[60] His early scientific work was stylistically unremarkable. As mentioned above, a trace of his now famous style can be found in a 1747 paper on burning mirrors, but this was not published until 1752. Appreciation for his style first came in 1749 with the publication of the first three volumes of the *Histoire naturelle*. Many reviewers remarked on his style at the time, variously describing it as noble, beautiful, elevated, forceful, clear, and elegant. The *Journal de Trévoux* pronounced Buffon a model for writers.[61] Sublime *tableaux* and *morceaux* were singled out and reprinted, in part for their style. The journalist Elie-Catherine Fréron, for example, praised the 'poetry' of

56. On the 'Discours' in general see Fellows and Milliken, *Buffon*, p.149-54; Gómez Torres, *Ideas de Buffon sobre retórica*; Charles Bruneau, 'Buffon et le problème de la forme', in *Œuvres philosophiques de Buffon*, p.491-99.

57. Gómez Torres, *Ideas de Buffon sobre retórica*, p.19-37.

58. Gómez Torres, *Ideas de Buffon sobre retórica*, p.37-59; Bruneau, 'Buffon', p.495-97. For more conservative estimates of the amount of individuality in the *homme* of Buffon's maxim see Fellows and Milliken, *Buffon*, p.153; Höhne, *Stil Buffons*, p.19-22; Emile Krantz, *Essai sur l'esthétique de Descartes. Rapports de la doctrine cartésienne avec la littérature classique française au XVIIe siècle*, 2nd edn (New York 1974), p.355-59.

59. See for example Bonnet to Haller, 31 January 1761, in *Correspondence Haller-Bonnet*, p.231; Louis-Sébastien Mercier, *Tableau de Paris, critiqué par un solitaire du pied des Alpes*, 6 vols (Nyon, Natthey 1783), ii.58. For a more cognitive interpretation of Buffon's dictum see *Mémoires inédits de Mme la comtesse de Genlis, sur le dix-huitième siècle et la Révolution française*, 10 vols (Paris 1825), viii.172. For an argument that Diderot actualised Buffon's prophecy about stylistic immortality see Aram Vartanian, 'Buffon and Diderot', in *Buffon 88*, p.131-32.

60. Hanks, *Buffon avant l'"Histoire naturelle'*, p.16, 19-20.

61. *Journal de Trévoux* (March 1750), p.595-96. For later development of the theme see H. Nadault de Buffon (ed.), *Correspondance générale [de] Buffon*, 2 vols (Geneva 1971), i.92n.; *Correspondance littéraire, philosophique et critique de Grimm et de Diderot*, 16 vols (Paris 1829), i.64.

Buffon's account of a newly created adult man's first day of existence. Fréron quoted this passage where Buffon's 'Adam' describes his discovery of 'Eve':

Je portai ma main sur ce nouvel être, quel saisissement! ce n'était pas moi, mais c'était plus que moi, mieux que moi. [...] Je la sentis s'animer sous ma main, je la vis prendre de la pensée dans mes yeux, les siens firent couler dans mes veines une nouvelle source de vie, j'aurais voulu lui donner tout mon être; cette volonté vive acheva mon existence, je sentis naître un sixième sens. Dans cet instant l'astre du jour sur la fin de sa course éteignit son flambeau.[62]

But comments on Buffon's style were relatively muted before 1753. Only with his admission to the Académie française and the publication of volume iv of the *Histoire naturelle* did Buffon's style begin to dominate reviews and assume mythic proportions. The *Journal des savants*, which did not mention his style in its long reviews of the opening volumes, praised it eight times in its reviews of volume iv. Buffon, said the *Journal*, wrote with 'une force et une majesté de style qui n'appartiennent qu'à lui'.[63] At the same time it became a cliché among critics to see Buffon's style as an *instrument dangereux* capable of misleading the gullible.[64] Though critical attention to the *Histoire naturelle* diminished after the 1750s, Buffon remained fixed for posterity in these stylistic stereotypes of 1753-1754.

Buffon's glorification of style in the 'Discours sur le style' was probably important in establishing his image as a stylist, but so too was his actual manner of writing in and after volume iv.[65] For in that volume, undoubtedly written before his nomination to the Académie française, Buffon began doing 'portraits' of individual animals, many of them opening with elegant high style. He did not simply switch from nondescript style to grand eloquence as Ernst Höhne has claimed.[66] As we have seen, the first three volumes were showy enough to win praise for their style, and reviewers were already quoting Buffon's *morceaux sublimes* et *magnifiques tableaux* before 1753. More importantly, as generations of specialists have pointed out in vain, most of the *Histoire naturelle* is written very plainly – at least three-quarters of the series in Jacques Roger's estimate.[67] Even the *Histoire des quadrupèdes* and the *Histoire des oiseaux* are for the most part stylistically unremarkable. What

62. See Fréron, *Lettres sur quelques écrits de ce temps* (15 December 1750), p.82-84; *HN*, iii.370.

63. July 1754, p.480; see also June 1754, p.329-39.

64. See the discussion of seduction below p.47-48. The cliché is anticipated in pre-1753 critiques by Malesherbes and Lignac. See Malesherbes, *Observations sur l'histoire naturelle*, i.221-22; [Joseph-Adrien Lelarge de Lignac], *Lettres à un Américain sur l'histoire naturelle, générale et particulière de M. de Buffon* (Hamburg 1751), pt 1, letter 1, p.7-8, pt 3, letter VII, p.10-11.

65. The reviewer is plainly conscious of Buffon's promotion to the Académie française in *Journal de Trévoux* (December 1753), p.2835-36. See also Roger, intro. to Buffon, *Epoques*, p.cxiv-cxv; Höhne, *Stil Buffons*, p.18. On the other hand the *Journal des savants* makes no mention of Buffon's style when belatedly reviewing volumes ii and iii in 1754. See *Journal des savants* (March 1754), p.148-59. Buffon later claimed that he only began to write and be noticed as an 'author' in 1751 or 1752. See *Mémoires inédits de Mme de Genlis*, ii.278-79.

66. Höhne, *Stil Buffons*, p.17-18, 26-27, 66-67.

67. *Buffon*, p.562.

changed in volume iv of the *Histoire naturelle* was not a whole style but rather the density and emphasis of an already established, elevated sub-style.

Buffon's 'style', conceived as unitary, has always been a myth, albeit an important and commercially useful one. It has been the project of critics since the 1750s to bolster the myth by sifting through the series and picking out set pieces for literati. Literary abridgements of the *Histoire naturelle* did not appear in massive numbers until the early nineteenth century,[68] but thanks to critics and word of mouth, eighteenth-century readers were well aware of the 'good parts' in the *Histoire naturelle.* Buffon himself was a willing accomplice in advertising his style. Not only did he define himself as a stylist with the 'Discours sur le style', he seems to have trumpeted if not exaggerated his stylistic standards.[69] His thematic index to the *Histoire des quadrupèdes* quotes and refers to all the series' stylistic high points: 'Il n'appartient qu'à M. de Buffon de tirer un aussi grand parti d'un ouvrage aussi peu fait pour plaire, qu'une table de matières.'[70] After the *Histoire des quadrupèdes* he corrected his collaborators' work to further their efforts to mimic his style.[71] High style became more and more a self-conscious project. Before launching a *tableau* of South American lowlands in the *Histoire des oiseaux*, for example, Buffon recalled and gave the location of an earlier *tableau*, his extraordinarily successful description of the desert in the article 'Chameau', thus priming his readers for another piece of bravura. The passage illustrates both his high style and his unabashed self-promotion:

Nous avons ci-devant peint les déserts arides de l'Arabie pétrée; ces solitudes nues où l'homme n'a jamais respiré sous l'ombrage, où la terre sans verdure n'offre aucune subsistance aux animaux, aux oiseaux, aux insects, où tout paraît mort, parce que rien ne peut naître. [...] Opposons ce tableau de sécheresse absolue dans une terre trop ancienne, à celui des vastes plaines de fange des savanes noyées du nouveau continent. [...] Des eaux stagnantes et répandues près et loin de leurs cours, couvrent le limon vaseux qu'elles ont déposé. [...] Et ces plages, alternativement sèches et noyées où la terre et l'eau semblent se disputer des possessions illimitées; et ces brossailles de mangles jetées sur les confins indécis de ces deux éléments, ne sont peuplés que d'animaux immondes

68. For a partial bibliography see Genet-Varcin and Roger, 'Bibliographie', p.527-29. Several editions of *Génie de M. de Buffon* were published in Buffon's lifetime, and encyclopedias, dictionaries, and other compilations regularly plundered the *Histoire naturelle* for its 'fine writing'. See Gunnar Broberg, 'The broken circle', in *The Quantifying spirit in the eighteenth century*, ed. Tore Frängsmyr, J. L. Heilbron and Robin E. Rider (Berkeley, Calif. 1990), p.68.

69. See Gómez Torres, *Ideas de Buffon sobre retórica*, p.14; Marie-Jean Hérault de Séchelles, 'Journey to Montbard', translated by John Lyon, in *From natural history to the history of nature: readings from Buffon and his critics*, ed. John Lyon and Phillip R. Sloan (Notre Dame, Ind. 1981), p.362-64, 369-73; Flourens, *Histoire*, p.1-2n.; Fuchs, 'Langue des sciences', p.587n. For testimony about Buffon's 'stylism' not necessarily encouraged by Buffon himself see Gómez Torres, *Ideas de Buffon sobre retórica*, p.88-89; *Mémoires de l'abbé Morellet de l'Académie française sur le dix-huitième siècle et sur la Révolution*, ed. Jean-Pierre Guicciardi (Paris 1988), p.126-28.

70. *Journal encyclopédique* 5.3 (August 1767), p.34-35. The index can be found in *HN*, xv.xlv-cccxxiv.

71. See Roger, *Buffon*, p.501-502; Pierre Flourens, *Des manuscrits de Buffon avec des fac-simile de Buffon et de ses collaborateurs* (Paris 1860); Nadault (ed.), *Correspondance générale [de] Buffon*, i.291n.; Höhne, *Stil Buffons*, p.10-13, 24-26.

qui pullulent dans ces repaires, cloaques de la nature, où tout retrace l'image des déjections monstrueuses de l'antique limon.[72]

Buffon's style has been described many times, mostly in its sporadic high-style form. As we have seen, eighteenth-century descriptions were typically vague, mentioning above all Buffon's elevation, evident in the vocabulary as well as the tone and rhythm of the high-style discourses. His enemies tended to turn elevation into amplification and stress Buffon's exaggeration and repetition, both typical fare in the *Histoire naturelle*. From Antoine Rivarol onward critics have dwelt on Buffon's ecclesiastical tone and compared him to the bishop and orator Jacques-Bénigne Bossuet.[73] Buffon's phrases and sentences have been lauded for their rhythm and symmetry. Among them Otis Fellows and Stephen Milliken have found 'some of the finest examples of classic Ciceronian periods' in modern times.[74] Buffon's favour for periods matches his preference for 'linear' prose in general, his ideas advancing sequentially under the direction of connecting words rather than piling up in cognitively difficult suspensions and hypotaxes.[75] More common than true Ciceronian periods are Buffon's carefully drawn antitheses and parallel, balanced, two- and three-member structures.[76] Lastly many commentators have evoked traces of oral style in the *Histoire naturelle*. These range from attention to sonority to Buffon's use of pronouns, verbs of saying, and rhetorical questions to create implicit dialogue.[77]

Jacques Roger has identified Buffon's style as old-fashioned for his time, particularly in the orally delivered preamble to the *Epoques*.[78] Certainly Buffon's intermittent use of formal periods and sublime generalities recalls not so much Voltaire, Montesquieu, or Diderot as the poets and grammarians of Louis XIV – whence the comparisons to Bossuet. But modernity and novelty can

72. *HNO*, vii.335-38. In theory Buffon rejected the analogy *ut pictura poesis*, but here and elsewhere he was happy to capitalise on his popular image as a 'painter' of nature. See Gómez Torres, *Ideas de Buffon sobre retórica*, p.72-76.

73. Fellows and Milliken, *Buffon*, p.157. Emile Faguet reproached Buffon for his supposed admiration for the prose of another predicant, Jean-Baptiste Massillon. See Höhne, *Stil Buffons*, p.28.

74. *Buffon*, p.155-56. See also Roger, intro. to Buffon, *Epoques*, p.cxv-cxvii, cxxv-cxxvi.

75. See Roger, intro. to Buffon, *Epoques*, p.cxviii-cxix; Hanks, *Buffon avant l'‘Histoire naturelle'*, p.229-30; Julie C. Hayes, *Reading the French Enlightenment: system and subversion* (Cambridge 1999), p.34-35; Jeff Loveland, 'Rhetoric and science in Buffon's natural history', Doctoral dissertation, Duke University, 1994, p.72. On Buffon's early use of connecting words see Hanks, *Buffon avant l'‘Histoire naturelle'*, p.85. Notice also Buffon's treatment of the long suspended sentences proposed by Bexon in Flourens, *Des manuscrits de Buffon*, p.31-32, 29-30, 148-51, 156-57.

76. See Höhne, *Stil Buffons*, p.33-39; Hayes, *Reading the French Enlightenment*, p.35-36. Buffon regularly sharpened and clarified Bexon's muddled antitheses. Compare for example Buffon's and Bexon's versions of the articles 'Aras' and 'Pélican' in Flourens, *Des manuscrits de Buffon*, p.140-41, 158-61.

77. See Roger, *Buffon*, p.564; Lepenies, *Ende der Naturgeschichte*, p.144-45n.; Hayes, *Reading the French Enlightenment*, p.34-35.

78. Roger, intro. to Buffon, *Epoques*, p.cxv-cxviii, cxxvii; Roger, *Buffon*, p.23. In contrast Roger considers Buffon's account of his newly awakened 'Adam' (*HN*, iii.364-70) a masterpiece of 'Louis XV' style. See Roger, *Buffon*, p.222. See also Bruneau, 'Buffon', p.495.

be found in Buffon's style as well. Malesherbes rebuked him for indulging in fashionably glib *style coupé* rather than the more deliberate language of traditional philosophy.[79] On a brighter note François-René de Chateaubriand wrote that 'Rousseau, et même un peu Buffon, dans un autre genre, ont créé une langue qui fut ignorée du grand siècle.'[80] In the end Buffon no doubt mixed innovation with tradition. With his sobriety, 'nobility', and emphasis on balance he could be seen as a holdover from neoclassical France. Especially in the provinces, where tastes ran toward the conservative and the rhetorical, this may have been an element of his attractiveness. But his style was also new. Not only was it lighter and less hypotactical than the *style soutenu* of Bossuet and his contemporaries, it created elevation outside the traditionally noble genres of neoclassical aesthetics.

As we have seen, Buffon's 'style' was only one of his styles, even in the *Histoire des quadrupèdes* and the *Histoire des oiseaux*. It became less and less his style toward the end of his life, though in a sometimes exaggerated form it remained the semi-official style of the series, thanks to the efforts of Guéneau and Bexon in the *Histoire des oiseaux* and Lacépède in his continuations of the *Histoire naturelle*. Anthologies of Buffon's 'masterpieces' from the early nineteenth century offer a useful index of the series' stylistic evolution. More than half the *morceaux choisis* typically come from descriptions of individual birds and quadrupeds in the *Histoire des quadrupèdes* and the *Histoire des oiseaux*. General discourses from the opening volumes and the sub-series on birds and quadrupeds account for most of the remaining 'masterpieces'. The *Suppléments* and above all the *Histoire des minéraux* are markedly under-represented, though both are almost always cited for a small number of passages. Regarding this stylistic ebb of the 1770s and 1780s Fellows and Milliken have argued that Buffon gave up on style once his literary reputation was established in order to ingratiate an increasingly professional scientific community.[81] This may be true in part, but the lack of style in Buffon's later works probably has something to do with his subjects as well. Geology and minerals, with few human qualities and little history in literature, are inherently harder to stylise than animals.

Within this stylistic evolution over series and decades, regular variations in stylistic level can be seen in volumes and articles. Buffon liked to begin volumes with gusto. 'Je fais cet arrangement dans la vue de commencer le neuvième volume par le bel article du cygne', he explained to Bexon in 1779.[82] Some volumes open with stylistic mediocrities, but many more feature high style

79. *Observations sur l'histoire naturelle*, i.188. On *style coupé* see Seguin, *Langue française au XVIIIe siècle*, p.143-48. Buffon himself expressed scorn for the 'asthmatic' quickness of *style coupé*. See Loveland, 'Rhetoric and science', p.70-75.

80. François-René de Chateaubriand, *Mémoires d'outre-tombe*, ed. Maurice Levaillant, 4 vols (Paris 1948), iii.279.

81. *Buffon*, p.162-68. On the style of the *Minéraux* see also Roger, *Buffon*, p.524.

82. Buffon to Gabriel-Léopold-Charles-Aimé Bexon, 24 December 1779, in Nadault (ed.), *Correspondance générale [de] Buffon*, i.447.

for the first several pages. Furthermore many articles begin with at least one stylish paragraph. Here, in his grandiose preambles, Buffon expresses general views of nature.[83] Later on in most articles, *grandes vues* give way to empirical detail and specific terminology, and the style deflates. The article 'Blaireau' exhibits a typical degeneration, both in its style and in generality. It begins with a poetic summary: 'Le blaireau est un animal paresseux, défiant, solitaire, qui se retire dans les lieux les plus écartés, dans les bois les plus sombres, et s'y creuse une demeure souterraine; il semble fuir la société, même la lumière, et passe les trois quarts de sa vie dans ce séjour ténébreux, dont il ne sort que pour chercher sa subsistance' (*HN*, vii.104). The vocabulary alone bespeaks elevation. All-purpose words such as 'demeure', 'séjour', and 'subsistance' could easily have been replaced with more specific ones. Likewise Buffon's superlatives ('les lieux les plus écartés' and 'les bois les plus sombres') seem to have been written more for their sublimity and ample ornateness than for any particular denotative function. Yet within a few pages the narration has lost nearly every hint of grandeur: 'Elle [l'espèce du blaireau] n'est point en Afrique; car l'animal du cap de Bonne-espérance décrit par Kolbe sous le nom de blaireau puant, est un animal différent; et nous doutons que la *fossa* de Madagascar, dont parle Flaccourt dans sa relation, *page 152*, et qu'il dit ressembler au blaireau de France, soit en effet un blaireau' (*HN*, vii.109).

Many other articles follow 'Blaireau' in their stylistic fall from grand openings to details, though some demonstrate a similar stylistic range without a high-style leader. The fact that high style appears in so many contexts, notably portraits of animals, justifies the vagueness of Buffon's prescription of its uses in the 'Discours sur le style'. As we have seen, the 'Discours' hesitates between making style a consequence of decorum and making it follow from personal perspective: 'Si l'on s'est élevé aux idées les plus générales, et si l'objet en lui-même est grand, le ton paraîtra s'élever à la même hauteur' (*HNS*, iv.10). In fact Buffon's high style correlates less with 'grandiose' subjects than with a certain generality of viewpoint and a tendency toward broad philosophical speculation. These aspects of Buffon's thought, well-known to his contemporaries, will be treated in chapter 4. For now it is sufficient to note that a penchant for *grandes vues* and casual 'system-building' marks much of the *Histoire naturelle* and regularly correlates with stylistic elevation.

Many eighteenth-century critics of Buffon's style were primarily concerned with these epistemological correlates, not style itself. Buffon's style was often deplored in the most general terms and in the same breath as his creativity and fertile imagination – a sign that many critics were less troubled by any specifically linguistic properties of the *Histoire naturelle* than by the link they perceived between 'systems' and 'poetry'. The complaint that Buffon 'dazzled' the public with 'magical representation' and 'splendid but delusive scenery presented by an ardent imagination' is typical in treating hypothesis-

83. See *Mercure de France* (March 1771), p.151.

ing as a 'stylistic' fault.[84] The hackneyed accusation that he had written a *roman de physique* was similar in adopting literature as a convenient but vague pretext for attacking speculation. Faced with such carping, Buffon found himself in a dilemma, for the same *grandes vues* that were condemned by the scientifically scrupulous were seemingly requisite for reaching the public. 'Tout le monde veut lire des systèmes', lamented Augustin Barruel toward the end of the century.[85] This was surely an overstatement, but Buffon himself recognised the importance of general discussions to his success as an author. When the series' sales declined in the 1760s he explained it as follows: 'Les volumes étaient trop faibles de substance et de matières qu'on pût lire.'[86] Already in 1757 Paul-Claude Moultou complained that the series was growing uninteresting thanks to its emphasis on minute details.[87] To remedy the problem Buffon inserted an apology at the head of volume xii: 'Comme les détails de l'histoire naturelle ne sont intéressants que pour ceux qui s'appliquent uniquement à cette science [...] nous avons cru que la plupart de nos lecteurs nous sauraient gré de couper de temps en temps le fil d'une méthode qui nous contraint, par des discours dans lesquels nous donnerons nos réflexions sur la nature en général, et traiterons de ses effets en grand' (*HN*, xii.i). Two general discourses followed in volumes xii and xiii, both conjoining *grandes vues* with verbal ostentation. This was the pairing Buffon chose for success, and the one at the root of many critiques of his 'style'.

Other critics considered Buffon's style scientifically improper in more immediate ways. Some, such as Condillac, found it imprecise in its striving for poetic effect: '[Il] substitue au terme propre celui qui frappe davantage.'[88] Buffon only encouraged this judgement with his 'Discours sur le style', which famously recommended the use of 'les termes les plus géneraux' (*HNS*, iv.8-9). Elsewhere, less spectacularly, he endorsed precise terminology (*HNS*, iv.8), but the lexical purism of the 'Discours' has dogged his reputation ever since. In some measure his association with neoclassical generality is deserved. Abstract, general vocabulary plays a major role in the *Histoire naturelle*, sometimes to the exclusion of specific terminology.[89] Buffon's quadrupeds find

84. 'Geology', in *Encyclopaedia Britannica*, 4th edn (Edinburgh 1810), ix.594. James Hutton is accused of being equally dazzling a few pages later. Since Hutton's literary style was anything but dazzling, hypothesising, not style, appears to be the real focus of both reprimands. Compare 'Geology', ix.597.

85. *Les Helviennes, ou lettres provinciales philosophiques*, 4 vols (Paris 1823), i.332.

86. Buffon to Louis-Jean-Marie Daubenton, 29 February 1764, cited in Roger, *Buffon*, p.298.

87. To Elie-Salomon Reverdil, 2 April 1757, in Leigh (ed.), *Correspondance complète de Jean-Jacques Rousseau*, iv.207.

88. Etienne Bonnot de Condillac, *Traité des animaux*, in *Œuvres philosophiques de Condillac*, ed. Georges Le Roy, 3 vols (Paris 1947-1951), i.353. See also Louis-Jean-Marie Daubenton, 'Histoire naturelle', in *Séances des écoles normales, recueillies par des sténographes et revues par les professeurs. Leçons*, new edn, 8 vols (Paris, Imprimerie du Cercle-social 1796-1798), i.288-89. For more on Condillac's attitudes toward Buffon's style see Downing Thomas, 'Philosophy, science, and style: Condillac on Buffon and figurative language', *Common knowledge*, forthcoming.

89. Houston, *Traditions of French prose style*, p.120-21. The passage in question can be found in *HN*, ii.518-19.

shelter in 'retraites', 'demeures', 'asiles', 'séjours', 'gîtes', 'manoirs', 'forts', 'abris', and 'domiciles souterrains' as often as they do in 'tanières', 'antres', 'terriers', or 'garennes'. General vocabulary for describing human anatomy, character, society, and emotion is regularly transferred to animal and even geological 'homologues' despite Buffon's theoretical opposition to anthropomorphising.[90] Buffon, like Bossuet, early psalmists, and many later writers, makes extensive use of biblical and religious vocabulary, seeking to capitalise on the sublime associations of words such as 'ténèbres', 'abîme', 'ensevelir', and 'firmament'. Periphrases appear again and again in the *Histoire naturelle*. The sun becomes an 'astre de lumière', the moon an 'astre de nuit', the sea the 'élément liquide' or the 'flots', air the 'substance aérienne' or the 'élément transparent', and water birds at one point 'le peuple ailé des eaux'. Most of these periphrases serve to avoid repetition of basic vocabulary. Others are more gratuitously poetic.[91]

As we have seen, Buffon was not alone among contemporary naturalists in using general, anthropomorphic, and even high-style vocabulary. He may have been an extreme case in matters of style, but he was not a singularity. Nor for all its abstractness and poetic resonance is his high-style vocabulary imprecise for the most part.[92] One of Buffon's periphrases, 'les sphères du monde' (*HN*, xii.vii), could refer to the planets or to some larger collection of 'celestial spheres', but in context the difference has little importance. As his friend Suzanne Curchod Necker pointed out, he wrote 'trente' in place of the cacophonous but more accurate 'vingt-sept' in one of his speeches for the Académie française, but the rounding had to do with an insignificant detail and the setting was literary, not philosophical.[93] Malesherbes rebuked him for calling powdered mica 'rotten glass' when he meant something subtler, but Buffon, to be fair, wrote 'en quelque façon pourri' and gave a full explanation beside his rough metaphor.[94] More seriously, Buffon's extremely frequent use of the word 'infini' and its derivatives to refer to large but finite quantities makes it hard to know what to think of the 'infinity' of organic molecules he evokes in volume ii.[95] For the most part, however, Buffon's contemporaries did

90. See *HN*, iv.3-110; Roger, *Buffon*, p.219-21, 370-77. For more on Buffon's humanisation of nature see chapter 2, p.69-71.

91. See for example the 'astre de la nuit' in *HN*, xii.ix. Periphrases are especially abundant in the high-style discourses. See for example *HN*, iii.364-70, xii.iii-xvi; *HNO*, vii.231-52. Jacques Roger, well aware of Buffon's periphrases, nonetheless argued that he avoided them 'as much as possible'. See Roger, intro. to Buffon, *Epoques*, p.cxxii.

92. Fellows and Milliken, *Buffon*, p.161.

93. *HNS*, iv.24. See Lepenies, *Ende der Naturgeschichte*, p.145. Paradoxically Buffon has also been accused of being 'pre-scientific' for *failing* to round his figures. See chapter 5, p.130-31.

94. Compare Malesherbes, *Observations sur l'histoire naturelle*, ii.69-70; *HN*, i.258-65.

95. *HN*, ii.20, 24, 40, 44. Buffon later stated that there was only a 'quantité déterminée de matière organique' (*HN*, xiii.viii), leading some historians to postulate an evolution in his thinking. See Roger, *Sciences de la vie*, p.550; Roselyne Rey, 'Buffon et le vitalisme', in *Buffon 88*, p.409; Jean Ehrard, *L'Idée de nature en France dans la première moitié du XVIIIe siècle*, revised edn (Paris 1994), p.221. But these conclusions are questionable. Not only did Buffon habitually apply the word 'infini' and its derivatives to finite quantities, he also rejected the notion of infinity as

not raise objections to his use of 'finite infinities'.[96] In the main those bemoaning his scientific imprecision offered few examples at the lexical level. Their complaints were probably less about loose terminology than about a pattern of hyperbole in Buffon's writing – a topic to which we shall turn in a moment.

In any case the *Histoire naturelle* includes hundreds of specific technical and scientific words alongside its lexically restricted, high-style passages.[97] Many are italicised or defined in footnotes, but many are presented with no explanation. In his article on the horse, for example, Buffon never explains the anatomical terms 'ganache', 'garrot', 'boulet', or 'fanon'. Ironically his reputedly more 'professional' assistant Daubenton ends up defining all of them in his own later treatment of the horse's anatomy.[98] Faced with unfamiliar natural objects, Buffon even coined words. Höhne provides a list of several dozen animal names that Buffon invented, many of them colloquial, compound names of the sort purists condemned: 'septicolor', 'rose-gorge', 'rouge-cap', and so on.[99] Most importantly his often extensive vocabulary helped Buffon communicate critical nuances. By studying his revisions to the *Epoques de la nature*, for example, Jacques Roger has shown how Buffon worked for precision and took pains to distinguish such near synonyms as 'dissolution' and 'délaiement'. If intelligibility and euphony concerned Buffon, so too did precision.[100]

Buffon's sexual vocabulary is also significant. The relative sexual explicitness of the *Histoire naturelle* was both scandalous and titillating in the eighteenth century. Marie-Jeanne Phlipon, the future Mme Roland, skipped parts of the series to preserve her moral purity.[101] Some alleged the series' eroticism as a main reason for its success, especially among women, and Buffon may have half-consciously cultivated it to humour his readers.[102] At the same time

physically unrealisable just pages after announcing the 'infinite' number of organic molecules (*HN*, ii.27). On his rejection of infinity see chapter 5, p.134. On his use of the oxymoron 'presque infini' see chapters 2 and 3, p.66, 80.

96. For two exceptions see Thomas, 'Philosophy, science, and style', forthcoming; Royou, *Monde de verre*, p.164.

97. For samples see Höhne, *Stil Buffons*, p.75-78; Fuchs, 'Langue des sciences', p.587-89.

98. Compare *HN*, iv.174-99; Daubenton, *HN*, iv.260-68.

99. Höhne, *Stil Buffons*, p.77-78. On composed words see Thomas, *Langue volée*, p.74-75. On the cultural connotations of Buffon's new names for hummingbirds see Spary, 'Codes of passion', p.111-14. For two of Buffon's neologisms outside natural history see Bruneau, 'Buffon', p.497.

100. Roger, intro. to Buffon, *Epoques*, p.cxx-cxxiv.

101. *Private memoirs of Madame Roland*, p.74-75, 169. See also Gita May, *Madame Roland and the age of revolution* (New York 1970), p.20, 42.

102. See 'Masculin/féminin. Discours sur le sexe et sexe du discours', in *Aimer en France 1760-1860*, ed. Paul Viallaneix and Jean Ehrard, 2 vols (Clermont-Ferrand 1980), ii.297-98; Fellows and Milliken, *Buffon*, p.33-34, 159-60; Bassy, 'Texte et l'image', p.153-54; [Lignac], *Lettres à un Américain*, pt 1, letter 1, p.11, pt 2, letter VI, p.54-56; Jean-Louis de Lanessan, intro. to Nadault (ed.), *Correspondance générale [de] Buffon*, i.iii; Antoine de La Salle, *Le Désordre régulier, ou avis au public sur les prestiges de ses précepteurs et sur ses propres illusions* (Bern [Auxerre] 1786), p.195-97; Royou, *Monde de verre*, p.59; Mornet, *Sciences de la nature*, p.181-82; Guillaume Raynal, 'Nouvelles littéraires', in *Correspondance littéraire, philosophique et critique par Grimm, Diderot, Raynal, Meister, etc.* (Paris 1877), i.336. On sexual suggestion and English polite science see Walters, 'Conversation pieces', p.126, 133-34.

his approach to sexuality was sufficiently tempered to strike many as compatible with *bienséance*. The *Journal de Trévoux*, for example, declined to discuss his racy 'De la puberté' but praised Buffon's tact in writing it.[103] Fortifying this impression of Buffon as tactful was his promise in 'De la puberté' to write with 'cette indifférence philosophique qui détruit tout sentiment dans l'expression' (*HN*, ii.479). The grave, moralistic tone of the article and Buffon's writing in general, different from the sexual lyricism of Maupertuis, Diderot, Julien Offray de La Mettrie, and others, probably went a long way toward legitimating his subjects.[104] This is not to say that the *Histoire naturelle* is without euphemism. Erections are described as a 'signe extérieur' and the clitoris, on one occasion, as 'certaine partie'.[105] But Buffon's treatments of sexuality are nonetheless clear, and for the most part they rely on specific anatomical terms such as 'testicule' and 'vagin'. This is not a domain where Buffon was imprecise.

The inaccuracy Buffon's critics saw in the *Histoire naturelle* was probably not so much lexical as general inaccuracy into which his style was thought to be luring the unwary. This much-lamented effect of 'seduction' or 'bewitching' could be understood in two ways. On the one hand the imposing, staid tone of the *Histoire naturelle* left little room for doubt. Barren of humour, stripped of obvious appeals to emotion, and written as a reference work in a style associated with unchallengeable genres such as panegyric and ecclesiastical discourse, the *Histoire naturelle* had the ring of the definitive.[106] But here Buffon was mostly just following the new sciences in general, which had long since abandoned flamboyant rhetoric in favour of a more powerful 'non-rhetorical' rhetoric. More often critics emphasised not the 'deceptive' objectivity of Buffon's style, but its active appeal to readers' imagination and feelings. Some suggested, almost charitably, that Buffon himself had been swept away by his style, notably by its grandeur.[107] He stood accused of practising rhetoric, in other words, and not without reason. For despite his dim view of

103. March 1750, p.582-83, 594-95.

104. Denise Brahimi, 'La sexualité dans l'anthropologie humaniste de Buffon', *Dix-huitième siècle* 12 (1980), p.113-26. See also Spary, 'Codes of passion', p.129-33.

105. See for example *HN*, ii.346, 485, 507, xii.3. See also Martinez-Contreras, 'Des mœurs des singes', p.565-66. Later the clitoris is evoked without euphemism. See *HN*, ix.124.

106. See Paul Lawrence Farber, 'Buffon and the concept of species', *JHB* 5 (1972), p.282.

107. See for example [Lignac], *Lettres à un Américain*, pt 3, letter VII, p.10-11; Thomas Jefferson, *Notes on the state of Virginia*, ed. William Peden (Chapel Hill, N.C. 1954), p.64; Barruel, *Helviennes*, i.148. For comments on Buffon's style as seductive of others see for example *Journal de Trévoux* (December 1753), p.2835-36; Lepenies, *Ende der Naturgeschichte*, p.151; Gerhard H. Müller, 'L'histoire d'un manuscrit allemand du Muséum: "La Diss[ertatio] contra Buff[onem]" de Georg Forster', in *Le Muséum au premier siècle de son histoire*, ed. Claude Blanckaert *et al.* (Paris 1997), p.588; Condorcet to Amélie Suard, October 1780, in Elisabeth Badinter (ed.), *Correspondance inédite de Condorcet et Mme Suard, M. Suard et Garat (1771-1791)* (Paris 1988), p.206; *HARS* 1788 (1791), p.65-72; Mornet, *Sciences de la nature*, p.203-206; Rousseau, 'Science books and their readers', p.221-22; Jeff Loveland, 'French thought in William Smellie's natural history: a Scottish reception of Buffon and Condillac', in *Scotland and France in the Enlightenment*, ed. D. Dawson, forthcoming; Félix Vicq d'Azyr, 'Eloge de Buffon', *Œuvres complètes de Buffon augmentées par M. F. Cuvier* (Paris 1829), i.li-lii; *Journal encyclopédique* 6.3 (September 1765), p.40.

eloquence in the 'Discours sur le style', his style was complemented in its seemingly non-rhetorical majesty with touches of traditional rhetoric, especially amplification in several forms.

Redundance, one element of *style ampoulé*, has long been a locus for disparaging suggestions that the *Histoire naturelle* is unscientific. As a *grand phrasier*, a pejorative sobriquet that referred to his prolixity as well as his pomp, Buffon could be written off as frivolous and unfocused, more interested in textual drama than 'simple' truth. Partly true to this image, the *Histoire naturelle* is filled with repetition:

> Tout ce qui a eu une origine, une naissance, un commencement, doit arriver à un but, à une mort, à une fin.
>
> Ministre de ses ordres irrévocables, dépositaire de ses immuables décrets, la nature ne s'écarte jamais des lois qui lui ont été prescrites; elle n'altère rien aux plans qui lui ont été tracés, et dans tous ses ouvrages, elle présente le sceau de l'éternel.[108]

In its extreme form Buffon's taste for paraphrase led the series toward preciousness, especially in the *Histoire des oiseaux*, where he often merely edited his assistants' overdone efforts to mimic his style. Working from Bexon's draft of the article 'Colibri', for example, he made it more precious with this poetic affirmation of a beauty evoked directly in following sentences: 'La nature, en prodiguant tant de beautés à l'oiseau mouche, n'a pas oublié le colibri.'[109] Though not usually so precious, Buffon's repetitions are often poetically striking thanks to their arrangement in symmetric two- and three-member groups. So configured, they no doubt 'seduced' the public by adding beauty and majesty to many parts of the *Histoire naturelle*. On a more concrete level they encouraged casual readers through their redundance, making the series' presentation slower and more digestible. Here again Buffon's critics came close to attacking vulgarisation with their complaints about style.

Buffon's repetitiveness extends beyond *elocutio* to higher levels of the *Histoire naturelle*. His denunciations of abstraction and classification recur again and again in the course of the series. Recapitulations are frequent, some stretching to several pages. Not only do they allow Buffon to update his thinking, they give readers a chance to review or catch up. Likewise certain natural objects, mostly those prominent in European culture, are mentioned as references throughout the series, not just in articles specifically devoted to them. In the *Histoire des quadrupèdes* the recurring elements are horses and dogs, lions and sheep, all evoked to identify alien species. The article on the guinea pig compares it not only to the domestic pig but also to the otter, the donkey, the fish, and the beaver (*HN*, xii.384-86). The following article, on the porcupine, pits

108. *HN*, ii.309; xii.iv. Notice the profusion of synonymy in the pastiche of Buffon's style in N. Chatelain, *Pastiches, ou Imitations libres du style de quelques écrivains du XVIIe et XVIIIe siècles* (Paris 1855), p.229-32.

109. *HNO*, vi.41. For Bexon's maladroit, less precious version see Flourens, *Des manuscrits de Buffon*, p.138-39. Notice the similar conceit in Pliny's notoriously precious description of the nightingale. See *Naturalis historia*, ed. Carl Mayhoff, 6 vols (Stuttgart 1967-1970), ii.243 (bk x, ch. 29).

its own subject against even more foils (*HN*, xii.402-408). Viewed philosophically, Buffon's many analogies may follow from his judgement that all knowledge comes from comparison,[110] but their practical importance is to facilitate assimilation of the *Histoire naturelle*. Redundance was essential to Buffon's natural history as a work of vulgarisation.

Those who saw Buffon's style as seductive tended to focus on what was implicitly another amplifying figure, namely hyperbole. A victim of his imagination if not his style, Buffon 'went too far', it was said, and this made him guilty of consistent overstatement. Like his contemporaries, Buffon was something of a hypocrite regarding hyperbole.[111] In theory he condemned it, though he was enough of a realist to urge his son to take advantage of it in his dealings with royalty.[112] In fact the *Histoire naturelle* is filled with hyperbole. Superlatives are frequent. Buffon's dog is the most pliant of animals (*HN*, v.186), his sloth the most miserable (*HN*, xiii.38-46), and his horse the 'plus noble conquête que l'homme ait jamais faite' (*HN*, iv.174). Later the camel will usurp the honour to become 'la plus utile et la plus précieuse de toutes les créatures surbordonnées à l'homme' (*HN*, xi.239). Human beings are extreme too, particularly in 'Variétés de l'espèce humaine'. The inhabitants of New Holland and Timor, for example, 'sont peut-être les gens du monde les plus misérables, et ceux de tous les humains qui approchent le plus des brutes' (*HN*, iii.408). Human beauty, on the other hand, is at its maximum in the temperate zones, notably in Circassia (*HN*, iii.421, 435, 528). Or are the Finns and the Danes the most beautiful people, as Buffon suggests later (*HN*, v.202-203)? With so many superlatives, inconsistencies are not hard to find.

But to focus on superlatives alone is to miss the better part of Buffon's hyperbole, for his exaggeration takes many forms. Emphatic adverbs pepper 'Variétés', making the world's peoples extremely ugly, very agile, supremely ignorant, or very well-built. A staggering number of animals in the *Histoire naturelle* are described as 'especially' amorous. Nature's movement is exalted as 'plus que perpétuel', an expression Bexon opposed as meaningless, but to no avail.[113] Categorical negations and affirmations add force if not credibility to many descriptions. Buffon's dog, for example, has 'nulle ambition, nul intérêt, nul désir de vengeance, nulle crainte que celle de déplaire' (*HN*, v.186), his guinea pig never drinks but urinates all the time (*HN*, viii.3-4), and the lazy inhabitants of Sierra Leone 'n'ont aucun goût que celui des femmes et aucun désir que celui de rien faire' (*HN*, iii.462). The negative 'ne

110. See Hanks, *Buffon avant l'"Histoire naturelle'*, p.226n. Descartes, Locke, and many of Buffon's contemporaries expressed similar judgements. See Stephen Gaukroger, *Descartes: an intellectual biography* (Oxford 1995), p.169; Hodge, 'Two cosmogonies', p.246.

111. See Peter France, *Politeness and its discontents* (Cambridge 1990), p.11-26.

112. Buffon to Georges-Louis-Marie Leclerc de Buffon, 10 June 1782, in Nadault (ed.), *Correspondance générale* [*de*] *Buffon*, ii.116-19. Implicit and explicit refusals of hyperbole can be found in *HNS*, iv.2-3, 8, 27-30.

113. *HNM*, ii.107; Flourens, *Des manuscrits de Buffon*, p.53.

... que', ideal for making overstatements, occurs regularly in the *Histoire naturelle*:

Le sable n'est que du verre en poudre (*HNS*, v.13).

[L'œil n'est] qu'un gros nerf épanoui (*HN*, iv.26).

La rumination n'est qu'un vomissement sans effort (*HN*, iv.461).

[Le goût] n'est qu'un instinct plus exquis (*HNO*, viii.20).

La chaleur n'est donc que le toucher de la lumière (*HN*, iv.355).

These reductive analyses present various theories, some with long histories in natural philosophy. What they all share is a certain literary snappiness and a pedagogical aspiration toward simplicity and disabusal. Many overstate Buffon's case to a degree. When he writes, for example, that a volcanic eruption 'n'est cependant que du bruit, du feu et de la fumée' (*HN*, i.503), this is not his best effort at characterising the process, but rather a provocative literary gateway to a theory of volcanoes detailed less dramatically in the following paragraph. So too his affirmation in the *Epoques de la nature* that 'un volcan n'est qu'un vaste fourneau' (*HNS*, v.137). Like his comparisons among natural objects in general, and like his comparison of volcanoes to cannons and tempests in other parts of the series, this is a disposable metaphor, albeit one cast as a transparent equivalence.[114] As in the maxims and epigrams of Buffon's literary contemporaries, overstatement here could be both instructive and entertaining. Writing natural history for the public at large, Buffon had good reason to use hyperbole when it promised both virtues.

Here we return to the theme of this section. Complaints about the scientific inappropriateness of Buffon's style were not what they might seem to be. Style for the most part was just an especially vulnerable target for critics opposed to two more general features of Buffon's natural history. First, it became a metonymy for imagination in the mouths of many critics, making deprecations of Buffon's style into deprecations of systematising and the philosophical *grandes vues* he so often expressed in high style. Since *grandes vues* were a key part of vulgarisation, amateurs were implicated in this critique of Buffon's style. Second, stylistic criticism provided a means of attacking a science that humoured literati, in Buffon's case with redundance, high style, and hyperbole among other things. As his enemies realised, Buffon's style was bound up not just with his epistemology but also, more importantly, with his readers and the very premises of scientific 'literature'. Faced with ostensibly stylistic reproaches, Buffon found himself in a quandary, for his whole natural history was being called into question. His style may not have been Buffon himself, as the 'Discours sur le style' appears to suggest, but it was certainly close to his identity as a popular writing naturalist.[115]

114. For comparisons of volcanoes to cannons and tempests see *HN*, i.502; *HNS*, v.138. For examples of 'comparisons' that threaten to become equivalences, though without the use of 'ne ... que', see Höhne, *Stil Buffons*, p.63-64.

115. On the relations between Buffon's style and thought see Roger, *Sciences de la vie*, p.772-73; *HARS* 1788 (1791), p.70-71; Lepenies, *Ende der Naturgeschichte*, p.150; Fellows and Milliken, *Buffon*,

There was no way to satisfy both his critics and his devotion to *grandes vues* and literary natural history, but Buffon did the best he could by diversifying the contents of the *Histoire naturelle*. Certain parts of the series – the *beaux morceaux*, the grand openings, and the general discourses – were self-consciously directed toward literati. Here Buffon relied on abstract vocabulary and created textual drama with repetition, hyperbole, and rhythmically organised parallel structures. Here too he secured the commercially advantageous myth of his 'style'. This style, we have seen, was only an idealised version of one of his styles. Much of the *Histoire naturelle*, written prosaically and filled with technical and empirical detail, was seemingly directed toward more serious readers. But this side of the *Histoire naturelle* was often hidden from direct public scrutiny, consigned to the *Suppléments* or to supplementary appendices following Buffon's main discourses.[116] This was not the *Histoire naturelle* that the public at large was expected to dwell on.

Buffon was not alone in appealing to different audiences with different styles. To a lesser degree both Perrault's *Mémoires* and the early *Histoire de l'Académie* balanced high-style generalities with dryer material, presumably in the interest of straddling the 'two cultures' of science and literature. Within the Académie des sciences style became more standardised during the eighteenth century as academicians withdrew from dependence on the culture of amateurs. Relative to academic colleagues of the mid-eighteenth century, Buffon was no doubt an anomaly with the numerous high-style declamations of the *Histoire naturelle*. But the contrast should not be made absolute. Even in official academic publications, high style subsisted in prefaces, dedications, eulogies, and descriptions of conceptually sublime phenomena. Nor was specialised scientific vocabulary an uncontested goal among academicians at any time in the eighteenth century. Dreams of 'polite science' – once encouraged by natural philosophers in need of protectors, now flattered by humanists, demonstrators, and popular authors – continued to exert a hold on the French public as well as on academicians long after many sciences had lost their immediate intelligibility. These dreams found an issue in vulgarisation, an application for which natural history was especially well-suited. Inside and outside the Académie des sciences many writers on natural history chose to address a broad public. Not all were stylish, but style helped many to reach what was essentially a literary public. Whatever its aesthetic virtues, Buffon's 'style' was just one among many in this domain.

p.169-70; John Lyon, 'The "Initial discourse" to Buffon's *Histoire naturelle*: the first complete English translation', *JHB* 9 (1976), p.134; Vicq d'Azyr, 'Eloge de Buffon', p.li-lii.

116. See Roger, *Buffon*, p.135, 514-16. As Roger notes, the sequestered 'empirical' sections of the *Histoire naturelle* sometimes served as a hiding place for unorthodox material. See Roger, *Buffon*, p.197.

2. Final causes

In the previous chapter we examined the role of literary style in creating a public for natural history. This chapter, too, will focus on a rhetorical tool aimed at drawing in readers, albeit a less familiar one – the textual invocation of final causes and teleology. As conceived by Aristotle and subsequent philosophers, a final cause was a cause acting toward an end. It was a necessary complement to Aristotle's three other causes – the efficient, the formal, and the material – in that none of the others could account for teleological or goal-orientated phenomena such as growth, heliotropism, or learning. At first glance it may not be obvious how final causes and teleology contribute to the project of vulgarised natural history. But final causes have always been a part of pedagogy and popular science. Like anthropomorphism, in which they participate, final causes help engage readers who are familiar with nature in its literary guises and who are looking for understanding, not piecemeal analysis. In fact teleology's ties to religion, literary culture, pedagogy, and everyday anthropomorphism made it more important than pleasing style in the rhetorical repertory of eighteenth-century popular naturalists. Nor were invoked final causes always very different from allusions, metaphors, and other traditional figures of rhetoric.

Before turning to the literary uses of final causes, however, we will look at debates about them in the Académie des sciences and elsewhere. Final causes were called into question by Aristotle's contemporaries and became a source of considerable dispute in seventeenth- and eighteenth-century France as mechanists squared off against Aristotelians, *cause-finaliers*, and pious common sense. These debates illustrate another way in which the question of final causes could be rhetorical. In seeking to persuade readers of the truth of their viewpoints, defenders and opponents of final causes defined their own rhetorics, both of them exaggerating and simplifying differences. The complex theological ramifications of beliefs about final causes only added to the rhetoric, inducing authors to express their ideas with particular, even hypocritical care.

In our own day philosophers continue to disagree about the place of teleology in the theory of evolution and biology in general.[1] Historians too differ in their approach to final causes. On the one hand Peter Gay has argued that 'the history of scientific progress is the history of liberation from teleology'.[2] For

1. Good overviews can be found in François Duchesneau, *Philosophie de la biologie* (Paris 1997), p.73-149; Richard J. Connell, *Nature's causes* (New York 1995), p.183-87.

2. *Enlightenment: an interpretation*, vol. i, *The Rise of modern paganism*, p.246. See also Gay, *Science of freedom*, p.161; Aram Vartanian, *Diderot and Descartes: a study of scientific naturalism in the Enlightenment* (Princeton, N.J. 1953), p.311-14.

Charles Coulston Gillispie on the other hand, 'providentialism and belief in fixity and divine design' have led to better scientific work than 'belief in process' as exemplified by Diderot and Goethe.[3] Many scholars with less categorical allegiances than Gay or Gillispie follow them in presenting final causes as a divisive philosophical issue for the Enlightenment. Accounts of attitudes toward final causes typically divide the Enlightenment into three groups. First come the credulous, notably William Derham, Bernhard Nieuwentyt, Pluche, and Bernardin de Saint-Pierre. Participants in the project of 'physico-theology' – uncovering divine providence in the design of living beings and the rest of the universe – they seem to have believed themselves capable of enouncing God's reasons for creating tides, the moon, and other natural objects. At the other extreme are scandalous free-thinkers such as Hume, La Mettrie, Diderot, and Buffon, who denounced the search for final causes as futile. In between are the moderates, always a majority in the Académie des sciences, who reproached their colleagues on both sides, insisting that final causes exist but are neither easy to discover nor necessarily scientific except as heuristics. Kant's view of finalism as regulating but not constituting knowledge represents this moderation at its most sophisticated.

This tripartite schema is satisfactory as a representation of consciously formulated attitudes toward final causes, but it fails to take account of the polarising effects of scientific rhetoric. Authors in the 'extreme' categories are almost always placed there because of forceful declarations, which may not reflect long-term or profound beliefs. Outside their manifestos disputants on both sides typically exhibited moderate, mixed presumptions about the existence of natural design. Beyond the inevitable inconsistency of those postulating final causes everywhere or nowhere, authors in all three categories appreciated and exploited final causes for their literary utility. As we will see, Buffon exemplifies all of these tendencies, for despite his reputation for vigorous anti-finalism, his attitudes were ambiguous and he invoked hundreds of final causes to satisfy his readers. Only by considering teleology's rhetorical aspects – its role in pleasing readers and its place in polemics – can we understand the terms in which Buffon and his contemporaries discussed final causes.

Philosophers and final causes

Final causes can be divided into several varieties. Internal final causes are those that direct from within, as in a developing organism or a human being with a plan. As these examples indicate, internal final causes can be conscious or unconscious. Attributing them to other living beings is a necessity of survival and everyday life, and they have naturally been a long-time concern of philosophers. Internal final causes were the ones theorised by Aristotle, at

3. *The Edge of objectivity: an essay in the history of scientific ideas* (Princeton, N.J. 1960), p.341. See also Emerson Thomas McMullen, *William Harvey and the use of purpose in the Scientific Revolution: cosmos by chance or universe by design* (Lanham, Md. 1998).

least for the most part. In a typical analysis he saw the 'form' of living beings as the final cause of their maturation. Taking finality to its limit in the way many others would, Aristotle held that every part of an organism's body functioned in the interest of maintaining its life, but in the case of the deer's horns – which he recognised as unsuitable for defence or attack – he admitted that finality could be more complicated.[4] Despite the concession, Aristotle retained his well-known motto that nature does nothing in vain. Consistent with his focus on internal final causes, the nature in question was primarily the nature of a particular thing and only occasionally a transcendental providence or Mother Nature.[5]

Other philosophers of antiquity were inclined to look for external final causes – purposes orchestrated from the exterior by a superior power. Aristotle's disciple Theophrastus, not as committed as his teacher to finding purpose in everything, saw nature as a whole, and consequently emphasised purposes that fostered harmony, not individual survival.[6] Where Aristotle made its form the goal of each organism, Galen looked for sub-goals at the level of organs, and counted on an external intelligence to make these sub-goals harmonise in a functioning unit.[7] External final causes became a cornerstone of Stoicism, and in an inevitable variation some Stoics claimed human welfare as the purpose of nature. Pliny, a moderate Stoic, asserted that all of nature was created for humans, and wrote his *Historia naturalis* as a survey of resources available to Romans. In practice his sense of providence went beyond humans, and he made no effort to find human value in certain insects, but compared to Aristotle's or Galen's, his finality was nonetheless anthropocentric.[8]

The competing visions of final causes explored in antiquity all remained operative in the Enlightenment. The expression 'final cause' most often referred to external final causes, which were sanctioned and encouraged by Christian apologetics and soon formed the object of a literary vogue. But however designated, internal final causes continued to attract attention, especially in the second half of the eighteenth century. Anathema to mechanists, they were implicated in the active matter, vegetative souls, *sensibilité*, and internal moulds invoked by vitalistic and Aristotelian philosophers. Whether internal or external, final causes could be sought everywhere or simply accepted when they provided a clear-cut explanation. The oft-repeated assertion that nature does nothing in vain could be interpreted strictly or loosely and with various conceptions of nature's goals. External final causes in particular could be conceived as directed toward natural harmony, plenitude, God's glory, or human welfare.

4. French, *Ancient natural history*, p.74-77.

5. French, *Ancient natural history*, p.15-16, 45-51, 68-69. For a dissenting point of view see Atran, *Cognitive foundations of natural history*, p.86, 101-103.

6. French, *Ancient natural history*, p.84-85, 92-103.

7. Roger, *Sciences de la vie*, p.74-76. See also French, *Ancient natural history*, p.186-95.

8. French, *Ancient natural history*, p.203-207, 248-52.

Christian monotheism both complicated and reinforced the notion of finality. External final causes found an obvious agent in divine providence, and God's wisdom and goodness could be sought and displayed in external final causes. This was a favourite pastime of eighteenth-century Protestants in particular, perhaps because everyday finality offered an alternative to the destabilising perils of miracles and 'enthusiasm', perhaps because their decision not to interpret things allegorically left Protestants wondering what nature meant.[9] But identifying God's purposes could be controversial, since theologians disagreed on the nature and visibility of divine providence. In a typical encounter Nicolas Malebranche and Antoine Arnauld debated the issue in the late seventeenth century, Malebranche contending that God worked through reasonable laws, Arnauld insisting that mortals could not pretend to make sense of God's reasons.[10] Neither position was theologically untouchable, but neither was patently heterodox either. Later free-thinkers such as Buffon and Diderot capitalised on the schism and took up Arnauld's argument in their own attacks on final causes.

Another force transforming final causes with respect to antiquity was the new natural philosophy of the sixteenth and seventeenth centuries. Though not universal, caution regarding final causes was widespread among the new philosophers. Francis Bacon considered final causes sterile in physics, if not metaphysics. Galileo and Thomas Hobbes found them practically unfruitful. William Harvey was led to his discovery of the blood's circulation by his faith in natural providence, but he refused final causes except as speculation,[11] a typical anticipation of Kant's approach to teleology. Robert Boyle urged 'Christian philosophisers' to recognise final causes, but he granted them a limited scientific role.[12] Descartes went especially far in denouncing final causes. His mechanical universe had no place or need for internal final causes, since passive matter under pressure accounted for everything. External final causes he refused as uncertain, justifying himself with the usual arguments about the weakness of reason. Traces of external finalism are not hard to find in his work on anatomy, and he sometimes compromised his principles for rhetorical advantage.[13] But Descartes's anti-finalism was as consistent as any, and his philosophy was more mechanical than anything previous.

9. See Mornet, *Sciences de la nature*, p.31; Lorraine Daston, 'Marvelous facts and miraculous evidence in early modern Europe', *Critical inquiry* 18 (1991), p.114, 123-24; Harrison, *Bible, protestantism*, p.161-62. Not surprisingly most natural histories were written by Protestants in the sixteenth and seventeenth centuries. See Roger French, *William Harvey's natural philosophy* (Cambridge 1994), p.45-50.

10. Roger, *Sciences de la vie*, p.400-404.

11. See McMullen, *William Harvey*, p.37-75; Jonathan Cohen, 'Teleological explanation', in *Purposive behaviour and teleological explanations*, ed. Frank George and Les Johnson (New York 1985), p.56; Phillip R. Sloan, 'The question of natural purpose', in *Evolution and creation*, ed. Ernan McMullin (Notre Dame, Ind. 1985), p.124n.

12. Roger, *Sciences de la vie*, p.225-27; Robert Boyle, *A Disquisition about the final causes of natural things* (London, John Taylor 1688).

13. See for example his description of semilunar valves and the nervous system as providentially designed in *Discours de la méthode*, in *Œuvres de Descartes*, ed. Charles Adam and Paul

A 'Cartesianism' much indebted to Malebranche and Christian Huyghens came to dominate the Académie des sciences toward the end of the seventeenth century, but in its approach to living beings it was no longer the natural philosophy Descartes had envisaged. The divergence was sharpest with respect to his posthumously published theories of reproduction, where true to his mechanism he attempted to explain sexual reproduction as a mechanical combination and fermentation of the two parents' 'seed'. This epigenesis convinced no one, yet Descartes's successful arguments for the passivity of matter made it hard to return to Aristotelian internal finalism as an explanation for reproduction. External teleology in the form of pre-existent germs became a necessary complement to Cartesian philosophy. In the Académie des sciences ovism was widely accepted by the last decades of the seventeenth century. Descartes's campaign against internal final causes had ironically subverted his effort to de-emphasise external final causes. Every aspect of the living world now risked being interpreted as reflecting God's design.[14]

Proponents of pre-existence remained nonetheless cautious regarding final causes. Leibniz, for example, encouraged the search for them, so much so that he was satirised as a naive teleologist by Voltaire and others, but outside metaphysics especially he warned against anthropomorphism and suggested coupling final causes with more useful efficient ones.[15] In the Académie des sciences Réaumur and Fontenelle embodied a typical circumspection. Though both Cartesian in a general sense, they took external final causes seriously and sometimes considered them in explanations. Réaumur, for example, wondered if certain animals were not 'given' the power of regenerating lost parts because of their susceptibility to being mutilated by predators. In the *Histoire de l'Académie* Fontenelle stressed the heuristic value of assuming natural providence and exulted in the design and equilibrium of nature. But despite the extravagant finalism of his *Entretiens sur la pluralité des mondes* and a few parts of the *Histoire* – a finalism we will turn to in the following section – Fontenelle was no enthusiast of the 'vulgar' final causes of popular literature. Réaumur, just as cautious, was uncomfortable with the 'precipitous' finalism of his friends Pluche and Bonnet.[16]

Tannery, 9 vols in 11 pts (Paris 1964-1974), vi.48; *Meditationes de prima philosophia*, in *Œuvres de Descartes*, vii.88.

14. Roger, *Sciences de la vie*, p.140-60, 267-69.

15. Compare Duchesneau, *Physiologie des Lumières*, p.65-102; Gottfried Wilhelm Leibniz, *Discours de métaphysique*, in *Opera philosophica quae exstant latina, gallica, germanica omnia*, ed. Renate Vollbrecht (Meisenheim am Glan 1974), p.825-26 (no. 19-22).

16. On Fontenelle see Marsak, *Bernard de Fontenelle*, p.26-29; Niderst, *Fontenelle*, p.199-200; Roger, *Sciences de la vie*, p.231-33, 236-37; Ehrard, *Idée de nature*, p.90-91; Patrick Tort, *L'Ordre et les monstres. Le débat sur l'origine des déviations anatomiques au XVIIIe siècle*, 2nd edn (Paris 1998), p.33. On Réaumur see Virginia P. Dawson, *Nature's enigma: the problem of the polyp in the letters of Bonnet, Trembley and Réaumur* (Philadelphia, Penn. 1987), p.27-28; Roger, *Sciences de la vie*, p.234-37, 248-49, 251, 392-95; Jean Torlais, *Réaumur. Un esprit encyclopédique en dehors de l'"Encylopédie"* (Paris 1961), p.45, 153-54; Roger, *Buffon*, p.109.

One early challenge to the Académie's metaphysics of finalism came in the debates over 'monsters' that culminated in the papers of Jacques-Bénigne Winslow and Louis Lémery in the 1730s. Aware that congenital monsters were not random in the forms they assumed, and committed to the dogma that nothing purposeful can develop unless guided by divine providence, first Joseph-Guichard Du Verney and then Winslow reached the conclusion that malformed beings, no less than other ones, were designed so by God. Lémery, among others, found this counter-intuitive and theologically troubling. Without openly questioning the assumption of pre-existence he suggested that monsters were due not to God's will but to accidents affecting well-designed 'germs' inside the womb. Explaining how accidents could create symmetric and seemingly half-purposeful monstrosities was difficult, and Lémery ended up professing a mild form of vitalism, claiming that nature made an effort to 'compensate' for accidents. Commonsensical though it was, this was not a solution that appealed to the Académie des sciences in the late 1730s.[17]

Nor did the Newtonian displacement of Cartesian mechanism in the 1730s and 1740s overthrow the Académie's position on final causes. Gravity, in principle, belied the strict mechanism which had led to attributing all final causes to God, but Newtonian philosophy remained generally mechanical and stressed matter's passiveness in its orthodox presentation. Newton's English admirers were inclined to glorify the providence built into his system, but the literature of English providentialism, sometimes non-Newtonian, had been familiar in France since the 1710s. In any case French Newtonians made less of Newton's finalism than did the English. In his tolerance of finalism Voltaire, for example, was no more or less liberal than the Académie des sciences. He defended everyday final causes, especially in the radical climate of the later eighteenth century, but he also ridiculed the extreme final causes of Pluche, Huyghens, and Nieuwentyt.[18] In the Académie des sciences early Newtonians were preoccupied with mathematical prediction, not final causes. Buffon and Maupertuis later cited attraction to justify their ascription of vital forces to matter, but both dismissed Newtonian physico-theology. Their goal-oriented matter, inspired by Leibniz as much as by Newton, was never widely accepted by other Newtonians.

But thanks in part to heterodox interpretations of monsters and gravity, the consensus governing final causes in the Académie des sciences was weakening in the first half of the eighteenth century. Further contributing to the demise of strict mechanism were any number of potentially vitalistic philosophies that haunted the margins of the Académie during the period: Spinozism, Cambridge Platonism, Stahlian animism, and others. Leibnizian philosophy,

17. See Roger, *Sciences de la vie*, p.405-18; Tort, *Ordre et les monstres*; Andrew Curran and Patrick Graille, 'The faces of eighteenth-century monstrosity', *Eighteenth-century life* 21 (1997), p.4-7; Michael Hagner, 'Enlightened monsters', in *The Sciences in Enlightened Europe*, ed. William Clark, Jan Golinski, and Simon Schaffer (Chicago 1999), p.188.

18. See Ehrard, *Idée de nature*, p.127-38; Roger, *Sciences de la vie*, p.732-48.

familiar to French readers via Jean-Henri-Samuel Formey and Du Châtelet, provided an especially important model for those seeking an alternative to the Académie's mechanism. Not only could monads be surreptitiously transformed into purposeful material atoms, as in the work of Buffon or Maupertuis, Leibniz's stress on *vis viva* as inherent in matter gave nature a certain general autonomy.[19] Above all, discoveries in natural history and anatomy undermined external providence and hinted at nature's internal power. Neither the teleology of pre-existence nor simple mechanism could explain the observed 'sensibility' and 'irritability' of organic tissue or the ability of certain animals to regenerate lost parts and reproduce by sectioning. The Académie's credo of cautious empiricism allowed such discoveries to persist unexplained alongside official mechanism, but the tension was not one that could be maintained forever.[20]

By the 1740s even the circumspect were expressing doubts about the tenability of the Académie's mechanism. Drawing on philosophers from Newton to Leibniz and on current 'anomalies' from monsters to sensibility, the more audacious inverted the Académie's policy on teleology, abolishing external in favour of internal final causes. La Mettrie published *L'Homme machine* in 1747, rejecting divine providence and making human beings not so much 'machines' as self-animated material systems.[21] An imprudent outsider, La Mettrie became a universal scapegoat in the manner of Spinoza, but similar ideas were being set forth in the Académie des sciences. Maupertuis, a long-time critic of physico-theology, attacked the pre-existence of germs in the 1740s and 1750s, proposing instead that corpuscles from both parents organised the embryo in virtue of their Leibnizian 'desire' and 'memory'. In 1748, building on Needham's prior observations, Buffon and Needham examined sterilised organic matter under a microscope and found that 'animalcules' like Antoni van Leeuwenhoek's could be generated spontaneously. Both proposed theories of reproduction dependent on active matter, not pre-existence, and Buffon no less than Maupertuis became famous for his attacks on external final causes.[22]

Buffon's theory of reproduction was in fact largely in place before his observations with Needham. His previous scientific work had given him ample opportunity for questioning pre-existence and the Académie's mechanism. As a former leader in the campaign for Newton in France, he constantly drew on gravity as an example of a non-mechanical power of nature. As a member of the Académie des sciences since 1734, he had first-hand knowledge of the

19. See Roger, *Sciences de la vie*, p.366-70, 418-39, 461-63; Sloan, 'Question of natural purpose', p.127-28; Sloan, 'Natural history', forthcoming.

20. See Vartanian, *Diderot and Descartes*, p.252-53; Dawson, *Nature's enigma*, p.5-9; Thomas L. Hankins, *Science and the Enlightenment* (Cambridge 1985), p.125, 130-34; Roger, *Sciences de la vie*, p.390-96.

21. See Ann Thomson, 'L'homme-machine, mythe ou métaphore?', *Dix-huitième siècle* 20 (1988), p.367-75; Roger, *Sciences de la vie*, p.490-94.

22. See Roe, 'Buffon and Needham', p.439-50; Roger, *Sciences de la vie*, p.494-520.

debates over monsters, parthenogenesis, and reproduction via sectioning. As a friend of Maupertuis and a reader of Du Châtelet, he was exposed to the dynamic possibilities of Leibnizian philosophy in the 1740s. In particular Maupertuis's writings of the early 1740s, which included a system of epigenesis and attacks on external teleology, furnished a clear precedent for the *Histoire naturelle*. Already in 1734 Buffon was familiar with Louis Bourguet's quasi-Leibnizian theory of reproduction via 'organic mechanism'. At the time Buffon refused it, among other reasons for its assumption of female spermatozoa, but Bourguet's ideas would return in the *Histoire générale des animaux*, notably his idea that internal 'moulds' are responsible for bodily development. Beyond his exchange with Bourguet, Buffon's early writings offer few clues about the evolution of his attitudes toward pre-existence and final causes. His 1735 translation of *Vegetable statics* omits a certain amount of Hales's rapturous finalism, but many in the Académie were probably equally wary of British teleology.[23] His critique of pre-existence and external finality probably solidified along with that of Maupertuis in the 1740s.

Unlike Maupertuis, Buffon did not explicitly postulate 'intelligent' corpuscles in his theory of reproduction, but he relied on his own kinds of internal finality. As commentators have noted from the eighteenth century onward, Buffon's internal moulds, charged with assembling offspring from organic matter, bear a conspicuous resemblance to Aristotle's forms, and their role is correspondingly teleological.[24] Internal finality also enters Buffon's theory at the level of organic molecules. Normally operating in tandem with internal moulds, these molecules can nonetheless spontaneously generate self-replicating micro-organisms in the absence of moulds – whence Buffon's repeated statement that nature tends to life and fecundity rather than to death.[25] 'Normal' reproduction thus appears doubly teleological as Buffon conceived it, but the two levels can be interpreted as large- and small-scale representations of the same ordering process, dependent in both instances on Newtonian micro-forces. Buffon's later efforts to reduce both moulds and molecules to by-products of heat make sense in this light.[26]

23. Hanks, *Buffon avant l'"Histoire naturelle'*, p.77, 82-87. On Bourguet and Buffon see Roger, *Sciences de la vie*, p.375-78; Roger, *Buffon*, p.181-83.

24. Sloan, 'From logical universals to historical individuals', p.121; Atran, *Cognitive foundations of natural history*, p.239.

25. For examples of this statement see *HN*, ii.37; *HNS*, v.458; *HNM*, i.421, iii.398. Buffon's finalistic statements about nature's fecundity shade into his equally finalistic statements about nature's drive to variety and plenitude. See for example *HNO*, i.xxiii, iii.188, iv.439. See also chapter 3, p.96-97, and Richard R. Yeo, 'The principle of plenitude and natural theology in nineteenth-century Britain', *British journal for the history of science* 19 (1986), p.263-82. On spontaneous generation without internal moulds see *HN*, ii.303-304, 320-24.

26. On Buffon's 'double' teleology see Roger, *Sciences de la vie*, p.556-58; François Duchesneau, 'Buffon et la physiologie', in *Buffon 88*, p.455-60. On his progressive naturalisation of internal moulds and organic molecules see Phillip R. Sloan, 'The history of the concept of the biological species in the seventeenth and eighteenth centuries, and the origin of the species problem', Doctoral dissertation, University of California, San Diego, 1970, p.321-27.

More interesting for our purposes than his implicit internal finalism is Buffon's rejection of external finality. His first attacks on external final causes came in the opening volumes of the *Histoire naturelle.* Already in volume i he dismisses Newton's explanation of planetary density as finalistic, and proposes instead a mechanical one.[27] Buffon's system of epigenesis in volume ii is naturally accompanied by aggressive repudiations of external teleology, which he associates with abstraction and rationalistic philosophers (*HN*, ii.31-33, 78-79). One of the most notorious of his attacks on final causes came in 1755 in an article on the pig. Here he concedes that teleology may exist, but doubts whether human beings can necessarily perceive it: '[Si la nature] ne fait rien en vain, elle ne fait rien non plus dans les desseins que nous lui supposons.' In particular he argues that the pig's vestigial internal phalanges are 'évidemment des parties inutiles, ou plutôt des parties dont il ne peut faire usage' (*HN*, v.103). A general diatribe follows (*HN*, v.103-104):

La nature est donc bien éloigné de s'assujettir à des causes finales dans la composition des êtres [...]. Peut-être y a-t-il, dans la plupart des êtres, moins de parties relatives, utiles ou nécessaires, que de parties indifférentes, inutiles, ou surabondantes [...]. Nous ne faisons pas attention que nous altérons la philosophie, que nous en dénaturons l'objet, qui est de connaître le *comment* des choses, la manière dont la nature agit; et que nous substituons à cet objet réel une idée vaine, en cherchant à deviner le *pourquoi* des faits, la fin qu'elle se propose en agissant.

Buffon goes on to identify other useless and even counter-productive organs in the *Histoire des quadrupèdes* and the *Histoire des oiseaux:* the allantois (*HN*, v.105-106), male nipples (*HN*, v.106-107), much of the sloth's body (*HN*, xiii.36-41), and several bird beaks (*HNO*, iii.449-50, vii.108-10, 136-39). Many of these passages are accompanied by tirades against the epistemological hubris of *cause-finaliers*, who confuse their reasons with God's. These two age-old arguments – the argument via useless organs and the argument via limitations on possible knowledge – were the favourite arms of anti-finalists in the Enlightenment.

But as Kant observed, among others, the arguments were contradictory. For if the human mind is too limited to discover final causes, as Buffon and other anti-finalists contended, how could they claim to know that something was useless? Knowing that an organ is useless – or, worse still, inimical to its possessor – would seem to require the same God's-eye view Buffon and others denied to teleologists. Like many appeals to epistemological humility, this one was hypocritical to a degree. Finalists too presented themselves as the party of intellectual restraint and their enemies as the party of reckless hypothesising. Both camps were in fact extrapolating from local common sense to philosophical absolutes – hardly a modest enterprise, and not one conducive to formal consistency. In the vocabulary of anti-finalists, terms

27. *HN*, i.145-46; Jean Seidengart, 'Le traitement du problème cosmologique dans l'œuvre de Buffon', in *Buffon 88*, p.320. For critiques of external teleology not mentioned below see *HNO*, i.41; *HNS*, ii.537.

such 'useless' and 'malformed' shifted symptomatically between informal meanings and more abstract ones. In a typical instance Buffon first describes the cross-bill's beak as a defect and an error of nature (*HNO*, iii.449-50), his judgement reflecting an immediate impression of dysfunctionality and aesthetic imbalance.[28] But then quickly turning to a more philosophical kind of usefulness, he grants that the bird's beak does have a use, like everything else, since 'il n'existe rien qui n'ait des rapports et ne puisse par conséquent avoir quelqu'usage' (*HNO*, iii.450). Shifts between everyday and abstract views of utility were characteristic of eighteenth-century polemics regarding external final causes.[29]

Despite the absoluteness of many of their pronouncements, anti-finalists mostly recognised that certain natural objects were marked by design. As we have seen, Buffon's article on the pig contrasts useless organs with useful ones. His articles on the toucan and the sloth, both malformed animals in his estimation, divide nature into two domains: one of regularity and beauty and one of imbalance and dysfunction (*HNO*, vii.109-10). Only regarding the latter group do the articles denounce the search for external final causes (*HN*, xiii.40). At the level of the foetus Buffon distinguishes between parts essential for development and those which 'semblent être faites autant pour l'ornement, la symmetrie et la perfection extérieure' (*HN*, ii.367). In his discussions of organs he regularly evokes their functions: seeing for the eyes (*HNS*, iv.421), protecting and lubricating the eyes for the eyelids (*HN*, ii.524), covering and balancing the body for feathers (*HNO*, iv.402), agitating spermatic fluid for spermatozoa (*HN*, ii.230), propulsion for webbed feet (*HNO*, vii.233), grasping and biting for beaks (*HNO*, vii.109), and so on. Functions are not the same thing as external final causes, but the line between the two has always been blurred.[30] In any case these assertions clash with Buffon's insistence that philosophers only ask *how*, never *why* – not to mention his one-time argument that the eye was *not* made for seeing (*HN*, ii.78).

In their conscious manifestos, anti-finalists no less than finalists generally avoided highlighting such ambiguities. Each side had its own battlefields of incontestable examples onto which the other side rarely dared venture. La Mettrie and Buffon may have boldly denied that the eye was for seeing, just as obstinate finalists managed to find divine providence in mosquitos and bicephalous animals, but these were momentary forays into hostile territory. Buffon and other anti-finalists concentrated their arguments on seemingly freakish animals and seemingly minor organs, as if to acknowledge that other

28. Aesthetics and normalcy outweigh functionality in many of Buffon's judgements about nature's errors. Thus, for example, while recognising the pangolin's armour as a superlative defence (*HN*, x.181-83), he nonetheless pronounces the animal malformed (*HN*, x.186; *HNO*, vii.136-37). For more on the role of aesthetics in Buffon's natural history see Geoffrey Bremner, 'Buffon and the casting out of fear', *SVEC* 205 (1982), p.82-83.

29. See Curran and Graille, 'Faces of eighteenth-century monstrosity', p.7, 12; Tort, *Ordre et les monstres*, p.41-42, 53-54.

30. Larry Wright, 'Functions', in *Purposive behaviour*, p.141-64.

natural objects exhibited purpose. Finalists, for their part, focused their case on plainly functional organs – especially the eye – and ecological balance, not 'anomalies' such as cannibalism or menstruation. In a sense these emphases merely reflected the empirical, commonsense motivations of each side's position, but they were also rhetorically motivated, designed as they were to highlight each position's strengths without delving too deeply into possible shortcomings.

When they converged on the same terrain through effort or accident, finalists and anti-finalists sometimes found their differences less remarkable than slogans might have it. The reactions of Haller and Bonnet to Buffon's treatment of the pig's 'useless' phalanges are suggestive in this light. Haller's response to Buffon's provocation was an equally truculent manifesto of finalism:

> [Buffon] raisonne en athée et n'est point physicien. Les osselets dont il parle font le même effet que le péroné; ils donnent origine à des muscles. [...] Les mamelons des mâles sont capables de devenir de véritables mamelles, et de suppléer au défaut d'une mère. Ils ne sont pas plus inutiles, que les mamelles des filles que Dieu prévoit devoir mourir vierges. Il doit arriver très rarement à un homme attentif d'ignorer l'utilité d'une partie.[31]

In contrast Haller's more empathetic friend Bonnet questioned Buffon's motives but agreed that organs need not have discoverable uses and that searching for relationships might be more profitable than looking for final causes.[32] This was the same Bonnet whose exuberant finalism worried Réaumur! Theology aside, 'extremists' on the issue of external teleology were not necessarily worlds apart.

The example of William Smellie, the pre-eminent eighteenth-century translator of the *Histoire naturelle* into English, is also significant. One of Smellie's friends tried to dissuade him from translating a series of which parts were 'atheistical', but Smellie decided not to heed the advice. In his resulting *Natural history* (1780-1785), however, he used his introduction and footnotes to express firm opposition to Buffon's dismissal of external final causes.[33] This was mostly a symbolic gesture – and perhaps a requisite piece of marketing in Protestant Britain – since Smellie himself admitted 'the absurdity of employing a final cause as a physical principle'.[34] In his *Philosophy of natural history* (1790-1799) Smellie mentions final causes but gives more attention to mechanical ones. The final cause of hibernation, for example, he identifies as avoiding starvation in a season of scarce food. But two pages later he complacently paraphrases Buffon's assertion that cold temperatures are the 'chief, if

31. Haller to Bonnet, 5 October 1761, in *Correspondence Haller-Bonnet*, p.246.
32. Bonnet to Haller, August 1761, in *Correspondence Haller-Bonnet*, p.244.
33. Loveland, 'French thought', forthcoming.
34. William Smellie (translator and ed.), *Natural history, general and particular, by the Count de Buffon*, 4th edn, 10 vols (London 1812), ii.416n.

not the only cause' of animals' 'torpor'.[35] Again, finalism and anti-finalism sometimes spoke the same language outside their wilfully provocative manifestos.

Along with his anti-finalism Smellie rejected Buffon's view of animals as thoughtless contrivances. The two issues had long been coupled, since animal cognition implied finality in one form or another. Regarding animals no less than external final causes, Buffon came close to Descartes's 'extremism' and parted company with common sense as well as academic 'moderates' such as Réaumur. It was Réaumur, in fact, who triggered Buffon's major attack on animal intelligence, the 'Discours sur la nature des animaux' of 1753.[36] As Hanks observes, Buffon attributed the field mouse's preparations for winter to 'foresight' in 1742, an attribution he explicitly repudiated in the 'Discours sur la nature des animaux'.[37] This may be a sign of evolution, but the polemical circumstances of the 'Discours' probably pushed him to exaggerate his position as well.[38] For with the 'Discours' Buffon sought vengeance for Réaumur's support of Joseph-Adrien Lelarge de Lignac's *Lettres à un Américain*, a blistering critique of the *Histoire naturelle*. This he accomplished by proving that insects are neither intelligent nor worthy of study, both major premises of Réaumur's *Mémoires pour servir à l'histoire des insectes*. More generally Buffon argued that animals are 'purement méchanique', human beings excepted (*HN*, iv.22-23). This assertion drew outcries from Smellie as well as French thinkers. But as Condillac noted, Buffon spent much of the rest of his 'Discours' retreating from the consequences of his animal-machine.[39] Just as Descartes conceded that animals possessed life and sentiment in the words' 'vulgar' acceptations, Buffon ended up giving animals 'material' forms of memory, feeling, imagination, and so on.[40] However convincing these distinctions may be, they were clearly a step back from his strongest pronouncements. Nor did Buffon describe individual animals as mechanical in subsequent volumes – quite to the contrary, as we will see. Yet Buffon's ephemeral, little-substantiated evocation of the animal-machine defined his position for critics. What fuelled quarrels over animal nature, as it did quarrels over external final causes, was less a lack of common ground than the symbolic engagement of authors on one 'side' or another.

35. Compare William Smellie, *The Philosophy of natural history*, 2 vols (Edinburgh 1790-1799), ii.405-12 and *HN*, viii.158-65, 224-25. Notice also Smellie's doubts about the final cause of poisonous animals in *Philosophy of natural history*, ii.480-81.

36. Roger, *Buffon*, p.317-20.

37. Hanks, *Buffon avant l'Histoire naturelle'*, p.187-88.

38. Certainly animal foresight achieved tacit recognition in subsequent articles of the *Histoire naturelle*. See for example *HN*, viii.293; *HNO*, iii.469-71; Philibert Guéneau de Montbeillard, *HNO*, ii.177-78, v.346-47. Notice also the refusal of implausible but not all animal foresight in Guéneau, *HNO*, ii.452-55, 459-60, 471.

39. Condillac, *Traité des animaux*, p.246.

40. See Roger, *Buffon*, p.320-29; Descartes, *Meditationes de prima philosophia*, p.426 ('Sextae responsiones'); Gaukroger, *Descartes*, p.166-67, 276-90, 324-25.

This is not to suggest that Buffon's or others' attacks on humanised animals and external final causes were scientifically meaningless. Similar attacks on the part of Descartes and others certainly contributed to the caution regarding external finality in the Académie des sciences. But the emphatic anti-finalists of the mid-eighteenth century exaggerated their differences with respect to conservatively speaking members of the Académie des sciences. Their highlighting of 'useless' organs and natural monstrosities betrayed their inability to overcome the commonsense perspective in which nature is ordinarily regular and purposeful. Likewise the model of the animal-machine resuscitated by Buffon and La Mettrie proved less than mechanical when subjected to close scrutiny. It was also irrelevant to these authors' descriptions of nature, which remained for the most part as anthropomorphic as those of their academic contemporaries. In the end what distinguished anti-finalists from their opponents was less their 'rejection' of external final causes or animal cognition than their replacement of pre-existence with internal finality, a far-reaching though always controversial solution.

Overstated as they were with respect to authors' beliefs, eighteenth-century attacks on external final causes were important symbolically, as demonstrated by the reactions that they elicited. Anti-finalism was not necessarily heterodox. To judge by its selection of 'reprehensible' passages from the opening volumes of Buffon's *Histoire naturelle*, for example, the Sorbonne did not consider his anti-finalism flagrantly or unproblematically heretical. But anti-finalism's classic theological basis, God's inscrutability, seems to have enjoyed less favour in the mid-eighteenth century than in the time of Descartes and Arnauld. Readers had been flocking to physico-theology since the 1710s. In spite of their considerable theological differences, writers as divergent as Fénélon, Pluche, and Guillaume-François Berthier embraced the vogue.[41] In this climate attacks on final causes risked raising ire. But it was also this climate that incited the attacks, most of which were protesting not so much against the moderation of academic finalism as against the extremes of physico-theology. Buffon, we shall see, both decried and participated in this literary tradition.

Using final causes and humanised animals

The above section summarises the philosophical context and substance of Buffon's critiques of external final causes and anthropomorphism. As just suggested, the context of literature, broadly defined, provides an equally rewarding framework for studying finality in the *Histoire naturelle*. Humanised animals have always been a popular literary motif, and religious moralists have long discovered providence in the phenomena of nature. But both of these tendencies were magnified in the Enlightenment, ironically at the very

41. As Ehrard notes, Christians and especially Jesuits grew warier of 'heavy' finalism as the century wore on. See Ehrard, *Idée de nature*, p.635-45.

time the Académie des sciences was urging restraint in assigning final causes. At a most general level the fashionableness of natural history from the late seventeenth century onward brought nature to the forefront of cultural awareness, creating opportunities for observing final causes and humanising animals. Intelligent animals became all the more current in eighteenth-century letters thanks to La Fontaine's spectacularly successful *Fables* (1668-1694) and their imitators.[42]

Meanwhile works displaying God's providence flooded Protestant Europe, especially Britain, in the late seventeenth century before arriving in France around 1710. The origins of this providential literature have been variously rationalised. It probably owed something to the christianised Galenism dominant in sixteenth- and seventeenth-century medical teaching.[43] It found support in mid-seventeenth-century Protestantism, which downplayed miracles, human sinfulness, and allegorical views of nature, stressing on the contrary nature's everyday benevolence.[44] The secular rationalism of Catholic France's natural philosophy also pushed Protestants to search for an alternative. But by the late seventeenth century the 'marvels of nature' revealed by the microscope had combined with the stalemate of pre-existence to make even French philosophers respectful of 'particular providence'. In the public at large, nostalgia for rural living and dissatisfaction with urban 'corruption' favoured the pastoral meditations of physico-theology.[45]

French interest in providential literature is often said to have peaked in the first half of the eighteenth century.[46] Not only were the classics of Protestant physico-theology translated into French then – John Ray's *Wisdom of God* (1691-1704) in 1714, Derham's *Physico-theology* (1713) in 1726, and Nieuwentyt's teleological proof of God's existence (1715) in 1725 – this was also the epoch of Fénelon's *Traité de l'existence de Dieu* (1713) and Pluche's *Spectacle de la nature*. But texts celebrating natural providence hardly died out in the second half of the century. The subject might be banal, as the *Journal encyclopédique* declared in 1760, but Nieuwentyt's treatise and especially Pluche's *Spectacle* continued to be printed while new studies of natural finalism were authored by Bonnet, Bernardin de Saint-Pierre, and many other authors.[47] Throughout the century the public had a huge appetite for final causes and providence.

42. Jean-Noël Pascal (ed.), intro. to *La Fable au siècle des Lumières, 1715-1815. Anthologie des successeurs de La Fontaine, de La Motte à Jauffret* (St Etienne 1991), p.7-12, 19-21, 28. The success of these fables was probably due in part to the decline of trope and allegory in scientific natural history, but as this chapter demonstrates, the boundaries between literary and scientific natural history remained ill-defined throughout the eighteenth century. See Harrison, *Bible, protestantism*, p.191-93, 270-72.

43. See Roger, *Sciences de la vie*, p.74-78; Sloan, 'Question of natural purpose', p.124.

44. Keith Thomas, *Man and the natural world: changing attitudes in England, 1500-1800* (Oxford 1983), p.20; Daston, 'Marvelous facts', p.114, 123-24; Harrison, *Bible, protestantism*, p.161-62.

45. See Roger, *Sciences de la vie*, p.240-49, 449.

46. See for example R. R. Palmer, *Catholics and unbelievers in eighteenth-century France* (Princeton, N.J. 1939), p.106; Norman Hampson, *The Enlightenment* (New York 1968), p.81-84; Mornet, *Sciences de la nature*, p.152; Ehrard, *Idée de nature*, p.635-45.

47. See Mornet, *Sciences de la nature*, p.29-36, 148-59.

Buffon is usually placed outside this tradition of providential literature because of his attacks on external final causes. Paradoxically, however, the *Histoire naturelle* often describes nature in finalistic language. Again and again Buffon evokes purposes for the existence of animals, organs, behaviours, and other phenomena. Most of the time he makes these assertions conjectural by qualifying them with expressions such as 'it seems', 'it appears', or 'so to speak'. In this mode, for example, he writes that nature 'appears' to have created the sheep so that every part of its body would be useful to humans (*HN*, v.6), that nature 'appears' to have designed the baboon for sexual debauchery (*HN*, xiv.135), that nature 'appears' to have made the seal's face as hydrodynamic as possible (*HNS*, vi.304), and that nature 'appears' to have stored up vast deposits of salt and coal to provide for humanity's present and future needs (*HNM*, ii.298; *HNS*, v.112). Buffon's use of verbs and adverbial phrases to attenuate the reality of his providential declarations recalls the grammarian César Chesneau Du Marsais's suggestion that hyperbole be accompanied by similar qualifiers.[48] It also recalls Buffon's use of the oxymoron 'almost infinite' to designate large quantities, seemingly an effort to weaken a beloved overstatement so as to make it more logical.[49] In each case we find a solution to the age-old dilemma of wanting to say something without really asserting it.

Occasionally Buffon proposes final causes without adding qualifiers to make them conjectural. This is often the case with his favourite form of internal finality: nature's favour for life and fecundity. More rarely he presents external final causes as matters of fact. Because the mole is blind, he writes for example, nature gave it huge genitals in 'compensation' (*HN*, viii.81). This notion of compensation takes on an ecological meaning in the article 'Rat', where Buffon sees fecundity as nature's compensation for small, defenceless animals. Once again the argument is purely declarative: 'Nous trouverons que la nature a su tout compenser.'[50] Similarly finalistic compensations and balances are mentioned elsewhere in the *Histoire naturelle*, as in natural histories since time immemorial, but mostly they are stated as mere possibilities.[51] The platitude of nature's fairness is pointedly refuted in the article on sloths: 'Pourquoi n'y aurait-il pas des espèces entières d'animaux créées pour la misère?' But at the last moment Buffon discovers a compensation for the sloth's many handicaps – its insensitivity to suffering – and this authorises

48. *Des tropes, ou des différents sens dans lesquels on peut prendre un même mot dans une même langue*, in *DES TROPES ou des différents sens, FIGURE et vingt autres articles de l'Encyclopédie, suivis de L'ABREGE DES TROPES de l'abbé Ducros*, ed. Françoise Douay-Soublin (Paris 1988), p.133.

49. In the 'Premier discours', typically, Buffon uses the word 'infinite' and its derivatives eight times without qualifiers (*HN*, i.8, 10, 11, 14, 17, 18, 61) and four times with 'almost' (*HN*, i.19, 47, 59). Other writers too used the expression 'almost infinite', but Buffon had especial reason to feel uncomfortable with it in view of his denunciation of the whole concept of infinity. See chapters 3 and 5, p.80, 134.

50. *HN*, vii.278. For a similar discussion see *HN*, iv.438-39.

51. See for example *HN*, vi.255, xv.22; *HNO*, vii.280.

his conclusion that 'dans ses productions les plus négligées, la nature paraît toujours plus en mère qu'en marâtre'.[52]

If taken seriously Buffon's more extravagant final causes are little different from the ones mocked in the works of the century's *cause-finaliers*. His suggestion that nature develops useful mineral compounds and deliberately avoids dangerous ones such as gunpowder and poisons (*HNM*, iii.398) may not be as ridiculous as Pangloss's contention that the nose was made to hold glasses up, but it could easily fit into one of Voltaire's satires. Likewise Buffon's suggestion that the stormy petrel was created to warn sailors of coming storms might be read as a corollary to Fénelon's credo that 'non seulement les plantes, mais encore les animaux, sont faits pour notre usage'.[53] Some of Buffon's final causes probably came from the same sources as those of his contemporaries: Pliny, Aldrovandi, British physico-theologians, and others. Nor is the *Histoire naturelle* always far from Christian apologetics. Several of Buffon's animals are assigned the purpose of making humans humble, albeit before nature rather than God.[54] Buffon also attributes nature's variety and plenitude to its desire to awe or confuse human beings. For Christians and non-Christians alike, this discovery of human humility as the purpose of the purposeless was the paradoxical terminus of extreme finalism.[55]

Read in light of the rest of the *Histoire naturelle*, few of the external final causes Buffon proposes can be said to reflect his definitive opinions. As we have seen, his implicit beliefs about purpose in nature were more mixed and moderate than his official declarations against final causes. No more than other anti-finalists could he look at living organisms without seeing design and function. Internal finality – which he cultivated, like other anti-finalists – helped account for life's purposiveness but in many contexts lacked the intelligibility of explanations appealing to external teleology. The resulting tension between his intuitions and his theoretical principles may have led Buffon to experiment with external final causes in a conjectural perspective. Certainly some of the goals the series proposes for transcendent nature – especially life, natural balance, and overall functionality in living beings – appear to have been seriously entertained by Buffon, whether as heuristics or as exploratory half-beliefs. It seems unlikely, however, that he placed any real credence in the series' many extreme or anthropocentric final causes. Why, then, are these final causes included at all?

Some are obviously no more than literary conceits. It is nearly impossible, for example, to see Buffon's suggestion that nature made deer to add beauty to the forest (*HN*, vi.63) as anything but a precious way of stating that deer are

52. *HN*, xiii.34-41, 45-46. Notice also Buffon's contention that 'l'on voit par les erreurs même de la nature qu'elle se méprend toujours le moins qu'il est possible' (*HN*, ii.371-72).

53. *HNO*, ix.328-29; François de Salignac de La Mothe-Fénelon, *Traité de l'existence et des attributs de Dieu*, in *Œuvres de Fénelon* (Versailles 1820), i.1-284, at p.32.

54. See for example *HNO*, viii.456, ix.298.

55. See for example *HN*, xii.408; *HNO*, i.397, vii.237, viii.115. See also Lepenies, *Ende der Naturgeschichte*, p.63-64; Harrison, *Bible, protestantism*, p.162.

beautiful and live in the forest. A comparable analysis holds for his statement that the chaffinch was made to be the minstrel of the woods (*HNO*, iv.114). Like the majority of Buffon's external final causes, these two are all the more poetic for drawing on the trope of personified nature.[56] Given their potential for creating literary effect, it is not surprising to find final causes plentiful in 'stylish' parts of the series: opening paragraphs, major discourses, and broad views of nature. Nor is it surprising to find final causes multiplied and abused by Buffon's assistants and successors, concerned as they were with showing their mastery of a mythical style. Many of the above examples were probably conceived by Bexon before being edited and approved by Buffon. One of Guéneau's first articles for the *Histoire des oiseaux* resuscitated the language of Buffon's article on the sloth to describe the dysfunctional dodo (*HNO*, i.480-81). Lacépède, for his part, used his first paragraph on an animal in the *Histoire des quadrupèdes ovipares et des serpents* to demonstrate his proficiency in Buffonian grandeur. Significantly this implied a reference to nature's compensations alongside the predictable noble vocabulary.[57] Allusions to nature's purposes were seen as an aspect of Buffon's style and thus as a contributor to his popularity.

While literarily ornamental to a degree, many of Buffon's final causes also have pedagogical value. Evoking nature for the uninitiated meant capturing its appearance as well as a possibly deeper reality, and this led Buffon to emphasise apparent design, regardless of its credibility to him or to others. In the case of the zebra, never seen by most readers, external teleology allowed him to recreate a probable spontaneous impression: 'Il semble que la nature ait employé la règle et le compas pour [peindre le zèbre]' (*HN*, xii.2). Sometimes his final causes are coupled with obviating efficient causes, a sign that their role is mainly provisional. Occasionally this relationship is spelled out, as when Buffon proposes a finalistic reason for the existence of mountains, pronounces it physically irrelevant, and goes on to explain the same phenomenon in mechanical terms.[58] More often the priority of efficient causes is left unstated. Buffon's suggestion that nature designed species to fit their climates, and climates to fit their species cannot be taken as his last word, for example, since he identifies a non-teleological reason for the correspondence in the same paragraph: the change-provoking influence exerted on organisms by their environments (*HN*, vi.56-57). But by recasting his correlation in the language of finality and poetic epigrams, Buffon stands a better chance of reaching his

56. On personification in general in the *Histoire naturelle* see Höhne, *Stil Buffons*, p.49-56. As Darwin pointed out, personification is also 'almost necessary for brevity'. On Darwin's anthropomorphism see Gillian Beer, '"The face of nature": anthropomorphic elements in the language of *The Origin of species*', in *Languages of nature: critical essays on science and literature*, ed. L. J. Jordanova (New Brunswick, N.J. 1986), p.212-43; Sloan, 'Question of natural purpose', p.132-41.

57. Bernard-Germain-Etienne de La Ville-sur-Illon de Lacépède, *Histoire naturelle des quadrupèdes ovipares et des serpents*, 2 vols (Paris, Hôtel de Thou 1788-1789), i.45-46. Lacépède begins four more articles in the same volume with the familiar conceit of provident nature. See *Histoire naturelle des quadrupèdes ovipares*, i.54, 188, 267-69, 309-10.

58. *HN*, i.308. See also *HN*, ii.160-61.

readers. More generally his many associations of 'well-adapted' organs with nature's providence might have been avoided in favour of mechanical explanations perhaps truer to his theoretical views, but the passages in question would have been longer and less comprehensible.[59] Even today, while professional biologists take pride in their theoretical extirpation of final causes from the theory of evolution and their discipline in general, popular biology retains older, more teleological explanations for their simplicity and intelligibility.

Nor should Buffon's attribution of human qualities to animals necessarily be read as a theoretical commitment. As we have seen, his claim that animals are mechanical was ambiguous from the start. While firm in his conviction that human beings are far superior to animals mentally, Buffon never returned to the stronger assertions of the 'Discours sur la nature des animaux'. His descriptions of animals regularly draw on an anthropomorphic vocabulary at odds with the 'Discours', but not all of this humanising should be taken at face value. Buffon, like all ethologists, probably had trouble not seeing human qualities in the animals he dealt with, but he also used anthropomorphism as a pedagogical and literary tool, sometimes in telltale conjectural constructions.[60] He was not alone in doing so. Humanised animals, like Mother Nature, have been a stock element of natural histories since antiquity. Avoiding them was especially difficult in the Enlightenment because of ingrained habits and reluctance to cultivate specialised jargon. As Fontenelle wrote of natural history in a review of Réaumur's *Mémoires pour servir à l'histoire des insectes*, 'On emploie malgré soi dans toute cette matière des expressions impropres de connaissance et de prévoyance.'[61] But writers on nature, including Fontenelle in his following comments, also humanised animals to be entertaining and intelligible. As the title indicates, for example, pleasure was a guiding principle for Linnaeus's supreme exercise in anthropomorphism, the *Deliciae naturae*.[62]

Among eighteenth-century naturalists Buffon remains especially stigmatised for his humanised animals, thanks in part to his fame and probably also to the legend that he made the lion 'king of the animals'. Already in January 1795, in a lecture on natural history at the Ecole normale, Daubenton denounced Buffon's lion and his anthropomorphism in general. Against Buffon he argued that the lion did not have courage, nobility, or a confident gaze. Even its roar, he insisted, was not terrible but 'plaintive', an anthropomorphic attribution he later retreated from. Daubenton concluded drama-

59. See for example *HNO*, vii.233-34, viii.302-303, 472.

60. See Roger, *Buffon*, p.219-20, 370-71, 376-77; Mengal, 'Psychologie de Buffon', p.607-608; Martinez-Contreras, 'Des mœurs des singes', p.566-67; Richard W. Burkhardt, Jr, 'Le comportement animal et l'idéologie de domestication chez Buffon et chez les éthologues modernes', in *Buffon 88*, p.572-73.

61. *HARS* 1740 (1742), p.8. See also chapter 1, p.44-45, and Henri Daudin, *De Linné à Jussieu. Méthodes de la classification et idée de série en botanique et en zoologie (1740-1790)* (Paris [1926]), p.107-108.

62. Lindroth, 'Two faces of Linnaeus', p.7-8.

tically: 'Le lion n'est pas roi des animaux: il n'y a point de roi dans la nature.'[63] Like his demonstration that bees have no queens and his critique of the 'kingdoms' of natural history, Daubenton's dismissal of the lion-king was welcome news to republicans in 1795. His students at the Ecole normale are said to have applauded for fifteen minutes in response to his dethronement of Buffon's lion, though at least one was not convinced that the lion's roar was plaintive.[64] Etienne Geoffroy Saint-Hilaire would later insist that Daubenton's 'lessons' were pronounced with no thought of their political connotations, but this seems implausible in light of his selective concern with royal anthropomorphism.[65] In any case the cliché of the lion-king predated Buffon, and nothing suggests that he took it very seriously.[66] Bonnet was both fairer and more perceptive than Daubenton when he dismissed the gross anthropomorphism of the *Histoire des oiseaux* as nothing but poetry.[67] This is not to imply that Buffon's anthropomorphism and allusions to natural hierarchy are uniformly superficial. His theory of degeneration presupposed a kind of ranking of living beings. But his anthropomorphism was neither as direct nor as naive as Daubenton and others later suggested, and it was hardly exceptional in the eighteenth century. Daubenton himself humanised animals in the same way. In the *Encyclopédie méthodique*, for example, he portrayed the hamster as a model warrior.[68] As we have seen, he even defended anthropomorphic vocabulary with regard to the horse (*HN*, iv.268-69), while Buffon adopted more technical terminology.

63. Daubenton, 'Histoire naturelle', in *Séances des écoles normales. Leçons*, i.290-93. For a later denial that he called the lion's roar plaintive see Louis-Jean-Marie Daubenton, 'Histoire naturelle', in *Séances des écoles normales, recueillies par des sténographes et revues par les professeurs. Débats*, new edn, 3 vols (Paris, Imprimerie du Cercle-social 1800-1801), i.93-95.

64. On queen bees and kingdoms see Daubenton, 'Histoire naturelle', in *Séances des écoles normales. Débats*, i.92-93, 96. For reactions to these statements see Daubenton, 'Histoire naturelle', in *Séances des écoles normales. Leçons*, i.291n.; Sébastien Gérardin, 'Cours élémentaire de l'histoire naturelle des animaux, ou leçons extraites des plus célèbres naturalistes', 2 vols, BMHN, Ms. 1537-38, i.8-9.

65. For Geoffroy Saint-Hilaire's argument see Frank A. Kafker and Serena L. Kafker, *The Encyclopedists as individuals: a biographical dictionary of the authors of the 'Encyclopédie'*, *SVEC* 257 (1988), p.90n. See also James Guillaume, 'Le berger Daubenton. Encore une légende contre-révolutionnaire', in *Etudes révolutionnaires* (Paris 1908-1909), i.195-96. Notice Daubenton's glorification of working bees in 'Histoire naturelle', in *Séances des écoles normales. Débats*, i.92-93. In addition, his lion of 1795 was noticeably less noble than previous versions. See for example 'Cours de zoologie (mammifères) par Daubenton, professé après 1786, soit à l'Ecole normale, soit au Collège de France, soit à l'Ecole d'Alfort', BMHN, Ms. 809, f.145*r*; Daubenton, *HN*, ix.26.

66. Roger, *Buffon*, p.10-11. As Roger observes, the reference to the lion-king is ironic in the article 'Lion' (*HN*, ix.4). As noted below, a reference to the lion-king in 'Tigre' is translated immediately into a statement about force.

67. Bonnet to Haller, 25 June 1771, in *Correspondence Haller-Buffon*, p.942.

68. Emma Spary, 'Political, natural and bodily economies', in *Cultures of natural history*, p.192-93. Even in his lectures at the Ecole normale, Daubenton insisted that the sciences be appealing to amateurs, a statement more in tune with his practice as a naturalist than his harsh condemnation of style and anthropomorphism in the lecture discussed above. See Daubenton, 'Histoire naturelle', in *Séances des écoles normales. Leçons*, iv.133-34.

But even when superficial with respect to Buffon's theories, anthropomorphism, like finalism, helped make the *Histoire naturelle* popular and understandable. Buffon's articles on the vulture, the wolf, and the pigeon, for example, would have been considerably less entertaining without their portraits of their subjects as cowardly criminals (*HNO*, i.146-47), bellicose sociopaths (*HN*, vii.41-42), and consummate lovers (*HNO*, ii.523), respectively. His article on the beaver would have been less picturesque and less comprehensible without its evocation of the lodge's 'tapis', 'fenêtres', 'balcon', and 'bain' (*HN*, viii.295). In concession to exacting readers, the series' anthropomorphic terminology is sometimes translated into less human language or subjected to caveats. The lion is king of the animals, Buffon writes in 'Tigre', only in the sense of being the most powerful.[69] Working under Buffon's supervision and perhaps from his notes, Guéneau warns readers of the *Histoire des oiseaux* that his language notwithstanding, birds' 'loves' should never be taken for anything more than reproductive behaviours (*HNO*, iii.256). Within chapters he is back to erotic, emotional avian love (*HNO*, iii.359).

Buffon, like Guéneau, seems to have felt no inconsistency in evoking final causes for literary and pedagogical purposes even as he rejected them on theoretical grounds. Nor does the discrepancy appear to have troubled his readers. Some took his anti-finalism as his last word and ignored his frequent recourse to finalistic language, but even the less than orthodox were not above enjoying poetic finalism. The *Journal encyclopédique*, for example, received Buffon's final causes and moralising anthropomorphism with fervent enthusiasm even while admitting that animals were alien to human values.[70] Other readers seem to have overlooked or ignored Buffon's rare attacks on finalism and appreciated his evocations of final causes as displays of true piety. The *Journal de Trévoux* balked at the 'Discours sur la nature des animaux' but exulted over Buffon's conjectural final causes, especially his suggestion that animal nuisances exist to remind humans of their fallen condition. Not coincidentally this was the same periodical that praised the engraving of God creating the Earth which accompanied Buffon's anti-biblical *Histoire et théorie de la terre*.[71] The Jesuit François-Xavier de Feller savoured Buffon's final causes and wrote off his anti-finalism as a momentary lapse.[72] Jean-L. Ferri de Saint

69. *HN*, ix.130. See also Guéneau's retreat from a reference to the mocking bird's 'soulful' singing in *HNO*, iii.326.

70. For praise of Buffon's finalism and humanisation see *Journal encyclopédique* 7.2 (October 1765), p.27-28, 4.3 (June 1772), p.342-53, 7.2 (October 1780), p.191-204, 8.3 (December 1781), p.420-29. For an admission that animals are not moral beings see *Journal encyclopédique* 8.3 (December 1781), p.421-23.

71. See *Journal de Trévoux* (October 1749), p.2232-35, 2240 (November 1749), p.2367-69 (December 1753), p.2841-45 (January 1754), p.144. The final cause of animal nuisances can be found in *HN*, iv.171. On the engravings for the *Histoire et théorie* see Bassy, 'A l'heure des grandes synthèses', p.176-80.

72. Mornet, *Sciences de la nature*, p.150-53. See also John Pappas, 'Buffon vu par Berthier, Feller et les *Nouvelles ecclésiastiques*', *SVEC* 216 (1983), p.26-28; John Pappas, 'Buffon matérialiste? Les critiques de Berthier, Feller et les *Nouvelles ecclésiastiques*', in *Etre matérialiste à l'âge des Lumières. Hommage offert à Roland Desné* (Paris 1999), p.243-45.

Constant's introduction to *Génie de M. de Buffon* (1778), an early abridgment of the *Histoire naturelle*, used paraphrases from the series to depict Buffon as a thoroughgoing finalist.[73] Such 'misreadings' support Alain-Marie Bassy's suggestion that the *Histoire naturelle* contained something for everyone, regardless of credo.[74]

In a more critical spirit Bonnet interpreted Buffon's attacks on external finalism as reflecting conviction, but he also remarked on Buffon's invocation of final causes, which in Bonnet's opinion had become scandalously frequent by the first volume of the *Histoire des oiseaux*. Bonnet may have been referring to anthropomorphism, which rose to new heights in the *Histoire des oiseaux* as a result of birds' musical, 'feminine', 'passionate' natures, their symbolic status, and their seemingly purposeful nests and migrations.[75] Or he may have been referring to conjectures about external providence, which remain as numerous as ever in the *Histoire des oiseaux*. Either way he should be taken seriously in his suggestion that final causes were a way of attracting and satisfying readers: 'Jamais Mr de Buffon n'avait fait un emploi plus fréquent des *causes finales*; c'est apparemment, qu'il les a jugées très propres à embellir ses tableaux.'[76] Neither Buffon nor his assistants probably thought about final causes with such lucid cynicism. Bexon, for his part, was a self-proclaimed finalist, quietly unhappy with his patron's view of the solar system as arising by chance.[77] In practice this divergence created minimal tension – less, for example, than Bexon's tendency to erudite prolixity – since Buffon no less than Bexon, Guéneau, or Lacépède recognised finalism as one of the series' more important stylistic and pedagogical tools.

Buffon's dual approach to final causes was not unique. As we have seen, Fontenelle was sceptical of the anthropocentric final causes characteristic of contemporary physico-theology, but they are staples of his *Entretiens sur la pluralité des mondes*, where the characters delight in discovering purposes and harmonies in the context of astronomy. Traces of this finalism even trickled into Fontenelle's *Histoire de l'Académie*, as in his reviews of Réaumur's *Mémoires*.[78] Many vulgarisers, likewise, were liberal with final causes relative to their sources. The close ties between final causes and vulgarisation are especially evident in popular abridgments of texts on nature, which regularly retain their model's final causes while dropping other material. Abridgments of the

73. [Jean-L. Ferri de Saint-Constant], intro. to *Génie de M. de Buffon* (Paris, Panckoucke 1778), p.xv-xvii. These passages are paraphrases of *HN*, xii.ix and perhaps also *HN*, i.308.

74. 'A l'heure des grandes synthèses', p.176-80.

75. On the meaning of birds in French culture see Spary, 'Codes of passion', p.123-33; Farber, *Discovering birds*, p.1-6.

76. Bonnet to Haller, 25 June 1771, in *Correspondence Haller-Bonnet*, p.942. See also Haller to Bonnet, 7 July 1771, in *Correspondence Haller-Bonnet*, p.947.

77. See Gabriel-Léopold Bexon, 'Discours sur les phénomènes de l'univers', BMHN, Ms. 804, discours ii, p.96-100.

78. See Alain Niderst, *Fontenelle à la recherche de lui-même (1657-1702)* (Paris 1972), p.278-81; Niderst, *Fontenelle*, p.93; Sutton, *Science for a polite society*, p.152-53; Roger, *Sciences de la vie*, p.237n. See also for example *HARS* 1736 (1739), p.8-37.

Histoire naturelle from Buffon's lifetime onward exemplify the tendency. Popular adaptations took even more liberties with respect to originals. A major source for many vulgarisations, Réaumur's *Mémoires* gave rise to texts that were considerably less circumspect in their approach to finality – Gilles-Auguste Bazin's *Abrégé de l'histoire des insectes* (1747-1751), for example, and Pluche's even more finalistic *Spectacle de la nature*.

It is important to remember the popularity of final causes in vulgarised natural history when examining the Enlightenment's most ridiculed finalists. Consider Pluche for example. In the footsteps of intellectuals since Pluche's lifetime it is easy to find comedy in the finalism of the *Spectacle de la nature*. This is a work asserting that nature was made for human beings in its entirety. More specifically Pluche states among other things that God created tides and made the sea salty to prevent it from putrefying, and that God designed atmospheric refraction to allow a little extra daylight immediately before sunrise and after sunset.[79] But despite his anthropocentric principles, many of Pluche's final causes are directed toward the well-being of plants and animals, not of human beings, and they were and remain more commonsensical than comic. His acceptance of 'providential' instinct makes his discussions of nesting and migration more intuitively reasonable than Buffon's in the 'Discours sur la nature des animaux', for example. Pluche also writes of the human 'uses' of natural objects as often as he does of their God-given 'purposes', making it unclear at times if he is listing final causes or simply the facts of human resourcefulness.[80] Several of his final causes – for example, the platitudinous ones concerning the 'uses' of mountains and the Earth's mineral 'stockpiles' – can be found in the *Histoire naturelle* too, albeit in conjectural form.[81]

Most of all, Pluche's final causes were part of a literary enterprise, not philosophical assertions thrown out in a vacuum. 'Il y a certainement plus que du badinage dans tout ce que nous venons d'entendre', observes Pluche's count after the *prieur* describes the donkey allegorically, but the count's words imply that not all is serious in the *prieur*'s gently mocked 'style' of doing natural history.[82] Here as elsewhere final causes were a means of attracting and entertaining an audience. In Pluche's case, in fact, they were part of a conscious campaign to satisfy readers. As Dennis Trinkle has shown, Pluche was under pressure from readers after volume i of the *Spectacle* to make nature more intelligible and providential, and he responded in later volumes with more final causes.[83] Like other finalists, Pluche was undoubtedly a

79. *Le Spectacle de la nature, ou entretiens sur les particularités de l'histoire naturelle, qui ont paru les plus propres à rendre les jeunes gens curieux, et à leur former l'esprit*, 9 pts in 8 vols (Paris, Etienne, 1763), ii.3, iii.190-93, 292-93.

80. See Hampson, *Enlightenment*, p.82.

81. Compare Pluche, *Spectacle de la nature*, ii.xi-xiii, iii.129-30; *HN*, i.308; *HNM*, ii.298; *HNS*, v.112. Pluche too occasionally uses conjectural final causes. This one, for the butterfly, could easily be mistaken for Buffon's or Bexon's: 'Il semble que la nature se soit fait un jeu d'étaler et de mélanger avec art tout ce qu'elle a de plus brillant' (*Spectacle de la nature*, i.62).

82. Pluche, *Spectacle de la nature*, i.353-58.

83. 'Noël-Antoine Pluche's *Le Spectacle de la nature*', p.131.

sincere believer in the final causes he proffered, but the finalism of his *Spectacle* had determinants going beyond simple inner conviction.

The same case can be made for Bernardin de Saint-Pierre. Like Pluche's *Spectacle*, Bernardin's enormously successful *Etudes de la nature* (1784-1788) represent providence as especially and sometimes exclusively concerned with humans: 'Toutes ces harmonies sont faites pour l'homme.'[84] Again like Pluche, Bernardin pushes this 'belief' to extremes on occasion. In one much ridiculed statement he suggests that nature divided melons into ribbed sections to facilitate their consumption by human families.[85] Elsewhere his final causes appear more promising in retrospect, as in his repeated argument that plants' and animals' colours are not accidental. Against Buffon he insists that the toucan's beak has a function, while like Pluche he accredits a number of Buffon's conjectural final causes.[86] All this might be expected of a naive finalist. But in other genres of writing Bernardin limits and alters his finalism. In his *Voyage à l'Ile-de-France* (1773) – a travelogue, not a celebration of natural harmony – he is mostly commonsensical about nature's purposes. While his *Etudes* claim that all plants were designed for human usage, his *Voyage* mentions plants and animals that are of no use to anyone.[87]

Perhaps Bernardin's ideas evolved between the *Voyage* and the *Etudes*. But the same sort of contrast can be drawn between the *Etudes* and his contemporaneous novel *Paul and Virginie*, which appeared in the third edition of the *Etudes* in 1788. In the novel Paul and Virginie grow up together in the wilderness of Ile-de-France (modern Mauritius), supervised by their mothers. Their budding love is frustrated by Virginie's forced departure for Paris. Some time later she returns unscathed from the corruptions of civilisation, but a hurricane kills her and a number of her shipmates in the harbour of Ile-de-France, causing many in the story to have doubts about providence. In the *Etudes* Bernardin proposes that nature uses hurricanes to clean the environment.[88] But this blithe suggestion is clearly inappropriate to the tragic conclusion of *Paul et Virginie*. The only providence the novel's narrator can discover in Virginie's death is abstract and contrary to the spirit of physico-theology: 'La mort, mon fils, est un bien pour tous les hommes.'[89] Even if heart-felt, the cheery, seemingly naive finalism of the rest of the *Etudes* was thus limited to a

84. Jacques-Henri Bernardin de Saint-Pierre, *Etudes de la nature*, in *Œuvres complètes*, ed. L. Aimé-Martin, 12 vols (Paris 1818), vol. iii-vii, at ii.154. On the success of the *Etudes* see Mornet, *Sciences de la nature*, p.155-56.

85. *Etudes de la nature*, ii.444.

86. On the toucan compare Bernardin de Saint-Pierre, *Etudes de la nature*, i.298-99; *HNO*, vii.108-12. For an example of common ground compare Bernardin, *Etudes de la nature*, i.292-95; *HN*, xiii.45-46.

87. Compare for example Jacques-Henri Bernardin de Saint-Pierre, *Voyage à l'Ile-de-France*, in *Œuvres complètes*, vol. i-ii, at i.94, 107; *Etudes de la nature*, i.64-65, ii.450. Practical as it is, the *Voyage* foreshadows the *Etudes* with its suggestion that the island's millipedes were created to keep people away from unhealthy locales. See *Voyage à l'Ile-de-France*, i.108.

88. Compare *Etudes de la nature*, i.46-47; Jacques-Henri Bernardin de Saint-Pierre, *Paul et Virginie*, in *Etudes de la nature*, iv.178-83.

89. *Paul et Virginie*, p.198. See also Mornet, *Sciences de la nature*, p.157.

specific context. Bernardin's extreme finalism was a literary stance, not just an abstract philosophical position.

The literary character of much extreme finalism becomes apparent on the margins of eighteenth-century natural history, where scientific pretensions gave way to erudite moralising in the name of religious instruction and bookish entertainment. One work outstripping Pluche and Bernardin de Saint-Pierre in this regard was Gaspard Guillard de Beaurieu's *Abrégé de l'histoire des insectes* (1764), a major goal of which was to make up for the 'lack' of moralising in popular natural history.[90] Beaurieu was not indifferent to scientific accuracy. While praising Johannes Goedaert's still popular, providential study of insects (1662-1669) for its 'tableaux instructifs', he lamented the work's evident errors of fact.[91] But the emphasis of Beaurieu's *Abrégé* is religious and humanistic: 'La partie physique n'est, pour ainsi dire, dans cet ouvrage que l'accessoire de la partie morale.'[92] Beaurieu frequently drifts into sermonic high style and includes quotations from La Fontaine, etymological studies, and interpretations of natural objects as moralistic emblems. Final causes and anthropomorphism are everywhere in the *Abrégé*, but they are no more obviously meant as scientific truths than the fables, emblems, and high-style ornaments surrounding and reinforcing them. Finalism here is part of a style and a genre as much as a philosophy.

Pluche, Bernardin, and Beaurieu were all successful authors. To judge by Daniel Mornet's list of the most represented books in eighteenth-century French libraries, many of the best-sellers of natural history were self-consciously finalistic.[93] There was clearly an audience for studies of nature's providence and symbolic meanings. As reactions to the *Histoire naturelle* show, even anti-finalists trafficking in conjectural final causes could tap into this market. Careless reading, it might be said, allowed some to miss the point of the *Histoire naturelle*. But this was a work that thrived on making different points on different levels. Nor did advocates of finalism necessarily consider teleology a scientific explanation. What they wanted in many cases was simple recognition of providence, which Buffon gave them. Furthermore, final causes were appreciated by most everyone, not just zealous finalists. As the example of Beaurieu's *Abrégé* indicates, they existed on a continuum with other pedagogical and literary devices. Final causes brought intuitively appealing *grandes vues* of nature. They functioned as intelligible abbreviations for more complex relations. They allowed authors to express commonsense impressions, if only as preludes to other explanations. And especially in the form of anthropomorphism, they were a time-honoured element of literary

90. [Gaspard Guillard de Beaurieu], *Abrégé de l'histoire des insectes dédié aux jeunes personnes*, 2 vols (Paris, Panckoucke 1764), i.iii, xxii-xxiii.

91. *Abrégé de l'histoire des insectes*, i.6. Johannes de Mey and later editors were in fact responsible for much of the heavy providentialism of Goedaert's *Metamorphosis naturalis*. See Peter W. van der Pas, 'Goedaert, Johannes', in *Dictionary of scientific biography*, v.439-40.

92. [Beaurieu], *Abrégé de l'histoire des insectes*, i.lxvii.

93. Mornet, *Sciences de la nature*, p.248-49.

culture. Both finalists and anti-finalists understood that final causes were important in ways that went beyond the 'merely' philosophical.

Final causes were certainly at issue in the Enlightenment – that is, they were the object of self-conscious manifestos and recriminations. Here, at the level of the formal dispute, the three traditionally identified attitudes toward external final causes – credulity, caution, and repudiation – enjoyed a clear-cut existence. Anti-finalists were those who presented arguments against final causes, finalists were those who claimed final causes were visible everywhere, and moderates were those who reproached both extremes. But beyond the clean lines of polemical self-fashioning, the three groups shared more than their stances would suggest. As we have seen, the arguments favoured by anti-finalists presumed purpose in nature. Nor were other anti-finalists more successful than Buffon in extirpating finality from their descriptions of the organic, let alone from their thinking and conjectures on the subject. Likewise, outside the genre of the celebration of providence, finalists recognised that not everything is useful, at least in a humanly meaningful way. Most importantly authors on both sides realised that final causes were a key part of vulgarisation. Eighteenth-century attitudes toward external final causes were unquestionably varied, but these attitudes need to be distinguished from formal pronouncements, which were in general a great deal more polarised.

3. The chain of being

IN *The Great chain of being* Arthur O. Lovejoy summarised what has become a truism of intellectual history: 'It was in the eighteenth century that the conception of the universe as a chain of being, and the principles which underlay this conception – plenitude, continuity, gradation – attained their widest diffusion and acceptance.'[1] As Lovejoy admitted, the concept of the chain of being meant different things to different thinkers, and this was perhaps a reason for its widespread appeal. For naturalists it was a means of approaching and organising nature, but it also had political and theological roles. Conservatives could cite it in defence of aristocratic privilege and the political status quo. The pious could invoke it to rationalise evil or celebrate providence. Materialists, for their part, could draw on it to demonstrate that humans were animals, that minerals had feelings, or that organic form was a function of continuity rather than providence. Not everyone was persuaded. But critics of the chain of being, among them Réaumur, Voltaire, d'Alembert, Daubenton, Samuel Johnson, and Johann Friedrich Blumenbach, could do little damage to a concept that thrived on vagueness, served myriad functions, and could be framed so as to sound like religious orthodoxy. Nay-sayers were drowned out by the chain's numerous proponents, among them such luminaries as Leibniz and Alexander Pope. Even Voltaire, at mid-century, succumbed to the chain's charm: 'Tout tend à nous faire croire qu'il y a une chaîne d'êtres qui s'élèvent par degrés.'[2]

Buffon plays a secondary role in discussions of the chain of being in the Enlightenment, with primary roles typically going to the chain's famous proponents or to less famous but more zealous ones such as Charles Bonnet and Jean-Baptiste-René Robinet. Still, we need read no farther than the 'Premier discours' to find Buffon proclaiming that nature is continuous and hierarchically organised from human beings to minerals: 'On peut descendre, par degrés presque insensibles, de la créature la plus parfaite jusqu'à la matière la plus informe' (*HN*, i.12). The abundance and salience of such declarations, particularly in the early volumes of the *Histoire naturelle*, have led some to associate Buffon with the chain of being. But most experts have been sceptical of Buffon's talk of continuity, judging nature as he saw it to have been significantly discontinuous.[3] Even in the 'Premier discours', alongside his

1. Cambridge, Mass. 1936, p.183.

2. Voltaire, *Eléments de la philosophie de Newton*, p.640. Compare with Voltaire, *Dictionnaire philosophique*, in Voltaire 35-36, at i.513-28; Voltaire, *Dictionnaire philosophique*, in M.xvii-xx, at iii.294-96.

3. Hanks, *Buffon avant l'"Histoire naturelle'*, p.124; Roger, *Buffon*, p.126-27; Robert Wohl, 'Buffon and his project for a new science', *Isis* 51 (1960), p.197; Jean Svalgelski, 'Buffon et les "intermittences de la nature"', in *Buffon 88*, p.165-74; Lorin Anderson, *Charles Bonnet and the order of the*

descriptions of a continuous chain of being, Buffon expresses doubts about the existence of key intermediaries such as lithophytes (mineral-plants) and zoophytes (animal-plants), both seemingly necessary for true continuity (*HN*, i.34-35). 'Cette loi de n'aller que par degrés nuancés paraît se démentir', he admitted at roughly the same time (*HN*, ii.263).

If Buffon was unconvinced of the existence of a continuous chain of being, why invoke it so frequently in the *Histoire naturelle*? It is not hard, at one level, to identify rhetorical interests behind his recourse to natural continuity. Like final causes, the image of the chain of being functioned rhetorically in two different ways in eighteenth-century natural history. First, as a representation of nature with a literary pedigree and a resonance with commonsense anthropocentrism, the chain of being provided a means of attracting, entertaining, and even instructing the public. Second, the chain of being was a useful piece of supporting evidence in certain debates, so useful that it was often cited by thinkers who cast doubt on it in other contexts. Most obviously the chain of being could be a devastating arm for attacking taxonomists such as Linnaeus. If natural forms change imperceptibly from one being to the next, what could be more foolish than to try to divide nature into species, genera, and other distinct taxa? Recognising the polemical and literary value of Buffon's pronouncements on continuity need not entail that they were no more than rhetoric.[4] So vague and flexible was the notion of a chain of being that one could hardly avoid believing in it within certain limits. But to acknowledge that the chain of being was only valid within limits was to eliminate its drama and polemical force. The more it was idealised through absolute principles of continuity and plenitude, the better the chain of being functioned as rhetoric. Here it becomes interesting to study strategies for easing tensions between nuanced, everyday knowledge about natural objects and a compelling if extreme image for expressing such knowledge. As it turns out, Buffon's attitudes toward the chain of being and natural continuity were no more contradictory than those of his contemporaries. Moreover his management of these problematic concepts followed patterns common in the Enlightenment and earlier.

Holding an ambiguous, impractical belief

In its usual eighteenth-century form, the chain of being was a linear hierarchy stretching from inanimate substances through plants, animals, and humans, to God.[5] Several of the more detailed representations of the chain were laid out

known (Dordrecht 1982), p.38-42; Rey, 'Buffon et le vitalisme', p.406-408; Robert Wokler, 'Tyson and Buffon on the orang-utan', *SVEC* 155 (1976), p.2314-19; James L. Larson, 'Linné's French critics', *Svenska Linné-Sällskapets Årsskrift* (1978), p.78; Larson, *Interpreting nature*, p.15.

4. Daudin, *De Linné à Jussieu*, p.166-69; Anderson, *Charles Bonnet*, p.41-42. For more cynical appraisals of Buffon's talk of continuity see Wohl, 'Buffon and his project', p.197; Lovejoy, 'Buffon and the problem of species', p.88-96, 100.

5. On God's place in the chain see Yvon Belaval, 'Leibniz et la chaîne des êtres', *Analecta Husserliana* 11 (1981), p.60-61.

by Bonnet, the first one appearing in his *Traité d'insectologie* (1745). Bonnet's 'échelle des êtres naturels' began with subtle matter and moved upward through the four elements of antiquity, the metals, and the minerals. Between the mineral kingdom and the plants, borderline entities such as asbestos, coral, and lichens assured a smooth transition. Between plants and insects – the lowest of the animals – Bonnet placed the 'sensitive' mimosa, the sea anemone, and the 'polyp', or freshwater hydra. From here the scale moved upward through shellfish, snakes, fish, birds, and quadrupeds to terminate with humans. As between lower groupings, exceptional cases facilitated the passage between each of these grades: snails appearing between shellfish and snakes, eels between snakes and fish, flying fish between fish and birds, the bat and the flying squirrel between birds and quadrupeds, and a number of primates between quadrupeds and humans.[6]

So presented, the chain of being seems a clear enough concept. But all through its history it was subject to philosophical and theological idealisation that at the same time contributed to its cultural importance and threatened to reduce it to esoteric nonsense. Chief among these destabilising forces of idealisation was the principle of plenitude, whose origins Lovejoy discovered in Plato.[7] In the *Timaeus* Plato describes the world's creation by a being whose consummate goodness necessitates creating everything possible, including all manner of imperfect beings. This vision of a necessarily exhaustive creation was a key inspiration for neo-Platonism and for subsequent conceptions of a 'full' chain of being. Adapted to Christianity, the principle of plenitude was implicated in a broad range of theological issues, notably in questions about the nature of universals and the extent of divine choice. A plenitudinous creation might seem a fit testament to God's power and glory, but if God 'merely' created everything possible, what of his role as a willing, choosing agent? Could the notion of possibility be prior to God's will? Opting decisively for divine reason over divine will, Spinoza and Leibniz became the most famous advocates of the principle of plenitude in the Enlightenment.

But was a plenitudinous chain of being really a chain? The idea of a chain entails discreteness, with every link connected to a predecessor and a successor, while the principle of plenitude requires links between links *ad infinitum*.[8] Kant rationalised the paradox elegantly, making the full chain a transcendental and thus unrealisable regulative model, but philosophers before and after wavered between a discrete chain and a continuum. On the one hand representations of the chain of being portrayed discrete links. Authors could identify intermediate species, but this could be carried only so far. On the other

6. Charles Bonnet, *Traité d'insectologie*, in *Œuvres d'histoire naturelle et de philosophie*, 18 vols (Neuchâtel, Samuel Fauche 1779-1783), vol. i, at xlix-li.

7. Lovejoy, *Great chain of being*, p.45-55. On the roles of Plotinus, Augustine, scholastic philosophers, and Spinoza, in defining and transmitting the principle of plenitude see Harrison, *Bible, protestantism*, p.163; Ehrard, *Idée de nature*, p.191-92; Atran, *Cognitive foundations of natural history*, p.173n.

8. See Michel Foucault, *Les Mots et les choses* (Paris 1966), p.158-60.

hand the chain's enthusiasts often invoked the principle of plenitude as a theoretical tenet. From antiquity onward the expression *natura non facit saltus* ('nature makes no leaps') was an axiom of natural history. Between any two links in the chain of being, it was asserted, another could be found. As Albertus Magnus wrote in the thirteenth century, 'Nature does not make [animal] kinds separate without making something intermediate between them.'[9]

Such rules about intermediaries masked the problem of discreteness versus continuity. Were the intermediaries themselves susceptible to being separated by more intermediaries, thus opening the way for perfect continuity and infinite regress? If not, at what level were intermediaries not separated by more intermediaries?[10] Despite the profusion of judgements regarding the necessity of intermediaries, these questions were almost always left unaddressed, and for good reason. Direct confrontation with infinite regress promised to be awkward. A single finite planet was plainly inadequate for housing an infinity of nested intermediaries. But if intermediaries were no more than dead-end exceptions between more legitimate, higher-level links, what became of natural plenitude and *natura non facit saltus*? Rules making intermediaries a uniform necessity hovered precariously between two unsavoury philosophical extremes. Emblematic in a small way of naturalists' unwillingness to choose one way or the other was the oxymoron 'presque infini', often used to characterise nature's abundance.[11] Ambiguity allowed the chain of being to be sublime and commonsensical at the same time.

Set against this backdrop of equivocal shallowness, Leibniz's view of the chain of being appears bold and ambitious. In asserting that God had created all possible beings and that intermediaries could be found between any two entities, Leibniz was only following a longstanding precedent. He was the first, though, to interpret the chain of being as a mathematical continuum.[12] As he wrote in a letter to Pierre Varignon in 1702, 'All the different classes of beings [...] are, in the ideas of God [...] only so many ordinates of a single curve so closely united that it would be impossible to place others between any two of them.'[13] The reference to sequential but inseparable points, contradicting Leibniz's faith in the omnipresence of intermediaries, is a little misleading. The analogy with mathematics makes it clear that these

9. Quoted in Lovejoy, *Great chain of being*, p.79.

10. See Belaval, 'Leibniz et la chaîne', p.64.

11. See for example Louis-Jean-Marie Daubenton, 'Botanique', in *Encyclopédie*, ii.344; Charles Bonnet, *Contemplation de la nature*, in *Œuvres d'histoire naturelle et de philosophie*, vol.vii-ix, at i.51; Bernardin de Saint-Pierre, *Etudes de la nature*, i.20, ii.153; Smellie, *Philosophy of natural history*, i.v, ii.3. For an argument that the almost infinite is a contradiction in terms see Fénelon, *Traité de l'existence et des attributs de Dieu*, p.118. In the same spirit Antoine-Laurent de Jussieu described his taxa as 'almost arbitrary' – a neat means of reconciling continuity with taxonomy. See Stevens, *Development of biological systematics*, p.48.

12. Sloan, 'History of the concept of the biological species', p.216-23; Daudin, *De Linné à Jussieu*, p.93-95.

13. Quoted in Lovejoy, *Great chain of being*, p.144-45. See also Sloan, 'History of the concept of the biological species', p.224n.

are infinitesimals, ideal objects outside the purview of ordinary logic. Continuity in the chain of being has become the continuity of geometers, and God's creation has been nuanced to the extreme of pure quantity. In a letter of 1686, significantly, Leibniz likens God's power for creating different 'Adams' to that of a geometer freely adjusting an ellipse's eccentricity.[14]

As Leibniz recognised in his posthumous *Nouveaux essais sur l'entendement humain* (1765), absolute natural continuity posed problems for taxonomy. If nature was an unbroken mathematical continuum, it stood to reason that taxa could only be arbitrary. Nor did the rest of his metaphysics offer any basis for classifying. His elemental 'monads' were essentially and necessarily different from each other, since God would have had no reason for creating two of the same one. Consequently natural objects, each one of them overseen by a dominant monad, would also end up being essentially different. Still, Leibniz refused to criticise the enterprise of taxonomy, insisting that his theory had no bearing on practice: 'On peut prendre l'espèce mathématiquement et physiquement. Dans la rigueur mathématique la moindre différence qui fait que deux choses ne sont point semblables en tout, fait qu'elles diffèrent d'espèce [...]. Mais les hommes établissant des espèces physiques ne s'attachent point à cette rigueur.'[15] As we will see in chapter 6, the harshest critique of taxonomy in the early Enlightenment came in Locke's *Essay concerning human understanding*, which ironically never claimed that the chain of being was infinitely continuous.[16]

Leibniz's refusal to mix 'rigour' and taxonomy was not an isolated act of lenience. Similar distancing of metaphysics and physics occurs in many of his discussions of natural continuity. At times it is reflected in subtle shifts of wording. In his famous letter to Varignon, for example, Leibniz first states that his law of continuity 'requires' the existence of the mythical zoophyte, but he lapses into tentative language when the context becomes more practical, telling us that the zoophyte 'may' be discovered and that its discovery 'would not surprise him'.[17] Familiar with contemporary research in the sciences of the living world, Leibniz was happy to signal agreement between his metaphysics and certain scientific discoveries – for example, Leeuwenhoek's discovery of 'spermatic animals' – but he also insulated his metaphysics from potentially falsifiable scientific results.[18] His implicit recognition of different orders of knowledge was especially marked when it came to the

14. Leibniz to Ernst von Hessen-Rheinfels, 12 April 1686, in Gottfried Wilhelm Leibniz, *Discourse on metaphysics, correspondence with Arnauld, monadology*, translated by George Montgomery (La Salle, Ill. 1902), p.80.

15. Gottfried Wilhelm Leibniz, *Nouveaux essais sur l'entendement humain*, in *Opera philosophica*, p.194-418, at p.313 (bk 3, ch. 6); Sloan, 'History of the concept of the biological species', p.216-24.

16. Locke's incremental chain of being is described in *An Essay concerning human understanding*, ed. Alexander Campbell Fraser, 2 vols (New York 1959), bk 3, ch. 6, no. 12.

17. Quoted in Lovejoy, *Great chain of being*, p.144-45.

18. Duchesneau, *Physiologie des Lumières*, p.65, 78-87; Roger, *Sciences de la vie*, p.370; Annie Ibrahim, 'La pensée de Buffon. Système ou anti-système?', in *Buffon 88*, p.178-79. On Leibniz's metaphysical use of corroborative physics see Jacques Roger, 'Leibniz et les sciences de la vie', in

principle of absolute continuity. Where others avoided trouble by leaving the chain of being in uneasy ambiguity, Leibniz idealised it and then pronounced it irrelevant in practical matters.

Later proponents of perfect continuity would adopt the same strategy. Leibniz's disciple Bonnet provides a good example. Lorin Anderson has written of the tension between Bonnet's practical knowledge and his oft-stated belief in a perfectly continuous chain of being. Typically Bonnet insisted there must be a link between vegetables and minerals even as his experimental work assured him of the contrary.[19] Rivals such as Robinet had no trouble seeing the inconsistency. Bonnet himself signalled a similar inconsistency in Pierre-Simon Pallas's treatment of continuity, yet regarding his own views he denied being contradictory.[20] Like Leibniz he cultivated different if related modes of thinking and wording in physics and metaphysics, and this surely contributed to his imperviousness to others' accusations that he was inconsistent. One of his remarks on the chain of being summarises his two roles as he conceived them: 'Le métaphysicien peut trouver dans la *loi de continuité* la raison de cette *progression*; le naturaliste se borne à l'établir sur les faits.'[21] The two tasks are intertwined and yet sufficiently different to give rise to different judgements about continuity. Speaking as a metaphysician Bonnet declares that intermediates can be found between any two species, but speaking as a naturalist he more commonly offers that intermediates *may* be found between given species, as in this suggestion about 'lower' animals: 'Peut-être qu'il en est plus encore entre ce polype et le dernier des animaux.'[22]

Buffon's approach to continuity also varied according to distance from practical specifics. In theoretical contexts he regularly and forcefully evoked the inevitability of intermediaries and natural plenitude: 'Tout ce qui peut être, est' (*HN*, i.11). In practical contexts he admitted discontinuities. As his assistant Guéneau put it in 1770, 'Les forces de la nature paraissent infinies lorsqu'on la contemple en gros et d'une vue générale; mais lorsqu'on la considère de près et en détail, on trouve que tout est limité' (*HNO*, i.401). Natural plenitude, in other words, was in the eye of the beholder. But if naturalists' changing attitudes toward the chain of being were a function of perspective, the perspectives they adopted were often strategically chosen. When attacking

Akten des internationalen Leibniz-Kongresses, Hannover, 14.-19. November 1966 (Wiesbaden 1969), ii.209-19, at p.209-14.

19. Anderson, *Charles Bonnet*, p.50-51. See also Peter Bowler, 'Bonnet and Buffon: theories of generation and the problem of species', *JHB* 6 (1973), p.262; Daudin, *De Linné à Jussieu*, p.103n., 179; Daubenton, 'Histoire naturelle', in *Séances des écoles normales. Leçons*, iv.3-5.

20. Bonnet to Haller, 25 November 1766, 29 April 1769, in *Correspondence Haller-Bonnet*, p.548-49, 816.

21. Charles Bonnet, *La Palingénésie philosophique, ou idées sur l'état passé et sur l'état futur des êtres vivants*, 2 vols (Geneva, Philibert and Chirol 1769), i.203.

22. Bonnet, *Contemplation de la nature*, ii.186. For similarly tentative judgements in practical contexts see Bonnet, *Palingénésie philosophique*, i.232-33, ii.111-12. Compare these with the more definitive metaphysical judgements in *Contemplation de la nature*, i.51-53; *Palingénésie philosophique*, i.245. Antoine-Laurent de Jussieu and Adanson predicted 'missing links' confidently if not specifically. See Stevens, *Development of biological systematics*, p.50-51, 179.

taxonomy in the 'Premier discours', for example, Buffon evoked nature from a general perspective to bolster his argument that intermediaries could be found between any established taxa. But when it came to justifying his own division of the three kingdoms of nature he assumed a closer perspective that allowed him to deny the existence of two supposed intermediaries, zoophytes and lithophytes. Limited in its influence, the chain of being served him with a minimum of unwanted consequences.

This, then, was one way of dealing with the useful but sometimes troublesome 'mathematical' chain of being: separate metaphysics from physics, or broad views from specific ones, and relegate the chain of being to the former, at least in its strict form.[23] Continuity may have been a matter of fact, not philosophy, in its commonsense form,[24] but commonsense continuity was hard to articulate except in forms redolent of ideal continuity. Nor did polemicists have much to gain from pointing out, for example, that nature appeared continuous in many particulars. Yet for all their convenience and elegance, laws of absolute continuity threatened to be paradoxical in practical contexts, whence the efforts to limit them to more philosophical discussions. Kant's 1781 'resolution' of the paradox of natural continuity would formalise and consummate this approach to the problem. Despite their tacit anticipation of Kant's position, however, proponents of plenitude could not bring themselves to separate metaphysics and physics explicitly or definitively. Uncomfortable with the implications of their absolute chains of being, they also recast older methods for making continuity sufficiently ambiguous to allow vacillation between metaphysical and practical interpretations. Leibniz's doctrine of 'compossibility' was one much imitated measure for giving the chain of being an unspecified slack. Since individually possible entities might not be possible in combination, Leibniz suggested that God created not all things possible but merely the best collection of compossible ones – that is, the best collection of compatible ones. Criteria for compossibility were never spelled out and presumably could not be. Interpreted liberally as an indistinct principle of ecological harmoniousness, compossibility offered a means of rationalising holes in the known chain of being. Interpreted narrowly it differed little from possibility, and the principle of plenitude remained in full force. Indeed Leibniz's distaste for emptiness led him to insist at his most metaphysical that any *vacuum formarum* arising from non-compossibility must be outside the parameters of the world as we know it, an insistence that effectively negated his and others' use of compossibility to eliminate the conflict between metaphysics and physics.[25]

Another way of making plenitude less contrary to common knowledge was to foist missing links on other times and places. Leibniz conceded that the Earth might not sustain a complete chain of being and suggested that the

23. See the discussion between Foucault and Bernard Balan in Michel Foucault, 'La situation de Cuvier dans l'histoire de la biologie', *RHS* 23 (1970), p.78-79, 81-83.

24. Stevens, *Development of biological systematics*, p.153-54, 260-61.

25. Lovejoy, *Great chain of being*, p.180-81.

balance could be sought on other planets. Closer to home, proponents of the chain of being pointed to unexplored regions as probable locations for missing links. For the more daring, the past and the future could function in the same way. Leibniz was ambivalent regarding the possibility of essential change in nature, but his suggestion that time could change organisms' outer 'envelopes', if nothing else, was a critical precedent for eighteenth-century transformism. While some versions of transformism retained the principle of plenitude as a synchronic truth, others distributed it over all of history.[26] Especially after the 1760s Buffon often claimed that nature's infinite possibilities were realised only serially in the course of millennia. His historical views on species were not designed to make plenitude compatible with observation, but others were. Maupertuis, for example, rationalised the incompleteness of the known chain of being by imagining that a comet destroyed the original plenum of Earth's chain of being.[27]

Thus far we have looked at reactions to the paradoxes of plenitude and perfect continuity. Another ambiguity in the notion of a chain of being concerns the number of qualities such a chain makes continuous. In *De generatione animalium* Aristotle ranked groups of animals from humans to shell-fish on the basis of the perfection of their offspring and eggs. While tagged as representing a single variable, this chain of being is usually interpreted as approximating his scale of overall perfection. Yet he acknowledged that perfection in one function did not imply it everywhere – a belief ruling out any monotone chain of being. Moreover his comparisons among animal attributes suggest mini-chains of being that are not always consistent, whether with each other or with the famous chain of *De generatione animalium*.[28] Even if measuring overall perfection, then, this latter chain did not propose continuity in all variables. Many later chains of being were also put forward as discontinuous in some ways. Mini-chains of being representing single variables were latent in the analogies of all eighteenth-century naturalists. Even Daubenton, a critic of the chain of being in its traditional form, recognised that organs and structures varied by nuances from species to species.[29] Alongside his more extended ones, Buffon proposed mini-chains of being based on specific features, as in this judgement regarding the pig: 'Il fait la nuance, à certains égards, entre les *solipèdes* et les *pieds fourchus*, et à d'autres égards entre les *pieds fourchus* et les *fissipèdes*' (*HN*, v.101). Likewise, under duress and in marginal caveats, nearly all the chain's proponents would admit that their versions of it were just possibilities based on a limited number of variables, not perfect truths reflecting all

26. Foucault, *Les Mots et les choses*, p.163-70. On Leibniz and transformism see Roger, 'Leibniz et les sciences', p.215-19; Lovejoy, *Great chain of being*, p.255-62.

27. Pierre-Louis Moreau de Maupertuis, *Essai de cosmologie*, in *Œuvres de Maupertuis*, 4 vols (Lyon, Bruyset 1768), i.xxvi-xxvii, 72-74.

28. See French, *Ancient natural history*, p.43, 59-60; Daudin, *De Linné à Jussieu*, p.91-92; Lovejoy, *Great chain of being*, p.55-59; Atran, *Cognitive foundations of natural history*, p.100-22.

29. See Daubenton, *HN*, v.25-26; Burkhardt, *Spirit of system*, p.21, 124-25. Daubenton's lectures at the Ecole normale were filled with arguments against the chain of being.

considerations. In all of these instances awareness of the multiplicity of possible chains of being anticipated a growing consensus in the late eighteenth century that a complete model of nature would have to be multi-dimensional rather than linear.[30]

But complete chains of being were too attractive to be given up. A chain based overtly on arbitrary criteria commanded less aesthetic and philosophical interest than a chain purporting to capture the hierarchy of the universe. From the standpoint of utility, only a universal chain of being could be used to predict properties of neighbouring species – long a desideratum of medicinal botany. Leibniz for one wrote in his letter to Varignon that 'the law of continuity requires that when the essential attributes of one being approximate those of another all the properties of the one must likewise gradually approximate those of the other'.[31] Many authors endorsed this Leibnizian ideal, however badly it fitted their ideas and however vulnerable it was to attack on specifics. Why, for example – one critic demanded – was the hydra, the last of the animals in many chains of being, so different from the tree, presumably first in the ranks of the plants?[32] More generally, attention to specifics pushed taxonomists and others to evoke natural relationships incompatible with seriality. So sophisticated was eighteenth-century taxonomy relative to visual models that it often obviated not only the linear chain of being but also multi-dimensional images of nature.[33]

Denying discontinuities could also be heretical. Situating humans in a universal chain of being was especially delicate. Robinet, an extreme case, sacrificed orthodoxy as well as common sense in the name of consistency, extending human attributes such as feeling and reason to even 'lifeless' matter.[34] Bonnet was more typical in undermining his supposedly universal scales of nature with assurances that different ranks enjoyed different properties. Human reason, notably, was denied to the rest of the chain by Bonnet and most

30. Giulio Barsanti, *La Scala, la mappa, l'albero: immagini e classificazioni della natura fra Sei e Ottocento* (Florence 1992), p.3-96; Giulio Barsanti, 'Buffon et l'image de la nature. De l'échelle des êtres à la carte géographique et à l'arbre généalogique', in *Buffon 88*, p.281-92. Many naturalists no longer entertained a coherent image of nature. See Stevens, *Development of biological systematics*, p.194-96.

31. Quoted in Lovejoy, *Great chain of being*, p.145.

32. [Lignac], *Lettres à un Américain*, pt 3, letter IX, p.24-27.

33. See Barsanti, *Scala, la mappa, l'albero*, p.ix, 105-107; Larson, *Interpreting nature*, p.55-59. For more examples of images being superseded by taxonomy see Daudin, *De Linné à Jussieu*, p.233; P. F. Stevens, 'Haüy and A.-P. Candolle: crystallography, botanical systematics, and comparative morphology, 1780-1840', *JHB* 17 (1984), p.79-80; P. F. Stevens, 'Metaphors and typology in the development of botanical systematics 1690-1960, or the art of putting new wine in old bottles', *Taxon* 33 (1984), p.175-77.

34. Robinet was nonetheless unable to overcome dualism according to Roselyn Rey, 'Les paradoxes du matérialisme de Robinet', *Dix-huitième siècle* 24 (1992), p.137-51. On the fortunes of Robinet's *De la nature* see Terence Murphy, 'Jean-Baptiste-René Robinet: the career of a man of letters', *SVEC* 150 (1976), p.200-201n. Even Robinet's apologists admitted that his ideas would seem absurd to the 'prejudiced multitudes'. See for example *Journal encyclopédique* 2.3 (March 1768), p.65, 3.1 (April 1768), p.61, 66, 71, 3.2 (April 1768), p.61, 3.3 (May 1768), p.99.

others – much to the outrage of unencumbered purists such as Robinet.[35] Likewise most naturalists admitted a gap between the living and the non-living. Even Needham, long intent on finding intermediaries between living beings and brute matter, ended up admitting that a 'sensitive principle' qualitatively distinguished the former from the latter.[36] Nor did Linnaeus's sometime principle of natural continuity deter him from sanctioning the traditional discrete kingdoms of nature: 'LAPIDES crescunt. VEGETABILIA crescunt et vivunt. ANIMALIA crescunt, vivunt, et sentiunt' ('Minerals grow. Plants grow and live. Animals grow, live and feel').[37] This discontinuous 'chain' of kingdoms would eventually be used against Linnaeus's theory of plant sexuality, since love without feeling was hard to conceive of.[38]

As their inventors often admitted in marginal caveats, chains of being announced as representing generic perfection usually emphasised just a few qualities, notably intelligence, external form, and spiritual excellence. At the same time many naturalists suggested that their chains of being reflected relations of dependence as well as of physical or moral similarity. Richard Bradley offered typically that 'all bodies have some dependence upon one another; and [...] if any one was wanting, all the rest must consequently be out of order.'[39] At one extreme, chains stressing dependence could become a metaphor for an interconnected, even deterministic universe. D'Alembert, for example, repeatedly described the universe and the knowledge that mirrored it as being chains.[40] Those endowing the chain of being with proto-ecological meaning rarely renounced its claim to represent morphological or moral continuity. When complete chains of being were actually drawn up, it tended not to matter what qualities or relations they were alleged to reflect. To the extent they were made specific, scales of natural perfection respected a commonsense anthropomorphism. The hierarchy of nature was almost always the hierarchy of biblical creation, passing upward in sequence through inanimate matter, small plants, trees, fish, birds, and land animals to humans.

All in all the chain of being was an ambiguous, flexible concept, well-suited for metaphor if not precision. In its shallow form it corresponded to commonsense views of nature and humans' place in the universe, but philosophy and

35. Anderson, *Charles Bonnet*, p.34-35, 48-49; Lovejoy, *Great chain of being*, p.269-77. On humans in the chain of being see Lovejoy, *Great chain of being*, p.195-98; Wokler, 'Tyson and Buffon', p.2304-305; Roger, *Buffon*, p.208-11; Gunnar Broberg, '*Homo sapiens*, Linnaeus's classification of man', in *Linnaeus*, p.156-74; Thomas, *Man and the natural world*, p.124-36.

36. Roger, *Sciences de la vie*, p.519-20. See also Burkhardt, *Spirit of system*, p.50-51.

37. Carl Linnaeus, *Philosophia botanica* (Stockholm, Kiesewetter 1751), p.1. See also Lindroth, 'Two faces of Linnaeus', p.16, 24-25; Gunnar Eriksson, 'Linnaeus the botanist', in *Linnaeus*, p.98-99.

38. François Delaporte, *Le Second règne de la nature* (Paris 1979), p.153-55.

39. *A Philosophical account of the works of nature* (London, W. Mears 1721), p.159.

40. Thomas L. Hankins, *Jean d'Alembert: science and the Enlightenment* (Oxford 1970), p.104-31. See also Paolo Casini, 'D'Alembert, l'économie des principes et la "métaphysique des sciences"', in *Jean d'Alembert, savant et philosophe. Portrait à plusieurs voix*, ed. Monique Emery and Pierre Monzani (Paris 1989), p.141; Voltaire, *Dictionnaire philosophique*, in Voltaire 35, p.522-28; Voltaire, *Dictionnaire philosophique*, in M.xx.294-96.

theology made it counter-intuitive by pushing it beyond its everyday limits.[41] Pressure to make the chain of being absolute and rigorous mounted with the rise of 'mathematical' philosophy in the seventeenth century. Theology added its own tensions. Like the chain, God's kingdom was presumably a hierarchy and a plenum of sorts, since anything short of that would be unworthy of omnipotence. The chain of being was also useful in rationalising evil, as in William King's *De origine mali* (1702), Leibniz's *Théodicée* (1710), Pope's *Essay on man* (1733-1734), and Albrecht von Haller's 'Über den Ursprung des Übels' (1734). But religious piety shirked from the consequence that humans and higher beings might not be qualitatively different from lower beings.

Perhaps more importantly the fullness and continuity of the Earth's chain of being were being 'confirmed' by the discovery of thousands of new plants and animals.[42] The number of recognised plant species grew more than tenfold between Andrea Cesalpino's *De plantis* (1583) and Ray's *Methodus plantarum nova* (1682) and continued to rise during the eighteenth century, leading Bernardin de Saint-Pierre to speculate that the Earth might support several million plant species.[43] The chimpanzee, a critical link in nearly all chains of being by the mid-eighteenth century, became familiar to Europeans only with Edward Tyson's *Anatomy of a pygmy* (1699). Other higher primates entered European consciousness in the next hundred years. These discoveries made the chain of being a crucial model for conceptualising nature, in part because they weakened alternative models. Lacking any obvious symbolic value, many new-found plants and animals did not fit easily into medieval, allegorical views of the universe and thus pushed the pious, particularly Protestants, to rely more and more on the chain of being as a means of rationalising nature.[44] Discoveries also provoked and reinforced platitudes about nature's plenitude and perfect continuity. In a world that was constantly offering unsuspected new objects, it may have seemed credible to assert that 'nature makes no leaps' and that 'all that can be, is'. If nothing else, such utterances were a means of affirming a form of control over the surprises of a bountiful planet.

More support for the chain of being came from research on microscopic and invertebrate life. The discovery of micro-organisms in the seventeenth century gave the chain of being another direction for expansion, a complement to the new worlds revealed by the telescope. Perhaps nature's continuity and infinite

41. See Lovejoy, *Great chain of being*, p.183, 251; Atran, *Cognitive foundations of natural history*, p.11-13, 206-11, 248-49; Thomas, *Man and the natural world*, p.60-61; Daudin, *De Linné à Jussieu*, p.105-106.

42. Atran, *Cognitive foundations of natural history*, p.121; Daudin, *De Linné à Jussieu*, p.159-60; Ehrard, *Idée de nature*, p.194-95; Maynard Mack, *Alexander Pope: a life* (New Haven, Conn. 1985), p.525-26, 542. Daubenton argued that discoveries merely complicated existing chains of being in 'Histoire naturelle', in *Séances des écoles normales. Leçons*, i.420, vi.11.

43. Bernardin de Saint-Pierre, *Etudes de la nature*, i.16-17; A. C. Crombie, *Styles of scientific thinking in the European tradition*, 3 vols (London 1994), ii.1263; Atran, *Cognitive foundations of natural history*, p.188; Ritvo, *Platypus and the mermaid*, p.10-11.

44. Harrison, *Bible, protestantism*, p.88-91, 161-63.

diversity, difficult to make out among human-size objects, were merely hidden by sheer disproportion. The idea of an organised, organic microcosm shocked common sense and posed problems for anthropocentric physico-theology, but worlds within worlds were taken seriously enough for pre-existence to become the official if uncomfortable doctrine of the French scientific community by 1705.[45] Studies of visible but previously neglected invertebrates by Réaumur, Bonnet, Abraham Trembley, and Needham also gave credence to the idea of a chain of being by providing intermediaries to fill out retrospectively critical lacunae. Proponents of the chain were especially quick to capitalise on the freshwater hydra, whose 'zoophytic' nature was only brought to light in the 1740s. With such surprises lurking so close at hand, who dared suggest that nature was limited?

In summary, the chain of being was a locus of tension in the Enlightenment, reinforced and contradicted by factors involving common sense, philosophy, theology, and new discoveries. What is interesting in this context is to examine how the chain of being was balanced or managed – that is, how proponents tried to reap its benefits without falling victim to its various perils. There were three common strategies for avoiding the chain's unwanted side effects. First, metaphysics and physics could be separated, an expedient that exempted ideal chains from their often disquieting practical consequences. Such separation could be accomplished by fiat or by ignoring absolute continuity when it risked being awkward. Second, ambiguity and vagueness allowed proponents of the chain of being to adjust it to different circumstances, making it alternately ideal enough for polemics and metaphysics and pragmatic enough for concrete natural history. Compossibility, for example, was used restrictively to rationalise lacunae in nature, but a more liberal interpretation allowed it to slip backward toward simple possibility – leaving plenitude intact for metaphysical purposes. Likewise missing links could be relegated to other times and planets, but naturalists went on invoking *natura non facit saltus* as a rule for here and now. Third, chains of being could be accompanied by conflicting signals about their meaning and scope. Detailed chains of being encouraged the inference that they were continuous in all variables, but when matters became delicate, authors retreated to safer positions, admitting that their chains were just possibilities and did not measure everything. Such caveats, rarely dwelt on, could be forgotten in moments, opening the door to grander presumptions.

45. Roger, *Sciences de la vie*, p.364. Bernardin de Saint-Pierre, among others, resisted the idea of microscopic plenitude since it would lack any 'purpose', even though he claimed elsewhere that 'la nature a fait tout ce qui était possible'. Compare *Voyage à l'Ile-de-France*, i.122; *Etudes de la nature*, ii.227. On 'vulgar' resistance to an organised microcosm see Pierre-Joseph Amoreux, *Notice des insectes de la France, réputés venimeux* (Paris, Hôtel Serpente 1789), p.19.

Using the chain of being

As with much of his philosophy, Buffon arrived early at his conceptions of continuity and plenitude. His career in mathematics familiarised him with continuity of an absolute kind. In a letter to Cramer in 1731, for example, he recognised that the function $y=(1+x)^{1/x}$ is discontinuous where y is negative, its curve being no more than 'une suite de points qui laissent des intervalles entre eux'.[46] By 1741 he was already wielding the 'laws' of continuity and plenitude that would reappear constantly in the *Histoire naturelle*. In a letter to Martin Folkes announcing the recently discovered peculiarities of the aphid and the hydra, Buffon declared that they demonstrated 'que toute la nature va par nuances et que tout ce qui peut être est'.[47] At roughly the same time his more pragmatic self was already cultivating a contrary rhetoric limiting possibilities. 'Tout a des limites', he asserted in 1747, commenting on the potential for creating more powerful burning mirrors.[48] Applied to the natural world this strand of his thought would vie with his principle of plenitude throughout his career, forcing recognition of 'l'étendue des limites dans lesquelles la nature semble se jouer' (*HNO*, iv.223).

How Buffon's ideas on continuity developed is unclear and perhaps unimportant. Not only was the chain of being a ubiquitous image by the early eighteenth century, the whole notion was sufficiently intuitive that it required little source beyond the need to give order to new findings about nature. As objects of study that defied classical taxa, the aphid and the hydra may have convinced Buffon of nature's continuity and plenitude, but nothing proves that his 1741 letter represents a conversion. While Buffon's pronouncements on the chain of being are often absolute in a way that recalls Leibnizian rather than Lockean philosophy, Phillip Sloan has pointed out similarities between the descriptions of the chain of being in the 'Premier discours' and those in Locke's *Essay concerning human understanding*, as translated into French by Pierre Coste.[49] The parallels between the two texts are not striking enough to suggest direct borrowing, but familiar as he was in eighteenth-century France, Locke could easily have influenced Buffon's views on the subject.

By 1741, in any case, Buffon was a proponent of the principles of plenitude and continuity. In 1749 the 'Premier discours' included his first published suggestion that living beings could be ordered on a linear scale. Giulio Barsanti has made the most thorough study of Buffon's 'image of nature' after 1749. According to Barsanti, Buffon regularly evoked the scale of nature both allusively and theoretically in the early volumes of the *Histoire naturelle*, but these references grew less detailed and serious, if not less numerous, after 1755. In

46. Buffon to Gabriel Cramer, 21 January 1731, in Françoise Weil, 'La correspondance Buffon-Cramer', *RHS* 14 (1961), p.109-10.

47. Buffon to Martin Folkes, 18 July 1741, quoted in Hanks, *Buffon avant l'"Histoire naturelle'*, p.189n.

48. Buffon, 'Invention de miroirs ardents', p.97.

49. Sloan, 'Buffon–Linnaeus controversy', p.362.

1755, concomitantly, Buffon introduced the image of nature as a map or a network – an image that came to dominate the *Histoire naturelle* between 1755 and 1766 and then once again after 1780. His 1755 diagram of relations of degeneration among breeds of dogs was presented not only as a map of lineage and habitat but also as a genealogical tree (*HN*, v.225-29). In 1766 and the several years afterward, such trees became Buffon's preferred metaphor for relations among species and groups of species, but trees gave way to maps and less specific networks in his final ten years. Typical among naturalists of the late eighteenth century in rejecting the chain of being for two-dimensional models, Buffon was the first, in Barsanti's judgement, to suggest visualising nature as three-dimensional – which he did somewhat explicitly from 1766 onward.[50]

Barsanti's analysis reinforces received views about Buffon's growing awareness of temporal change. Scholars used to reading Buffon's 'Nomenclature des singes' and 'Dégénération des animaux' as the climactic expression of a long path toward historical natural history will have no trouble understanding his adoption of multi-dimensional images of nature, though his abandonment of the genealogical tree may seem anomalous. Just as interesting as the evolution Barsanti charts is its ambiguity. Long after finishing the *Histoire des quadrupèdes*, Buffon resorted to simple chains and scales as metaphors for nature. None of these linear images are defended theoretically after the 1760s, and most do not purport to describe all of nature, but a small number seem to. In 1779, for example, Buffon suggested that the parrot be placed immediately after humans in the 'chaîne du grand ordre des êtres', and in 1782 he rationalised the bizarrely conformed manatee as a necessary link in the 'chain of being'.[51] Buffon's inability to renounce the chain of being cleanly reflected a common recidivism. Well into the nineteenth century, long after its 'replacement' by non-linear networks, the chain of being continued to show up in the writings of naturalists.[52] Easily written off as vestigial or poetic simplifications, such relapses hold limited interest for those solely interested in officially theorised knowledge, but they show that the chain of being could function in other ways than as a serious, precise model. Not only was it a heuristic with a common-sense grounding, the chain of being could also be a rhetorical device.

To explore this idea further, consider a prime accessory to Buffon's image of nature – the principle of plenitude. Barsanti has associated the phrase 'tout ce qui peut être, est' with Buffon's abandonment of chains in favour of multi-dimensional images of nature.[53] Jacques Roger, on the other hand, has proposed that 'tout ce peut être, est' is characteristic of the early years of the

50. Barsanti, 'Buffon et l'image de la nature', p.255-96. See also Svalgelski, 'Buffon et les "intermittences"', p.166-67.

51. *HNO*, vi.80; *HNS*, vi.381. See also *HNS*, iv.340; *HNO*, vi.321.

52. Stevens, *Development of biological systematics*, p.3-4, 173-75, 184; Ritvo, *Platypus and the mermaid*, p.30-31; Yeo, 'Principle of plenitude', p.263-64.

53. 'Buffon et l'image de la nature', p.278.

Histoire naturelle.[54] In fact this and other evocations of plenitude are made with equal seriousness throughout the series. We have already noted Buffon's use of 'tout ce qui peut être, est' in 1741. The same phrase, used in identical form by Descartes and others, can be found in volumes of the *Histoire naturelle* published in 1749 (*HN*, ii.319), 1755 (*HN*, v.102), and 1765 (*HN*, xiii.40). While this distribution seems to support Roger's association of plenitude with the series' early years, differently worded affirmations of plenitude can be found in nearly every volume, though many formulations after the 1750s suggest that plenitude is realised only in time: 'Tout ce qui peut être a été amené par le temps, et se trouve ou s'est trouvé dans la nature.'[55]

Here it seems clear that we are dealing with something other than the neutral expression of a firmly held conviction, for the principle of plenitude is also contradicted throughout the series, often within pages of being advanced. Shortly after declaring that 'tout ce qui peut être a été amené par le temps' – a declaration meant to bolster his claim that fertile 'half-mules' once existed – Buffon denies that bulls and mares could ever produce hybrids (*HN*, xiv.340-41, 347-48). Likewise his exclamation regarding the pig's useless organs – 'tout ce qui peut être, est' – belies his assertion in the same article that there are no intermediaries between the pig and neighbouring species (*HN*, v.99-100, 102). More generally he places strict limits on nature's creative powers, even in his later, 'historical' works. On several occasions he asserts that extant species were inevitable and would spring up in the same form on other planets.[56] He concedes that organic molecules may join together randomly to create a potential *plenum formarum* – albeit one limited by the availability of molecules – but little of this spontaneous 'life' will be capable of surviving or reproducing.[57] Such limitations on viable life are difficult to square with his principle of plenitude, not to mention his use of it to argue for the existence of an unknown variety of fertile half-mules. As Malesherbes observed: 'Malgré ce principe, il est certain qu'on peut imaginer des espèces dont l'existence ne renferme en elle-même aucune contradiction, et qui cependant ne se trouvent point dans la nature.'[58]

There remains room for reconciliation among Buffon's opinions regarding continuity and plenitude. Following Leibniz, he might have argued that the logically imaginable was not coextensive with the possible. In one form or another he resorted to all the standard strategies for easing the conflict between opposed views of plenitude. But for all his potential and actual strategies for smoothing out his ideas there remains tension in his writings, if not the

54. *Buffon*, p.567. For a similar judgement see Lovejoy, 'Buffon and the problem of species', p.88-93.

55. *HN*, xiv.341. For other affirmations of historical plenitude see *HN*, vi.59-60; *HNM*, ii.107, iii.260-61. For differently worded evocations of simple plenitude see for example *HN*, ii.22, 402.

56. John H. Eddy, Jr, 'Buffon's *Histoire naturelle*: history? a critique of recent interpretations', *Isis* 85 (1994), p.653-57; Bowler, 'Bonnet and Buffon', p.268-75; Roger, *Buffon*, p.190-91, 433-34, 566.

57. Bowler, 'Bonnet et Buffon', p.272-73; *HNS*, iv.339-44, 357-64.

58. Malesherbes, *Observations sur l'histoire naturelle*, i.22.

blatant contradictions his critics supposed. Unlike his models for describing nature's shape – chains, trees, and so on – this tension changed little in the course of his career. Throughout the *Histoire naturelle* he switched between absolute and everyday continuity, between 'tout ce qui peut être, est' and the realistic plenitude of the following passage: 'Quoique la marche de la nature se fasse par nuances et par degrés, souvent imperceptibles, les intervalles de ces degrés ou de ces nuances ne sont pas tous égaux' (*HN*, iv.383-84). The question arises of why again and again he returns to his principles of perfect continuity and plenitude when they create tension in the series and are constantly contradicted by his practical assertions. No doubt his principles are heuristic and regulative, but principles of near-plenitude or near-continuity might have served the same purpose. Above all, we will see, his absolute principles were a rhetoric, a means of making the *Histoire naturelle* appealing and persuasive.

We have already noted one use of continuity which is of critical importance in the *Histoire naturelle*, namely casting doubt on Linnaean taxonomy. As mentioned above, Leibniz managed to reconcile continuity and taxonomy to his own satisfaction. Nor did their work as taxonomists stop Linnaeus and fellow classifiers from giving voice to the truism that nature makes no leaps. Talk of nature's continuity may have been ignored in large measure, like much scientific rhetoric,[59] but naturalists were aware of the potential for conflict between taxonomy and continuity.[60] Antoine-Laurent de Jussieu, for example, understated his belief in continuity in classificatory works to make his taxa appear better-defined.[61] Michel Adanson justified his own taxonomy by suggesting that nature was only continuous from God's perspective.[62] Following Locke and others, Buffon invoked natural continuity as an argument against classifying, and offended taxonomists responded by rejecting or adjusting his principle of continuity, a principle they saw as implicated in his attack.[63] The incompatibility of taxonomy and natural continuity is a main theme of the 'Premier discours'. Because 'il se trouve toujours dans chacune de ces méthodes un certain nombre de plantes anomales dont l'espèce est moyenne entre deux genres' (*HN*, i.14), botanical taxonomy is judged a doomed enterprise. All through the series, but especially early on, Buffon evoked continuity in

59. See Giulio Barsanti, 'Linné et Buffon. Deux visions différentes de la nature et de l'histoire naturelle', *Revue de synthèse* 113-14 (1984), p.88-89.

60. Daudin, *De Linné à Jussieu*, p.ii, 110-16, 121, 229-30; Dirk Stemerding, 'How to make oneself nature's spokesman? A Latourian account of classification in eighteenth- and early nineteenth-century natural history', *Biology and philosophy* 8 (1993), p.199, 201; Burkhardt, *Spirit of system*, p.53-54; Lindroth, 'Two faces of Linnaeus', p.24-25; Pierre Bulliard, 'METHODE botanique', in *Dictionnaire élémentaire de botanique*, 3 vols (Paris, Didot *et al.* 1783), ii.109.

61. 'Jussieu was convinced, in theory at least, that nature was continuous', but his presentation conceded nature's discreteness on many occasions. See Stevens, *Development of biological systematics*, p.28, 40-41, 57-58.

62. Stevens, *Development of biological systematics*, p.95.

63. See for example Roger, *Sciences de la vie*, p.687-88; Malesherbes, *Observations sur l'histoire naturelle*, i.10-11, 21-28, 33-43; Jean-Etienne Guettard, 'Description de deux espèces de nids singuliers faits par des chenilles', *MARS* 1749 (1753), p.200-205.

the same way to attack classification.[64] Sceptical of exactly intermediate lithophytes and zoophytes by 1749, aware of 'intervals' between species by the start of the *Histoire des quadrupèdes*, he never believed literally in the continuity he brandished against Linnaean taxonomy, but it was an excellent weapon. Furthermore it merely exaggerated his enduring belief that nature was larger and more complicated than taxonomists assumed.

Less documented is the role of Buffon's principle of plenitude in adding spurious weight to his theoretical positions. After suggesting, for example, that the domestic goat, the ibex, and the chamois have different male forms but identical females, Buffon brings in plenitude to corroborate his assertion: 'Ces rapports, comme tous les autres rapports possibles, doivent se trouver dans la nature des choses' (*HN*, xii.143). Above we encountered a similar instance. To support his claim that fertile half-mules existed long ago, Buffon brings in a law of historical plenitude: 'Tout ce qui peut être a été amené par le temps, et se trouve ou s'est trouvé dans la nature' (*HN*, xiv.341). As noted above, Buffon often found himself on the other side of the issue within a short time, denying that certain entities had ever existed. When this happened, his critics sometimes drew on a similar rhetoric. Pierre-Ambroise-François Choderlos de Laclos, for example, attacked Buffon's assertion that solitary, asocial humans had never existed by stressing human ignorance and the possibilities of unknown lands: 'Qui sait, qui osera dire que tout cela n'existe pas?'[65] The plainly tactical character of such appeals to plenitude leads one to wonder how seriously they were taken by readers and authors. Perhaps very casual readers, in whom conviction verged on literary pleasure, were the primary targets of this rhetoric of plenitude.

Whether or not the law of plenitude was convincing in arguments for particular natural objects, it may have promoted the historical aspect of Buffon's natural history. Before coming to terms with the limited transformism he would later endorse, Buffon offered nature's non-plenitude as a reason for being sceptical about organic change: 'S'il [...] y eût, je ne dis pas plusieurs espèces, mais une seule qui eût été produite par la dégénération d'une autre espèce [...] il n'y aurait plus de bornes à la puissance de la nature' (*HN*, iv.382). This was an odd argument coming after declarations that 'tout ce qui peut être, est', and Buffon would soon drop it, but it shows how plenitude could be implicated in the question of change. Buffon's later invocations of the powers of time to create all things possible were even more pertinent to beliefs about change. Readers probably read them in the same spirit he wrote them, as exaggerated affirmations of a real if restricted creativity in nature, but this may have been sufficient to dispose them more favourably toward historical natural history. In the same way Buffon's evocations of intermediaries between intermediaries *ad infinitum* may have helped readers appreciate

64. See for example *HN*, ii.260-63, 324, x.200-202, xiv.16-17, 21-24; *HNO*, vii.325-26; *HNM*, i.242.

65. *Des femmes et de leur éducation*, in *Œuvres complètes* (Paris 1979), p.412-13.

nature's abundance even if they balked at literal plenitude. Like final causes, continuity had considerable pedagogical value.[66]

Sometimes the principle of plenitude serves not so much to legitimate as to motivate and provide moralising comment on the ideas presented in the *Histoire naturelle*. Introducing mercury (*HNM*, iii.213-14), Buffon refers to the scale of minerals and nature's need to fill it: 'Pour compléter la suite de ses opérations n'a-t-elle pas dû produire aussi des eaux métalliques telles que le mercure?' Introducing the puffin and its unusual beak, he takes the same approach: 'Toutes les figures possible ont été tracées, et toutes les formes remplies; et pour que dans cette suite il ne reste rien à désirer ni même à imaginer, l'extrême de toutes ces formes s'offre dans le bec en lame verticale de l'oiseau dont il est ici question' (*HNO*, ix.360). In his 'Nomenclature des singes' Buffon invokes plenitude twice to introduce intermediaries among the apes:

> Mais, comme la nature ne connaît pas nos définitions, qu'elle n'a jamais rangé ses ouvrages par tas, ni les êtres par genres, que sa marche au contraire va toujours par degrés, et que son plan est nuancé partout et s'étend en tout sens, il doit se trouver entre le genre du singe et celui du babouin, quelque espèce intermédiare qui ne soit précisément ni l'un ni l'autre, et qui participe des deux.[67]

In all of these instances Buffon's principles of plenitude and natural continuity are rhetorically valuable in facilitating transitions and amplifications, but they lack any serious philosophical content. Taxonomists had long used parameter-specific 'principles of plenitude' to predict possibilities for new species and genera, but mostly without claiming to predict with any certainty whether these possibilities were actually realised. Buffon's 'predictions' in the examples above are of a different order, more conversational than theoretical. If nature had to produce a liquid metal to complete the scale of minerals, why not a liquid semi-metal? Is it impossible to imagine a more extreme beak than the puffin's? Likewise, despite the confidence of his invocations of continuity in 'Nomenclature des singes', not all the primates discussed in the article are set off by intermediaries. Stripped of their absoluteness, plenitude and continuity played a role in Buffon's thinking. He knew that every species included varieties, that species were often similar, and that many natural objects remained to be discovered. These are the more common pragmatic interpretations suggested by contemporaries for his and others' absolute claims regarding the chain of being.[68] Any of these 'principles' of partial continuity could have replaced his more ambitious ones in the *Histoire naturelle*, but Buffon preferred his denser, more dramatic overstatements.

A clue to the logic of these exaggerated principles can be found in Buffon's recourse to a comparable principle in his discussion of sexual culture in 'De la

66. Notice also the pedagogical value of the traditional chain of being in Stevens, *Development of biological systematics*, p.174-75.

67. *HN*, xiv.6-7. For more examples of the principle of plenitude as a means of introduction see *HN*, xiv.12; *HNO*, viii.456-57; *HNM*, i.14-15; Guéneau, *HNO*, vi.437.

68. See Malesherbes, *Observations sur l'histoire naturelle*, i.24-25; Burkhardt, *Spirit of system*, p.124-25; [Lignac], *Lettres à un Américain*, pt 3, letter IX, p.22-23.

puberté': 'On ne peut rien imaginer de bizarre et de ridicule sur ce sujet que les hommes n'aient mis en pratique' (*HN*, ii.481). No more than in the natural world is this principle of plenitude designed to be read literally. More interesting is how it verges on the proverbial in this example. Like the statements Buffon makes about natural continuity, the sayings and commonplaces of everyday oral culture are typically exaggerated and serve less to express truth than to give the apparently singular a ceremonial if superficial intelligibility. Buffon's principles of continuity and plenitude should be compared with such sayings as 'there's always a first time', 'when it rains, it pours', 'you learn something every day', or 'it's a small world'. What matters about such sayings is less their literal meaning – generally unverifiable and furthermore irrelevant to the context of utterance – than their ability to give listeners a sense of order and control in the face of the novel. Like folk wisdom reflecting on unusual situations, the principle of continuity was often just an elegant utterance for rationalising surprises. As John Ray implied in *Methodus plantarum nova*, it was pre-eminently something naturalists said: 'Cum enim *natura* (ut dici solet) *non faciat saltus*' ('since nature, as the saying goes, does not make leaps').[69]

Not only did his principles of plenitude and continuity give Buffon a 'philosophical' way of presenting quirks of nature, they embellished and elevated his natural history. Whether favourable or unfavourable toward the chain of being, everyone agreed that the idea was pleasing. A plenitudinous chain of being was also sublime, since it dramatised God's omnipotence, linked humans to God, and depicted the universe as a place of miraculous interdependence. This 'rêve sublime', as Jean-Claude-Izoard Delisle de Sales called it, inspired awe even in Cuvier, an opponent of the chain of being: 'This idea has grandeur and seizes the imagination.'[70] As William Smellie put it, the chain of being 'gives us exalted ideas of that variety and progression which reign in the universe'.[71] Likewise a review of Robinet's *De la nature* (1761-1768) in the *Journal encyclopédique* suggests that aesthetics and feelings could be more important than facts in the reception of visions of absolute continuity: 'Quand il ne serait pas évidemment prouvé que les plantes sont animées, que ce sont autant d'animaux, vivant, respirant et sentant à leur manière, ne serait-ce pas du moins une idée bien flatteuse et bien satisfaisante que celle de penser [...] au fond d'un bois, dans les lieux les plus isolés, qu'on n'y est jamais seul?'[72]

Throughout the *Histoire naturelle* Buffon used striking images, stylish writing, and sublime views of nature to attract a broad readership. In this

69. London, Faitborne & Kersey 1682 ('Praefatio ad lectorem').

70. Jean-Claude-Izoard Delisle de Sales, *De la philosophie de la nature*, 3 vols (Amsterdam, Arkstée and Merkus 1770), i.6; Anderson, *Charles Bonnet*, p.49. For more on Cuvier and the chain of being see Stevens, *Development of biological systematics*, p.67-72.

71. *Philosophy of natural history*, i.522. See also Bonnet, *Traité d'insectologie*, p.xliv-xlvii; Voltaire, *Dictionnaire philosophique*, in Voltaire 35, 513-15; Daubenton, 'Histoire naturelle', in *Séances des écoles normales. Leçons*, i.419-20, iv.10.

72. *Journal encyclopédique* 3.1 (April 1768), p.69. Notice also the evocation of Robinet's sublimity in *Journal encyclopédique* 2.3 (March 1768), p.50.

light it is no accident that he capitalised on the chain of being and nature's continuity and plenitude, since these were all images conducive to poetic development and breath-taking overviews. In parts of the series – the long review of apes in 'Nomenclature des singes', for example – his invocations of continuity constitute the only high-style passages and the only *grandes vues* for pages and pages. Given their superficialness in certain cases, one has to wonder if these invocations were not deliberately added to enliven rote discussions in accordance with the policy announced at the head of volume xii. Conversely many of the series' *beaux morceaux* evoke the chain of being or the principle of plenitude. In his celebrated preamble to 'Soufre', for example, Buffon conjures up a vision of an all-powerful nature whose 'mouvement plus que perpétuel, aidé de l'éternité du temps, produit, entraîne, amène toutes les révolutions, toutes les combinaisons possibles' (*HNM*, ii.107). This is not to say that he drew on the chain of being merely as entertainment or to enhance his reputation as a literary stylist. As shown above, he believed in continuity in a limited sense, but he also realised the advantages of presenting it in absolute form.

Hints can be found that Buffon recognised his principles of plenitude and continuity as heuristic, literary overstatements rather than truths. Sometimes he introduced them conjecturally as psychic appearances: 'Il semble que tout ce qui peut être, est' (*HN*, xiii.40). As we saw in the last chapter, this was one of several strategies he employed for invoking final causes, which he found too useful and interesting to eliminate from his writing even though, officially, he denounced them as falsehoods. The comparison between plenitude and final causes can be drawn even farther, since he often wrote of plenitude as a purpose of nature. Likewise his talk of continuity is often couched in language that personifies nature. The following passage illustrates both tendencies: 'Mille autres productions de figures non moins étranges ne nous prouvent-elles pas que cette mère universelle a tout tenté pour enfanter, pour répandre la vie et l'étendre à toutes les formes possibles?'[73] All of these practices, bending the philosophical toward the literary, the pedagogical, and the propaedeutic, suggest that perfect continuity was not something Buffon believed in very firmly or literally.

Buffon's personification of nature as the author of plenitude shades into his deployment of God in the same role: 'Il semble que tout ce qui peut être, est; la main du Créateur ne paraît pas s'être ouverte pour donner l'être à un certain nombre déterminé d'espèces' (*HN*, i.11). Since he was very possibly an atheist when he made this declaration, it may have been a conceit or a ploy to show his piety. Plenitude had long been a theologically important concept. From Plato onward philosophers linked nature's plenitude with God's goodness and power. In the Enlightenment devout thinkers linked it less often with God's omnipotence than with his providence. Richard Bradley, for example,

73. *HNO*, viii.449. For other associations of plenitude with nature's goals see *HN*, x.200; *HNO*, vii.237.

used the chain to emphasise what a reasonable, harmonious unity God had created.[74] While not obviously consistent in his presentation of the chain of being in the *Essay on man*, Pope too stressed the providential interdependence of the universe.[75] Likewise Pluche alluded to God's omnipotence in the *Spectacle de la nature*, but his emphasis was on well-designed natural harmonies.[76] By the mid-eighteenth century nature's plenitude was often being cited against final causes, since a world including everything included objects without purpose. As we saw in the last chapter, Buffon's attacks on final causes adduced plenitude frequently and with a certain impunity, since they could take God's omnipotence as their orthodox premise. For Buffon as for others, a full chain of being could serve as a locus for canonical tributes to God and nature even as it undercut final causes and providence.

In the second half of the eighteenth century heterodox thinkers such as La Mettrie and Robinet began using continuity to argue that human beings were not essentially different from apes, other animals, or even minerals.[77] Buffon was probably sympathetic to much such materialism, but in this matter his pronouncements were not obviously unorthodox, except to those predisposed to read them as such.[78] While recognising that 'l'homme [...] doit se ranger lui-même dans la classe des animaux, auxquels il ressemble par tout ce qu'il a de matériel' (*HN*, i.12), he generally made humans an unrationalised exception to his oft-expressed rule of absolute gradation, insisting that an infinite distance separated humans from other animals, just as it did living from non-living matter (*HN*, ii.443):

> La nature marche toujours et agit en tout par degrés imperceptibles et par nuances; cette vérité, qui d'ailleurs ne souffre aucune exception, se dément ici tout à fait; il y a une distance infinie entre les facultés de l'homme et celles du plus petit animal, preuve évidente que l'homme est d'une différente nature, que seul il fait une classe à part.

This contradictory statement was smoothed out and approvingly summarised in the *Journal des savants*, but others were quick to seize on Buffon's antinomy.[79] Paraphrasing the 'Discours sur la nature des animaux' in the *Encyclopédie*, Diderot pushed Buffon's principle of continuity to a more logical conclusion regarding human beings. Buffon's 'incomplete' application of *natura non facit saltus* also provoked criticism from Robinet.[80] No doubt the *Histoire naturelle*

74. *Philosophical account*, p.159-63.

75. See Mack, *Alexander Pope*, p.522-44; Colin Manlove, 'Parts and wholes: Pope and poetic structure', in *Alexander Pope: essays for the tercentenary*, ed. Colin Nicholson (Aberdeen 1988), p.145-47; F. E. L. Priestley, 'Pope and the great chain of being', in *Essays in English literature from the Renaissance to the Victorian age*, ed. Millar Maclure and F. W. Watt (Toronto 1964), p.213-28.

76. For a rare reference to plenitude see Pluche, *Spectacle de la nature*, i.261.

77. Ehrard, *Idée de nature*, p.195-96.

78. Roger, *Buffon*, p.220-22; Wokler, 'Tyson and Buffon', p.2314-19. For a reading of the 'Premier discours' that makes Buffon's chain of being seem brazenly materialistic see *Nouvelles ecclésiastiques* (6 February 1750), p.21.

79. *Journal des savants* (March 1754), p.148, 155.

80. On Diderot see Denis Diderot, 'Animal', in *Encyclopédie*, i.470; Roger, *Sciences de la vie*, p.599-600; Ehrard, *Idée de nature*, p.226-27. On Robinet see Roger, *Sciences de la vie*, p.645; Rey,

contains suggestions that humans and animals are not essentially different. As we saw in chapter 2, Buffon also retreated from his once sharp dichotomy between life and non-life and asserted that heat could engender new organic molecules. Yet unlike other philosophers he never supported his potentially materialistic reductions by citing the chain of being or perfect continuity.

Plenitude, continuity, and the *scala naturae* played four roles as rhetoric in the *Histoire naturelle*. First, Buffon used them to cast doubt on taxonomy, since the 'continuum' of nature could only be divided arbitrarily. Second, invocations of continuity and plenitude served to bolster – and prepare readers for – a number of his arguments, mostly small-scale ones, but also his vast, implicit argument for historical natural history. Persuasive or not, these invocations created inconsistencies which some critics highlighted. Third, Buffon used the chain of being to introduce and provide moralising comment on natural objects. Without it, his discussions would have been narrower and less poetic. Lastly, he drew on the plenitudinous chain of being for orthodox imagery, even as he wielded it against final causes. The fact that he used continuity and plenitude rhetorically does not mean he did not also believe in them somehow. As we have seen, he believed in the continuity of nature up to a point, just as he believed in a finite, everyday 'plenitude'. But his beliefs never measured up to his frequently absolute pronouncements. To account for these pronouncements and their placement in the series, we have to examine their functions in polemical and literary context.

Buffon was not alone in taking polemical advantage of concepts of continuity he only halfway endorsed. As we saw previously, belief in the chain of being could be hard to uphold, notably in specific and practical contexts. At a textual level plenitudinous chains of being engendered tension and inconsistency. Under these circumstances, what pushed authors to invoke continuity in its problematic, absolute form? Simple faith clearly motivated many thinkers, from Leibniz to Robinet, but even here rhetorical interests should not be ruled out. At the minimum, philosophers tended to bring up continuity only when it promised not to appear too implausible. More significantly they often used it in the same ways that Buffon did. Critiques of taxonomy based on natural continuity were common before and especially after the 'Premier discours'. Arguments of all kinds found sometimes dubious support in idealised chains of being. Fontenelle, for example, cited nature's plenitude as a token rationale for Maupertuis's analysis of hypothetical 'disc stars': 'L'immensité et la variété de l'univers demandent qu'il y ait des soleils de toutes les figures possibles.'[81] The theological overtones of plenitude and the chain of being were understood and cultivated by both the orthodox and the heterodox. As

'Paradoxes du matérialisme de Robinet', p.146-47. See also Louise-Florence-Pétronille Tardieu d'Epinay to Ferdinando Galiani, 3-10 November 1770, in Eugène Asse (ed.), *Lettres de l'abbé Galiani à Mme d'Epinay, Voltaire, Diderot, le baron d'Holbach, Morellet, Suard, d'Alembert, Marmontel, la vicomtesse de Belsunce, etc.*, 2 vols (Paris 1903), i.170-72.

81. *HARS* 1732 (1735), p.91. See also *HNS*, ii.522.

for the chain's use as a catalyst for high style and intelligible overviews, many naturalists were at least sensitive to the image's intuitive and aesthetic appeal.

'Confirmed' by new discoveries even as it was being displaced by more complex continuities, the chain of being was a widespread image in the Enlightenment. Pushed by philosophers and others beyond its commonsense moorings, it was perpetually in danger of becoming paradoxical and hard to believe in. The fact that the chain of being was at the same time so idealised and so widely endorsed by the early eighteenth century raises the question of how people reconciled it with their intuitions, experience, and practical assumptions. Here we have looked at textual strategies for minimising the conflict between mundane beliefs and a chain of being idealised with respect to continuity and plenitude. A second key question concerns the reasons for the chain's popularity in the Enlightenment. Traditional answers have pointed to its role as a principle of order, providential or scientific, in a world opened up by discoveries in the natural world. This is illuminating but requires a supplement, for the chain of being functioned rhetorically too. For Buffon as for others, the chain was simultaneously an exaggerated belief, a regulative heuristic, and a carefully calibrated rhetoric, useful, among other things, for confounding taxonomists and creating high-style natural history for literati.

4. Systems and theories

HENCEFORTH we will concentrate on the rhetoric of scientific polemic, largely abandoning the rhetoric of literary seduction figuring prominently in the three preceding chapters. Chief among the polemical tools for making scientific texts persuasive is what can be termed the rhetoric of methodology. Debates over method – over how to discover, present, or verify knowledge – sprang up alongside the new sciences of the sixteenth and seventeenth centuries. In the Enlightenment the most famous of the methodological debates centred on the validity of hypotheses and 'systems', the latter term denoting a wide range of theoretical constructions. Like the Quarrel of the Ancients and the Moderns, the quarrel over systems simmered uneventfully for much of its long existence, but it attained a measure of focus in the 1740s and early 1750s, thanks to a series of manifestos and replies. In 1740 Du Châtelet included an endorsement of hypothetical reasoning in her *Institutions de physique*, perhaps the first such endorsement from a quasi-Newtonian.[1] In 1748 Mairan read a defence of certain kinds of systems in the semi-annual public meeting of the Académie royale des sciences.[2] Striking back, d'Alembert declared in his *Recherches sur la précession des équinoxes* (1749) that systems were indefensible except when mathematical. The charges would be amplified in his 'Discours préliminaire' (1751) to the *Encyclopédie*.[3] Condillac's *Traité des systèmes* (1749) seems to have been welcomed by both sides as a vindication of their viewpoint – a curious reception, and one that suggests that the point of the arguing was less clear than disputants implied.[4]

The opening volumes of the *Histoire naturelle* constituted an important point of reference in these debates, not so much for their explicit methodology

1. Gabrielle-Emilie Le Tonnelier de Breteuil Du Châtelet, *Institutions physiques. Nouvelle édition (1742)*, ed. Sonia Carboncini (Hildesheim 1988), p.78-93. Du Châtelet changed the title from *Institutions de physique* to *Institutions physiques* in the second edition. See also Phillip R. Sloan, 'L'hypothétisme de Buffon: sa place dans la philosophie des sciences du dix-huitième siècle', in *Buffon 88*, p.211. Vague, one-line defences of hypothetical reasoning can be found earlier among Newtonians, as in Pierre-Louis Moreau de Maupertuis, 'Sur les lois de l'attraction', *MARS* 1732 (1735), p.348.

2. As secretary of the Académie, Mairan championed systems as early as 1740. See *HARS* 1740 (1742), p.83.

3. Jean Le Rond d'Alembert, 'Introduction aux recherches sur la précession des équinoxes, et sur la nutation de l'axe de la terre dans le système newtonien', in *Œuvres de d'Alembert*, 5 vols (Paris 1821-1822), i.437; Jean Le Rond d'Alembert, 'Discours préliminaire', in *Encyclopédie*, i.xxxi.

4. For reactions to Condillac compare d'Alembert, 'Discours préliminaire', p.xxxi; Voltaire, *Dictionnaire philosophique*, M.xviii.177; 'Hypothèse', in *Encyclopédie*, viii.417-18. See also Roger, *Sciences de la vie*, p.468. For summaries of the debates over systems of the 1740s and 1750s see Hankins, *Jean d'Alembert*, p.78-80; Ellen McNiven Hine, 'Dortous de Mairan and eighteenth-century "systems theory"', *Gesnerus* 52 (1995), p.54-65; Lesley Hanks, 'Buffon et les fusées volantes', *RHS* 14 (1961), p.138-40.

(which was ignored) as for their content. Within two years Buffon's theories of the Earth and reproduction were being widely deplored as unfortunate hypotheses.[5] Mairan, for his part, expressed hope that Buffon's theory of reproduction would vindicate his own arguments in favour of systems.[6] Ironically Buffon, a longtime critic of *esprit de système*, had taken great pains to set himself apart from system-builders and hypothesisers. His *Histoire et théorie de la terre* in particular is filled with methodological statements distinguishing theories – legitimate entities in Buffon's estimation – from their stigmatised cousins, hypotheses and systems. Why was this methodology so unsuccessful in warding off accusations of system-building?

Answers can be sought at two levels. First, it is important to recognise the rhetorical dimension of talk about systems in the Enlightenment. This is not to suggest that natural philosophers did not differ in their practices (experimental standards, use of instruments, modes of reporting, etc.) or that their methodologies always misrepresented these practices in a serious way. But methodologies did have considerable indeterminacy, thanks in part to the ambiguity of terms such as 'system', and the value of methodology as propaganda led writers of the period to take maximal advantage of its several possible connections to their forms of practice. Professions of method were regularly interested even when not positively hypocritical. Second, when Buffon's particular engagement in the quarrel over systems is examined in detail, it becomes clearer in retrospect why his attempts to distinguish his theories from others' systems were unsuccessful. Not only are his pronouncements on method in the *Histoire et théorie de la terre* diffuse and enigmatic, they also promise more than he can offer in the way of mathematised induction. The texts and endeavours which Buffon's methods purport to oversee are regularly more complex than these methods suggest.

Rhetoric and ambiguities in the quarrel over systems

At their most simplified, histories of systems and hypotheses in the seventeenth and eighteenth centuries recount the following episodes. Descartes, on the French side, provides an obvious starting point. Sceptical of scholastic 'dogma', he was more credulous with regard to his own world-building systems, and his foible for speculation was shared by many thinkers of the next generation, among others Huyghens, Boyle, Robert Hooke, and the Port-Royal theorists.[7] By the mid-seventeenth century the tide was turning.

5. See Gay, *Rise of modern paganism*, p.136; [Lignac], *Lettres à un Américain*, pt I, letter II, p.16; Malesherbes, *Observations sur l'histoire naturelle*, i.226-27, 260-61, ii.17, 26-27, 304; Hanks, *Buffon avant l'‘Histoire naturelle’*, p.150n.; Abraham Trembley to Willem Bentinck, January 1750, in Maurice Trembley (ed.), *Correspondance inédite entre Réaumur et Abraham Trembley* (Geneva 1943), p.330n.; Raynal, 'Nouvelles littéraires', i.336-38; Albrecht von Haller, *Tagebuch seiner Beobachtungen* 2 vols (Bern, in der Hallerschen Buchhandlung 1787), i.67.

6. Hanks, 'Buffon et les fusées volantes', p.139.

7. Larry Laudan, *Science and hypothesis* (Dordrecht 1981), p.9.

Emblematic of the change, the Royal Society and the Académie royale des sciences were founded in the 1660s, both proclaiming their determination to pursue sciences based on facts instead of hypotheses.[8] The downfall of systems was secured by Newton, whose celebrated dictum *hypotheses non fingo* ('I do not frame hypotheses') appeared in 1713 in the General Scholium of the second edition of his *Principia*. In the following three decades hypotheses fell into further disrepute in the Académie des sciences and France in general, and Newton was cited alongside his Dutch disciples and his 'precursor' Francis Bacon to confirm the new reign of facts, experiments, and mathematical analysis.

Much of the above could have been gleaned from eighteenth-century intellectual histories, for example from d'Alembert's 'Discours préliminaire'. But where d'Alembert contemplated only more of the same for the rest of his century, modern histories of methodology have made the 1740s a significant watershed, a 'return to hypotheses'. In the French 'sciences of life' this was the period in which the mechanism of the early eighteenth century was replaced by a new, vitalistic philosophy inherently attached to philosophical speculation.[9] Elsewhere in the physical sciences a number of disparate theorists were discovering the utility of postulating unobservable, hypothetical fluids and particles to explain large-scale effects such as gravity and reproduction.[10] Not surprisingly this was the period in which the quarrel over systems flared into its greatest public prominence with the manifestos of Du Châtelet, Mairan, Condillac, and d'Alembert.

Like all general schemata, this history of methods demands qualification. All its alternating periods of systematising and empiricism can be said to have coexisted with contrary tendencies. Sceptics such as Gassendi attacked Descartes's systems during his lifetime, though without stigmatising systems in general to the extent the eighteenth century would. Hooke, Boyle, and Huyghens, besides being hypothesisers, were first-generation members of the Royal Society and the Académie des sciences, both of which were premised on avoidance of hypotheses. Indeed Paul Wood has argued that the empiricism of the early Royal Society, as described in Thomas Sprat's *History of the Royal Society* (1667), was less a reflection of members' methodologies than propaganda aimed at making the Society appear politically and religiously trustworthy.[11] Similarly Sébastien Le Clerc strategically exaggerated the empiricism of the early Académie des sciences in his official portraits. In practice the Académie's empiricism often meant little more than a refusal to decide between members'

8. Hahn, *Anatomy of a scientific institution*, p.15-16, 31-34; McClellan, *Science reorganized*, p.48-51.

9. Roger, *Sciences de la vie*, p.457-68, 749-50, 755.

10. Laudan, *Science and hypothesis*, p.12-15. See also John Bender, 'Enlightenment fiction and the scientific hypothesis', *Representations* 61 (1998), p.6-28.

11. P. B. Wood, 'Methodology and apologetics: Thomas Sprat's history of the Royal Society', *British journal for the history of science* 13 (1980), p.1-26. Uneasiness with official empiricism is obvious in many publications of the early Royal Society – for example in Robert Hooke, *Micrographia: or some physiological descriptions of minute bodies made with magnifying glasses* (London, J. Martyn and J. Allestry 1665), p.A2*v*, b1*r* ('To the Royal Society', 'The Preface').

rival systems.[12] Even in the early eighteenth century members of the Académie continued to defend and propose conjectures, hypotheses, and systems. Réaumur found the empiricism of the Académie overstated and defended a kind of hypothetico-deductive method in his *Art de convertir le fer forgé en acier* (1722).[13] Fontenelle, likewise, made it amply clear that his cautions on systems in the first (1699) volume of the *Histoire de l'Académie* had not been intended to eradicate them. Conversely the second advent of hypotheses around 1740 left no shortage of much more empirical natural philosophers to whom detractors could compare Buffon and other alleged system-builders. Despite the 'return' of hypothesis around 1740, declamations against *esprit de système* became even more common in the second half of the century.[14]

Conflict was both a cause and a consequence of such methodological diversity. The battle between Newtonian and Cartesian natural philosophy that accompanied the quarrel over systems during the first half of the eighteenth century illustrates the process. As Newton became familiar to French natural philosophers, his methodological differences with respect to Descartes assumed mythic proportions. To cast Descartes as a hypothesiser indifferent to experience was unjust, but the idealisation of Newton pushed the bounds of the credible even further. Undoubtedly hypothesis was less central in Newtonian than Cartesian philosophy. Newton was also loath to publicise his more fanciful ideas. Still, as I. Bernard Cohen has pointed out, he did use hypotheses, even under that name in the *Principia*.[15] Ironically, in France, where the Dutch view of Newton as an empiricist was reinforced by his assimilation to Lockean and other scepticism,[16] readers were great enthusiasts of his explicitly conjectural 'Queries' at the end of the *Optics*. In the end only the doctrinaire zeal of the Newtonian movement can explain the ease with which its partisans in the 1720s and 1730s adopted *hypotheses non fingo* as an absolute motto.

The fact that French Newtonians were the harshest critics of *esprit de système* in the 1720s and 1730s makes Du Châtelet's defence of hypothesis in 1740 a noteworthy happening. By 1740, of course, Cartesian natural philosophy had been routed for the most part, leaving Newtonians with a diminished need to advertise their methodological superiority. This may have encouraged a new liberality regarding hypotheses. At the same time the triumph of Newtonian attraction added momentum to Newtonianism as a cultural movement, thus

12. Stroup, *Company of scientists*, p.5-8; Hahn, *Anatomy of a scientific institution*, p.31-34.

13. Roger, *Sciences de la vie*, p.204; *Réaumur's memoirs on steel and iron*, p.xxiv, 210.

14. Mornet, *Sciences de la nature*, p.103; Bensaude-Vincent, *Lavoisier*, p.177.

15. I. Bernard Cohen, 'Hypotheses in Newton's philosophy', *Physis* 8 (1966), p.163-84; N. R. Hanson, 'Hypotheses fingo', in *The Methodological heritage of Newton*, ed. Robert E. Butts and John W. Davis (Toronto 1970), p.14-33. See also Maurizio Mamiani, 'The rhetoric of certainty: Newton's method in science and in the interpretation of the apocalypse', in *Persuading science*, p.157-72.

16. Common perception in the French Enlightenment notwithstanding, Locke's attitudes toward hypotheses were not wholly negative. See Larry Laudan, 'The nature and sources of Locke's views on hypotheses', *Journal of the history of ideas* 28 (1967), p.211-23; Kenneth Dewhurst, *John Locke (1632-1704), physician and philosopher: a medical biography* (London 1963), p.310-11.

contributing to the popularity of empirical and anti-rationalistic Newtonian slogans in French culture at large.

If early French Newtonians found unity in their condemnation of systems and hypotheses, Cartesians were more fragmented. Fontenelle, typically, stressed caution and moderation when dealing with systems. In his 1699 'charter' for the newly reorganised Académie des sciences he recommended 'nul système général, de peur de tomber dans l'inconvénient des systèmes précipités'. Later in the essay he speculated that systems might some day be possible, pending more discoveries, and he defended conjectures, only insisting that the Académie as a whole retain its 'sage Pyrrhonisme'.[17] In subsequent issues of *Histoire de l'Académie* he periodically defended conjectures and systems, but his respect for experience makes him a disappointing Cartesian when compared to the phantoms conjured up by Newtonian polemics.

Even aside from the rising prestige of Newtonian natural philosophy, the polemical value of speculative truths as opposed to truths of experience had been declining since the Renaissance, and by the later seventeenth century authors could not afford to defend systems or attack experience with anywhere near the righteous bluster of those who did the opposite.[18] Sensationistic philosophy, which required ideas to be analysed and justified via their origin in experience, exerted such a strong hold on French intellectual life by the mid-eighteenth century that arguments could hardly afford to do without the 'authority of experience'.[19] Experience, of course, was a flexible concept. It could refer, among other things, to grandiose theories perceived as facts, to what everyone knew intuitively from accumulated memory, or to what had been seen in specific experiments. Thanks to such ambiguity, French natural philosophy regularly managed to pack a great deal more theory into its 'experience' than British natural philosophy of the same period. Vagueness, however, did not prevent experience from being touted more and more as an ideal of knowledge. This primacy of experience vis-à-vis speculation made Enlightenment quarrels over systems and hypotheses inherently lopsided. It is noteworthy that even those defending hypothesis in the 1740s premised their arguments on the foundational nature of experience. Not only did they require new hypotheses to be consistent with existing experience, they also emphasised the value of hypotheses as gateways to previously unimagined observations and experiments. As Fontenelle put it, 'Ce que l'on ne connaît

17. *HARS* 1699 (1702), p.í2*r*-í2*v*.

18. Ricken, *'Gelehrter' und 'Wissenschaft' im Französischen*, p.103-14; Roger, *Sciences de la vie*, p.184-85, 195. See also Elizabeth L. Eisenstein, *The Printing press as an agent of change*, 2 vols (Cambridge 1979), i.151.

19. John C. O'Neal, *The Authority of experience: sensationist theory in the French Enlightenment* (University Park, Penn. 1996), p.1-2, 14, 245. Visual observation in particular wielded enormous authority, a fact that always jeopardised the 'empirical' image of the notoriously myopic Buffon. See Emma Spary, 'Le spectacle de la nature. Contrôle du public et vision républicaine dans le Muséum jacobin', in *Le Muséum au premier siècle de son histoire*, p.459-62; Hanks, *Buffon avant l''Histoire naturelle'*, p.23n.

point, est assez mal observé, il faut savoir à peu près ce que l'on voit, pour le bien voir.'[20]

Eighteenth-century pronouncements on systems and hypotheses were thus rhetorical in two ways. Vigorous condemnations of *esprit de système* and stern evocations of *hypotheses non fingo* associated their authors with the Newtonians and, after the 1730s, with an apparently victorious programme of thought. More generally, disparagement of systems and insistence on the primacy of facts and experience fell into a longstanding and increasingly effective sceptical mode going back to the Renaissance. Thanks to the vagueness of 'system', 'fact', 'experience', and similar terms, such truisms of method could be used in all sincerity even by hypothesisers, and the prestige of method and experience guaranteed that they would be. A. I. Sabra goes too far in suggesting that 'such tools (so-called methodological rules) [...] only serve as *machines de guerre* which one can always direct against the adherents of rival theories'.[21] It is hard to believe that methodology never serves to express the general principles which helped to guide research. On the other hand complete cynicism regarding methodology is hardly a precondition for appreciating its rhetorical nature. In various proportions and senses the Enlightenment's pronouncements on method were both representational and manipulative.

Many historians have remarked on the rhetorical aspect of the quarrel over systems, noting the coexistence, especially in France, of hypothetical reasoning and self-serving protests of empiricism. As critics from Diderot onward have noted, Condillac was something of a systematiser, despite his often biting attacks on *esprit de système*.[22] Antoine Lavoisier hardly incarnated *esprit de système*, but his chemistry was often speculative, a fact which leads Bernadette Bensaude-Vincent to denounce his occasional broadsides against hypotheses as opportunistic. When it suited his needs, as in 1777, Lavoisier could just as well denounce the empirical approach.[23] Commenting on Bonnet's *Considérations sur les corps organisés* (1762), where the author insists he is not making systems, Jacques Roger argues that Bonnet is deluding himself. Elsewhere in the *Considérations* Bonnet both denigrated and defended *esprit de système* as his needs varied.[24] Wilda Anderson has argued that Pierre-Joseph Macquer presented hypotheses as facts in the second edition (1778) of his *Dictionnaire de chimie*. The century's obsession with empiricism encouraged such artifice, but like many dictionary-makers of the time, Macquer was also

20. *HARS* 1732 (1735), p.14.

21. *Theories of light from Descartes to Newton* (London 1967), p.279n.

22. See Denis Diderot, *Lettre sur les aveugles*, in DPV, iv.65-66; Isabel F. Knight, *The Geometric spirit: the abbé de Condillac and the French Enlightenment* (New Haven, Conn. 1968), p.16; Roger, 'Diderot et Buffon en 1749', p.224-25.

23. Bensaude-Vincent, *Lavoisier*, p.169-96.

24. Roger, *Sciences de la vie*, p.717. Compare for example Charles Bonnet, *Considérations sur les corps organisés*, 2 vols (Amsterdam, Marc-Michel Rey 1762), i.109-10, 139, ii.145-46.

committed to creating a 'self-service' science suitable for amateurs, a goal best realised by a dictionary of facts, not ready-made systems.[25]

Another example of anti-rationalism as propaganda comes from the dispute over 'monsters' between Winslow and Louis Lémery in the 1730s and 1740s. As noted in chapter 2, Lémery contended that life-bearing germs were all perfect when created by God and that monsters resulted from accidental deformities occurring prior to birth. Winslow, on the contrary, argued that the 'system' of accidental causes was empirically inadequate and that monsters existed preformed by God. The dispute took place on at least three different levels – the empirical (where Winslow won convincingly), the theological (where there was stalemate), and the methodological (where Winslow accused Lémery of systematising and Lémery did his best to ward off the charge). So important was this last level that Lémery and Winslow both took time off in their papers to interpret, in their favour, Fontenelle's moratorium on 'general systems' in the 1699 *Histoire de l'Académie.*[26] In retrospect Winslow's recourse to Fontenelle's 'Pyrrhonism' to justify his position comes across as a little hypocritical. He was, after all, imputing most monsters directly to Providence! This was perhaps a system to end all systems, as Roger describes it, but it was nonetheless a system,[27] and Winslow's presentation of himself as a Pyrrhonist was clearly self-serving.

Already by the mid-eighteenth century, observers of the ongoing quarrel over systems were aware of its rhetorical aspect. Malesherbes described his century as one in which 'le nom seul de système fait la critique d'un ouvrage'.[28] In the published version of his famous 1748 defence of systems, Mairan wrote ironically: 'Système ou chimère semblent être aujourd'hui termes synonymes [...]. *C'est un système*, fait souvent la critique entière d'un livre; se déclarer contre les systèmes et assurer que ce qu'on va donner au public n'en est pas un, est devenu un lieu commun des préfaces.'[29] Bazin remarked that 'le système présent, en matière de physique, [est] de n'avoir

25. See Wilda C. Anderson, *Between the library and the laboratory* (Baltimore, Md. 1984), p.53-69; Richard Yeo, 'Reading encyclopedias: science and the organization of knowledge in British dictionaries of arts and sciences, 1730-1850', *Isis* 82 (1991), p.42-43, 49. For more examples of dubious pretensions of empiricism see Mornet, *Sciences de la nature*, p.94-96; Gaston Bachelard, *La Formation de l'esprit scientifique. Contribution à une psychanalyse de la connaissance* (Paris 1993), p.30, 87; Lawrence Brockliss and Colin Jones, *The Medical world of early modern France* (Oxford 1997), p.423-24.

26. Jacques-Bénigne Winslow, 'Observations anatomiques sur un enfant', *MARS* 1740 (1742), p.598; Louis Lémery, 'Remarques sur un nouveau monstre', *MARS* 1740 (1742), p.609-10; Winslow, 'Remarques sur deux dissertations touchant les monstres', *MARS* 1742 (1745), p.110-11. See also Roger, *Sciences de la vie*, p.409-18; Curran and Graille, 'Faces of eighteenth-century monstrosity', p.3-7.

27. Roger, *Sciences de la vie*, p.383-84. See also Tort, *Ordre et les monstres*, p.95, 157, 200-201.

28. *Observations sur l'histoire naturelle*, ii.140.

29. Jean-Jacques Dortous de Mairan, *Dissertation sur la glace* (Paris, Imprimerie royale 1749), p.v. Mairan's complaint was taken up nearly word-for-word by Nicolas Beauzée in 1767. See André Joly, intro. to James Harris, *Hermès, ou Recherches philosophiques sur la grammaire universelle*, translated by François Thurot (Geneva 1972), p.52. I am grateful to Julie Hayes for bringing this passage to my attention.

aucun système', a paradox that summarises Winslow's stance against Lémery.[30] Of course both Mairan and Bazin highlighted the rhetorical nature of the quarrel over systems in order to prepare the way for their subsequent efforts to rehabilitate systems, and the interested nature of their 'observations' should not be forgotten. Still, the fact that both felt obliged to bring in meta-methodological social commentary in order to defend systems is an indication of how difficult they considered their task. All through the Enlightenment systems could be attacked with carefree passing comments, but defending them required strategy and preparation.

Thus far we have treated systems and hypotheses as equivalent and the passage from systems to *esprit de système* as unproblematic. Relations between these expressions were in fact very close in the Enlightenment, but given the role of rhetoric in the quarrel over systems, and given the instability and subtlety of rhetorically charged vocabulary, it is critical now to be more precise. A history of the vocabulary of the quarrel over systems is important in that historians have typically cast the quarrel in the terms of one idiosyncratic participant. In his 'Discours préliminaire' d'Alembert wrote: 'En effet, plus on diminue le nombre des principes d'une science, plus on leur donne d'étendue [...]. Cette réduction, qui les rend d'ailleurs plus faciles à saisir, constitue le véritable esprit systématique, qu'il faut bien se garder de prendre pour l'esprit de système avec lequel il ne se rencontre pas toujours' (p.vi).

Following d'Alembert, historians such as Ernst Cassirer have spoken of an opposition between two forms of rationalism in eighteenth-century France – *esprit de système*, a moribund tendency to dogmatic world-building in the tradition of Descartes, and *esprit systématique*, a more acceptable tendency to organise and order within the sciences. Such an opposition may well have existed at the level of concepts. As we shall see, nearly every author taking a detailed stand in the quarrel over systems ended up distinguishing two forms of rationalism resembling d'Alembert's in key respects. On the other hand the terminology of d'Alembert's distinction misrepresents eighteenth-century usage. 'Esprit de système' and 'esprit systématique' were consistently used as synonyms from at least the 1730s onward. No one clearly and consistently distinguished them in the twenty years before d'Alembert, and few if any of his contemporaries took up the distinction.

The word 'système' was first used in French as a technical term in music and astronomy, but references to philosophical systems of all sorts had become current by the second half of the seventeenth century.[31] The system of Descartes was constantly referred to, but discussions also alluded to systems of the soul, the system of Malebranche, the system of Saint Augustine, the system of the Moderns, and even Aristotle's system of the elements. Founded in 1666, the Académie des sciences frowned on systems, but as mentioned

30. Quoted in Roger, *Sciences de la vie*, p.465.

31. On the early usage and ambiguities of 'système' in French see Hayes, *Reading the French Enlightenment*, p.22-57. As Hayes demonstrates, the quarrel over systems went well beyond science and was never resolved as cleanly as d'Alembert and others may have wished.

already, it was no more successful in excluding them than it was in excluding Cartesians. By the early eighteenth century academicians were applying the word 'system' to a large number of seemingly acceptable scientific theories. Those dealing with reproduction were almost always called systems, probably because of their broadly philosophical character. But the *Mémoires* of the Académie countenanced other kinds of systems too, among them the system of plant sexuality, systems of magnetic and electrical current, and systems for explaining foetal blood flow.

Increasingly, however, systems were seen as dogmatic and irrelevant intellectual whimsies. Specific systems had been mocked in the seventeenth century for being wrong and misguided, but attitudes toward systems in general had become noticeably less favourable by the early eighteenth century, so much so, for example, that Jean-Antoine Nollet felt obliged to apologise in advance for his electrical system of 1745.[32] A small role in the transformation might be claimed for the collapse of John Law's fiscal scheme in 1720. Like financial and taxational plans of the seventeenth century, Law's plan was known as a system, or simply the System, and contemporary reactions show that *faiseurs de système* in economics and natural philosophy were seen as kindred spirits. Having developed in part as pedagogical tools, systems were also associated with teaching, schools, pedantry, and scholasticism.[33] In the Enlightenment, a period in which scorn for traditional education and its institutions was the order of the day, the word 'system' came to have many of the negative connotations of the word 'academic' today.

Still, the overriding determinant of the new view of systems was the sceptical Newtonianism of the early *philosophes*. For most French Newtonians, Newton's ideas on universal attraction were more than a system. Thanks to his method, Newton's results were truths, not metaphysical possibilities of the sort propounded by systematisers. Fair or unfair, this view of Newton and his break from the past steadily gained ground, and not just in the sciences in which he had worked. In medicine, typically, 'Newtonian' thinking stigmatised systems and largely ended their heyday – the medical 'age of systems' – by 1760.[34] Perhaps through inadvertence, Newtonians sometimes applied the word 'system' to Newton's own discoveries during the eighteenth century, especially in less polemical contexts. D'Alembert, for example, regularly referred to the Newtonian 'system' in individual articles in the *Encyclopédie*, but in his more focal 'Discours préliminaire' he insisted that Newton had made a theory, not

32. Jean-Antoine Nollet, 'Conjectures sur les causes de l'électricité des corps', *MARS* 1745 (1749), p.107-108. See also Sutton, *Science for a polite society*, p.301-303, 313; Licoppe, *Formation de la pratique scientifique*, p.166-69.

33. Mornet, *Sciences de la nature*, p.73-80; Edward G. Ruestow, *Physics at seventeenth and eighteenth-century Leiden* (The Hague 1973), p.135-36, 147-49; Sutton, *Science for a polite society*, p.93-95.

34. Brockliss and Jones, *Medical world*, p.414, 431-34. Describing the early to mid-eighteenth century as an age of medical systems is traditional. See for example Fielding H. Garrison, *An Introduction to the history of medicine*, 4th edn (Philadelphia 1929), p.310-11; Arturo Castiglioni, *A History of medicine*, translated by E. B. Krumbhaar, 2nd edn (New York 1947), p.578-82.

a system.[35] 'Vulgar', qualitative Newtonians such as Voltaire were more careful than d'Alembert in separating Newton from the notion of system. In his *Eléments de la philosophie de Newton*, for example, Voltaire reserves 'system' almost exclusively for non-Newtonian theories.[36]

If Newtonians were intent on heightening the contrast between their empirical theories and earlier systems, Fontenelle and other Cartesians sought to diminish the contrast by applying the word 'system' indiscriminately to all higher-level scientific results. In 1735, for example, Fontenelle wrote of the 'Galilean system', Galileo's law relating time to the distance fallen by an object in free fall. In part this reflected a common Cartesian concern with finding precursors and rivals for Newton's theory of gravity, but it was also an effort to validate systems. More vexing still for the Newtonians, Fontenelle regularly treated Newton's theory as a system, and one among several for dealing with gravity.[37] In his *Théorie des tourbillons cartésiens* (1752) Fontenelle and his editor reversed Newtonian rhetoric, opposing a Cartesian theory to the 'abstract', 'metaphysical' Newtonian system.[38] To describe what Fontenelle termed the Newtonian and Galilean 'systems', Newtonians preferred terms such as 'theory', 'law', or 'truth'.[39] Over the next thirty years this is the usage that would slowly win out. In 1748 Grandjean de Fouchy, Fontenelle's second successor as secretary of the Académie, suggested that the word 'system' was inappropriate for certainties – a clear rejection of Fontenelle's usage.[40]

Like other non-Newtonians, Fontenelle acknowledged the dangers of systems. Even his more direct efforts to defend the hypothetical conceded that the hour of worry-free system-building was not yet at hand: 'Ce sont là de petits systèmes précipités, qui ne sont pas encore trop de saison, si ce n'est peut-être parce qu'on se fait mieux une idée des faits, quand on imagine une cause, quelle qu'elle soit, qui les lie.'[41] But attitudes had evolved so far by the 1730s that caution regarding systems was no longer what it used to be. In 1736, the same year Fontenelle excused the *petits systèmes* above, the *philosophe* André-François Boureau Deslandes defined systems as simple dogmatism, and this

35. Within the *Encyclopédie* compare for example d'Alembert, 'Discours préliminaire', p.xxvi; 'Astronomie', i.793; 'Attraction', i.854; 'Attraction des montagnes', i.856; 'Newtonianisme', xi.122-25; 'Physico-mathématiques', xii.537.

36. For two possible exceptions see Voltaire, *Eléments de la philosophie de Newton*, p.233, 359. See also Voltaire to Claude-Nicolas Le Cat, 15 April 1741, D2463; *Dictionnaire philosophique*, in M.xvii.550, xx.467-68.

37. See for example *HARS* 1727 (1729), p.158; *HARS* 1733 (1735), p.94-95; *HARS* 1734 (1736), p.88-89. On the 'Galilean system' see *HARS* 1735 (1738), p.82. See also Sutton, *Science for a polite society*, p.236.

38. See especially Camille Falconet, intro. to Bernard Le Bovier de Fontenelle, *Théorie des tourbillons cartésiens; avec des réflexions sur l'attraction* (Paris, Hippolyte-Louis Guérin 1752), p.iii-xxxi.

39. On the 'Galilean system' see for example Voltaire, *Eléments de la philosophie de Newton*, p.413-17; Du Châtelet, *Institutions physiques*, p.319-20.

40. *HARS* 1748 (1752), p.35. Notice, in contrast, how Fontenelle slips indifferently from 'system' to 'theory' in *HARS* 1732 (1735), p.1, 13, 16.

41. *HARS* 1736 (1739), p.5.

was a definition many agreed with.[42] The possibility of dogmatic systematisers had long been recognised. This, more than anything, had motivated Fontenelle's caution regarding systems in the 1699 *Histoire*. But for their supporters, systems were essentially just linked, ordered truths open to use as well as abuse.[43] In this meaning, 'system' continued to play a limited role in designating scientific theories into the 1750s and beyond, but it thrived scientifically in only a few senses, notably as a synonym for classification. As we will see in chapter 6, taxonomical 'systems' such as Linnaeus's were seen as a special case throughout the eighteenth century. Above all they were relatively safe from carping about *esprit de système* in that their authors typically billed them as purely conventional. Outside taxonomy the preferred designations for theoretical constructions came to be 'theory' (for near certainties, especially mathematical ones) or 'conjecture' and 'hypothesis' (for possibilities) rather than 'system'.

If Fontenelle made a minimally successful attempt to rescue the word 'system' from its detractors, he apparently recognised the futility of trying to defend the related but even more stigmatised expression 'esprit de système'. The first attested references to 'esprit de système' date from the 1710s.[44] Nothing indicates precisely how novel the expression was at the time, but it was certainly rare before then. From this time onward, in any case, 'esprit de système' would remain stable in its contours, referring to dogmatism and adherence to abstract theories. Its connotations were almost always negative. A few courageous or marginal figures did attempt to defend various forms of rationalism under its aegis, making 'esprit de système' a principle of order equivalent to d'Alembert's 'esprit systématique',[45] but most of those seeking to redeem hypothetical thinking chose to forgo the inherently derogatory 'esprit de système' and defend an alternative such as 'theory', 'hypothesis', or even 'system', the connotations of all of which were less pejorative. As we will

42. See Pierre Brunet, *Les Physiciens hollandais et la méthode expérimentale en France au XVIIIe siècle* (Paris 1926), p.110.

43. See for example the definitions in Jean-Louis-Samuel Formey, *La Belle Wolfienne*, 6 vols, vol. iii (The Hague 1743), p.148; Sutton, *Science for a polite society*, p.94-95. Fontenelle's characterisation of his theory of mathematical infinity as a system is also revealing. See Bernard Le Bovier de Fontenelle, *Eléments de la géométrie de l'infini*, ed. Michel Blay and Alain Niderst (Paris 1995), introduction, p.14.

44. Roger, *Sciences de la vie*, p.205; 'Système', in *Dictionnaire historique de la langue française*, ed. Alain Rey, 2 vols (Paris 1993), ii.2069-70. Two earlier instances dating from 1714 can be found in Antoine Houdar de La Motte, *Discours sur Homère*, in *Œuvres complètes*, 10 vols in 11 pts (Paris, Prault 1754), ii.1-137, at p.103; Anne Lefèvre Dacier, *Des causes de la corruption du goût* (Paris, Rigaud 1714), p.306. A variant that never took hold, 'esprit à système' is mentioned in Antoine Furetière, 'Système', in *Dictionnaire universel*, 4th edn, ed. Henri Basnage de Beauval and Jean-Baptiste Brutel de La Rivière, 4 vols (The Hague, Husson *et al.* 1727), iv.Ggg3*r*.

45. For exceptional positive references to 'esprit de système' see Fontenelle, *Eléments de la géométrie*, introduction, p.24; Raynal, *Nouvelles littéraires*, i.340-41; *HARS* 1740 (1742), p.83; Claude-Adrien Helvétius, *De l'homme* (London, Société typographique 1773), p.243; Hine, 'Dortous de Mairan', p.62; *Journal encyclopédique* 4.2 (June 1759), p.99. As noted above, Bonnet alternated between condemning and praising 'esprit de système' within single texts. Other 'defenders' of 'esprit de système' were similarly inconsistent.

see, 'theory' in particular enjoyed a Newtonian pedigree and was thus an ideal if surreptitious vehicle for bringing back hypotheses.

Like 'esprit de système', 'esprit systématique' became current in the early eighteenth century. Paraphrasing Pierre Bayle, Leibniz used it in his *Essais de théodicée* (1710) to describe Cornelis Jansen.[46] Here the expression is not obviously pejorative, and it might be said to come close to d'Alembert's sense. By the 1720s, however, 'esprit systématique' had acquired the same meanings and functions as 'esprit de système'. The two expressions had merged.[47] Both could be defined positively as generic principles of ordering, simplifying rationalism, as was 'esprit systématique' in the 'Discours préliminaire', but they usually referred to dogmatic systematising, as in Condorcet's evaluation of Buffon: 'En s'abandonnant trop facilement peut-être à des systèmes spéculatifs [...] jamais il n'étendit cet esprit systématique sur des objets immédiatement applicables à l'usage commun.'[48]

D'Alembert's definition of 'esprit systématique' as a tendency to rational simplicity with little relation to 'esprit de système' was thus mostly wishful thinking in 1751. The fact that his terminology is the received one for discussing eighteenth-century epistemology exemplifies the way in which the history of the Enlightenment remains to a large extent the story told by the *philosophes*.[49] In fact, so pervasive and well-established was the synonymy of 'esprit de système' and 'esprit systématique' in d'Alembert's time that neither his collaborators nor he himself managed to uphold his distinction in subsequent pages of the *Encyclopédie*. The unsigned article 'Philosophie', for example, considers a definition of 'esprit systématique' vaguely like d'Alembert's but ends up with a more usual characterisation – 'celui qui bâtit des plans et forme des systèmes de l'univers, auxquels il veut ensuite ajuster, de gré ou de force, les phénomènes'.[50] A mere twenty-five pages after distinguishing the two, d'Alembert himself uses 'esprit de système' for what he has just called 'esprit systématique'.[51] Similarly, in his article 'Mars', d'Alembert associates 'esprit systématique' with conjecture and loose thinking – precisely the qualities he tried to foist off on 'esprit de système' in the 'Discours préliminaire'.[52]

46. Gottfried Wilhelm Leibniz, *Essais de théodicée*, in *Opera philosophica*, p.610.

47. 'Système', p.2069-70. Notice also Voltaire's use of the expressions as pejorative synonyms in *Lettres philosophiques*, ed. René Pomeau (Paris 1964), letter XIII, p.83, letter XIV, p.94.

48. *HARS* 1788 (1791), p.53.

49. Ironically even critics of the Enlightenment from Theodor Adorno and Max Horkheimer to Michel Foucault have taken d'Alembert's and others' pronouncements on systems at face value as proof that the epoch can be characterised by its obsession with an ultimately repressive *esprit systématique*. But if eighteenth-century systematising was less simple than rhetoric would have it, these critiques are as misleading as certain *philosophes*' idealisations of the Enlightenment. See Julie C. Hayes, 'Sequence and simultaneity in Diderot's *Promenade Vernet* and *Leçons de clavecin*', *Eighteenth-century studies* 29 (1996), p.291-92; Hayes, *Reading the French Enlightenment*, p.22-57.

50. 'Philosophie', in *Encyclopédie*, xii.515.

51. Compare d'Alembert, 'Discours préliminaire', p.vi, xxxi. See also Hine, 'Dortous de Mairan', p.57n.

52. Jean Le Rond d'Alembert, 'Mars', in *Encyclopédie*, x.150. See also d'Alembert, 'Fluide', in *Encyclopédie*, vi.890.

The fact that d'Alembert chose synonyms to designate what he and many subsequent historians of the Enlightenment have insisted are quite different ways of thinking is curious – even granted that his usage may have reflected earlier standards. Curious as well is the fact that Condillac's *Traité des systèmes* was welcomed both by 'enemies' of hypothetical thinking such as d'Alembert and Voltaire and by the anonymous 'friend' of hypothesis who wrote the main article 'Hypothèse' for the *Encyclopédie.* In fact, as Ellen McNiven Hine has convincingly argued, there was more common ground in the quarrel over systems of the 1740s and 1750s than most of the participants were willing to acknowledge.[53] In particular Hine has demonstrated that although d'Alembert and more recent scholars have not hesitated to classify Mairan as a defender of systems, the positions he assumed in his 'defence' of systems are almost identical to those of Condillac in his *Traité* – which d'Alembert endorsed wholeheartedly! By the late 1740s, Hine suggests, the quarrel over systems was more about cliques and reputations than methodology. Mairan, on this view, was and remains cast as a defender of systems because he was seen as an archaic, Cartesian system-builder, whereas Condillac was cast as an enemy of systems because of his ties to d'Alembert and the *philosophes*. Given the near agreement between the two 'sides' in the quarrel, the synonymy of terms in d'Alembert's dichotomy makes perfect sense, as does the ambiguity of Condillac's *Traité*. To the extent that it existed, disagreement was a matter of subtle differences magnified in the interest of promoting or discrediting rival lines of research.

Re-examining the manifestos on systems of the 1740s and 1750s in light of Hine's argument, we find much common ground, not only between Mairan and Condillac but also between Du Châtelet and d'Alembert.[54] All four attribute past successes in science to systems and hypotheses. All four distinguish between good and bad forms of rationalism, and recommend the good form as a necessary complement to a science of pure facts. Finally all four describe their good form of rationalism as deriving from or explaining facts, and their bad form as arbitrary, chimerical, or vague. Naturally there are differences in the four authors' attitudes. Mairan does not specify, as Du Châtelet and Condillac do, that hypotheses and systems must be explicitly marked as such and not passed off as facts. Although he too allows 'conjectures', d'Alembert emphasises reductive rationalism rather than the others' expansive, explicit hypothesising. Still, the amount of common ground is remarkable in a quarrel that has been presented by d'Alembert and subsequent historians in such polarised terms.

The same common ground can be found in other manifestos of method on either side of the quarrel over systems. For example, discussions of Fontenelle's

53. Hine, 'Dortous de Mairan', p.54-65.

54. The following comparison concerns the texts mentioned above, namely Mairan, *Dissertation sur la glace*, p.iii-xxix; Etienne Bonnot de Condillac, *Traité des systèmes*, in *Œuvres philosophiques de Condillac*, i.118-217; Du Châtelet, *Institutions physiques*, p.78-93; d'Alembert, 'Discours préliminaire', p.i-xxxiv.

1699 'charter' for the Académie des sciences have emphasised its stand against systems since the early eighteenth century, an emphasis reinforcing the Académie's historiographical role as an agent of empiricism. Like Condillac's 'condemnation' of systems, however, Fontenelle's charter allows for good forms of rationalism. Conjectures are explicitly authorised, provided they are marked as such, and Fontenelle envisages a possible return to systems once enough facts are gathered.[55] When texts are examined in their own right alongside their reputations, many of the more elaborate eighteenth-century methodologies appear less extreme than hearsay might have it.

Was the quarrel over systems no more, then, than a pretext for making myths out of institutional and personal reputations? This would be going too far. First, the fact that there is much common ground in many 'opposed' eighteenth-century statements of method need not imply that scientific practice was not polarised. As we have seen, Winslow and Lémery both professed to be respecting Fontenelle's 'ban' on systems in their work on monsters, yet their means of respecting it differed enormously, as their results show. Even if researchers all tried to follow their professed methods, methodological terms such as 'system', 'experiment', and 'hypothesis' are so subject to individual interpretation that it would not be surprising if a relatively centralised collection of verbal methods gave rise to a more dispersed collection of practical methods. In other words, a relatively silent but pivotal quarrel over systems may have been taking place at different levels of practice even as eighteenth-century methodologists engaged in what was often a pseudo-quarrel of method through manifestos.

Second, the broad agreement that can be found in more elaborate eighteenth-century methodologies is absent in what are undoubtedly the period's most numerous declarations of method – short, not necessarily consistent 'barbs' of method emitted in passing to shore up an argument, to attack a rival theory, or simply to moralise in a vague way. Here, instead of common ground, one finds *ad hoc* one-sidedness, usually skewed toward the rhetorically more powerful empirical point of view. In this spirit, articles in the *Mémoires de l'Académie* regularly note in passing that facts, observations, and experiments are the sole basis for science. Charles Le Roy, for example, prefaces an attack on Nollet's theory of electricity with the usual empiricist's *captatio benevolentiae*: 'La voie des expériences est la seule par laquelle nous puissions espérer d'en découvrir les causes, mais encore [...] en suivant cette voie nous ne pouvons marcher avec trop de circonspection.'[56] A less placid micro-methodology appears in the inconsistent but consistently belligerent article 'Phénomène' in the *Supplément à l'Encyclopédie*: 'On doit proscrire et éliminer de la physique toute les hypothèses et les conjectures.'[57] At this level the quarrel over systems

55. *HARS* 1699 (1702), p.i1*v*-i2*v*.

56. Charles Le Roy, 'Mémoire sur l'électricité résineuse', *MARS* 1755 (1761), p.264. For more on the rhetorical aspect of eighteenth-century quarrels over electrical theory see Sutton, *Science for a polite society*, p.291; Licoppe, *Formation de la pratique scientifique*, p.161-94.

57. Fortuné-Barthélemy de Félice, 'Phénomène', in *Supplément à l'Encyclopédie*, ed. Jean-Baptiste Robinet, 5 vols (Amsterdam, Rey 1777), iv.320.

featured a radical one-sidedness usually lacking in better-developed methodologies.

In sum, there are at least three levels on which to analyse the eighteenth-century quarrel over systems.[58] First, regarding practice, one can look at what concepts thinkers chose and allowed, what instruments they used and how they used them, their standards of relevance concerning different forms of information, and any number of other rarely verbalised factors. Judging mainly by the increased acceptance of hypothetical, unobservable entities, Jacques Roger and Larry Laudan both identify a shift at this level from empiricism to rationalism around 1740, and – allowance made for the inevitable exceptions and counter-tendencies – this appears plausible as a general movement. Second, one can look at in-depth treatment of systems and hypotheses in treatises, prefaces and other sustained works. Once again, Laudan and Roger note a surge in the number of defences of the hypothetical around 1740. The shift is indubitable, but its suddenness and sharpness should not be exaggerated. Hypotheses and systems may not have been defended much before 1740, but conjectures and other nearly equivalent forms of rationalism frequently were. Furthermore, as Hine has shown, defenders and enemies of systems were defined as such by reputation and image as well as by their actual ideas about systems, a fact that casts doubt on highly polarised presentations of the quarrel over systems even outside the small group of authors Hine treats in detail. Finally, one can look at brief, passing invocations of method in non-methodological texts. Here Sabra's view of methods as *machines de guerre* is often justifiable, and it is here that the quarrel over systems was most polemical.

Buffon in the quarrel over systems

As mentioned above, Buffon came to be considered a prime example of systematising after the publication of the first three volumes of the *Histoire naturelle* in 1749. Lesley Hanks has argued that already from the 1730s onward, Buffon's scientific work demonstrates a proclivity to rationalism and grand explanation. In 1734, for example, he presented a paper on the causes of aurora borealis to the Académie des sciences, a paper which Hanks judges daring in that new members of the Académie were not expected to theorise about causes at this time. In his work on arboriculture in the 1730s and 1740s Buffon was also prone to theorising and drawing conclusions from just a few observations.[59] In the preface to his 1735 translation of Hales's *Vegetable statics*, Buffon glorifies Hales's empiricism and criticises systems: 'En fait de physique, l'on doit

58. For a similar distinction of three levels of method see John A. Schuster, 'Methodologies as mythic structures: a preface to the future historiography of method', *Metascience* 1/2 (1984), p.18-20.

59. Hanks, *Buffon avant l'"Histoire naturelle'*, p.103, 160-61, 169-71, 180-82, 195-96, 203, 209-10, 213.

rechercher autant les expériences, que l'on doit craindre les systèmes.'[60] Regardless of such protests, Hanks maintains, Buffon yearned for systems and showed greater interest in Hales's theories than in his experiments.[61]

In all of this there is a measure of truth. When Buffon praises the experimental method and derides *esprit de système* in his *Statique des végétaux* it is hard not to see a cliché of the time, the rallying cry of so many Newtonians and simple opportunists of the early eighteenth century. So banal and indeterminate were attacks on systems that Hales's critics at Trévoux simply inverted Buffon's rhetoric, claiming with some justice that the whole *Statique* was based on a system, that of attraction.[62] Likewise it is significant that Buffon's preface to the *Statique* only pronounces systems unattainable for the moment – the same line taken by Fontenelle in the Académie's 1699 'charter' – and that it proposes Stahl, the inventor of phlogiston, as a model empiricist.[63] Ultimately it would not be surprising to find Buffon, an eighteenth-century Frenchman, less interested in the experimental side of the *Statics* than Hales, an Englishman. Still, if Buffon was a systematiser before 1749, his rationalism was not so blatant as to raise outcries and satires in the way that it later would. Condorcet's division of Buffon's career into a pre-1749 empirical period and a post-1749 rationalism certainly corresponded to many people's perceptions in the eighteenth century.[64]

Within just a few years of 1749 Buffon was identified as a systematiser by both detractors and admirers, and despite a handful of dissenting views this would be his reputation to the end of his life. His own rhetoric took a long time to adjust to public perception. The fact that he continued to insist on the sanctity of facts and experience is unremarkable. Virtually everyone did so, especially when 'empiricism' offered a means of pointing out weaknesses in rival doctrines. Nor was it exceptional for an obvious systematiser such as Buffon to malign others' hypotheses and systems as such, as he did many times in the *Histoire naturelle*. Still, such attacks risked being turned on the attacker, and so they often were in Buffon's case. Finally, in 1783, he came to terms with his reputation. Frustrated by 'empirical' critiques of his *Epoques de la nature*, he wrote his first explicit defence of systems and hypotheses in the *Histoire des minéraux*, berating strict empiricists for lacking in genius.[65]

60. Georges-Louis Leclerc de Buffon, preface to Stephen Hales, *Statique des végétaux*, translated by Buffon (Paris, Debure 1735), p.iv.

61. Hanks, *Buffon avant l'*'*Histoire naturelle*', p.74-76, 93-100.

62. *Journal de Trévoux* (October 1735), p.2054-73.

63. Buffon, preface to Hales, *Statique des végétaux*, p.vi.

64. See *HARS* 1788 (1791), p.53-59. See also *HARS* 1748 (1752), p.44-45; Roger, *Buffon*, p.50n.; Stephen Milliken, 'Buffon and the British', Doctoral dissertation, Columbia University, 1965, p.168-74.

65. *HNM*, ii.344-46; Roger, *Buffon*, p.524-25. For earlier pledges of empiricism and rejections of the hypothetical see for example *HN*, i.28, 66, ii.65, 77, vii.16; *HNM*, iv.172; *HNS*, i.520. Buffon treated the term 'system' non-pejoratively in passing comments long before 1783. See the discussion of 'slips' below, p.124, and Nadault (ed.), *Correspondance générale* [*de*] *Buffon*, i.54, 88-89, ii.140-41; *HNS*, iv.6.

Before 1783 many of Buffon's contemporaries saw hypocrisy in his preaching of empiricists' values. 'Lui qui parle tant contre les hypothèses', grumbled Haller, 'il en a sans fin à lui-même.'[66] As we have seen, such double-dealing was as much the rule as the exception in the Enlightenment. Still, the tension between Buffon's reputation for systematising and his efforts to cast himself as a 'Newtonian' empiricist remains interesting in exemplifying a frequent occurrence. How could Buffon and other systematisers reconcile their rhetoric and their theories? And assuming Buffon managed to convince himself that he was different from other systematisers, why were his arguments so ineffective in persuading the public? To answer these questions we will focus on a text where the tension between Buffon's rhetoric and his readers' perceptions reached its highest level, his 1749 *Histoire et théorie de la terre.* Along with his 'system' of reproduction via organic molecules and the *Epoques de la nature*, the *Histoire et théorie* was seen as a notorious example of systematising, even though it features Buffon's most careful effort to ward off the accusation.

Early in the text Buffon contrasts theories with systems and hypotheses, pronouncing the former acceptable and the latter unacceptable in scientific work. This is one basis on which he will subsequently castigate three British authors treating the Earth's development – Thomas Burnet, William Whiston, and John Woodward, all systematisers from Buffon's point of view (*HN*, i.66-67). As we have seen, this was a typical strategy in the quarrel over systems. On both 'sides' of the quarrel nearly everyone recognised good and bad forms of rationalism. Buffon's choice of 'theory' to cover his good form was also felicitous. Adapted from Ephraim Chambers' *Cyclopaedia*, the article 'Théorie' in the *Encyclopédie* equates theories with systems and hypotheses, but this is misleading.[67] The word 'theory' had been gaining favour with respect to 'system' and 'hypothesis' since early in the century, in part because it had Newton's authority to back it up. For as Buffon surely knew, Newton too had distinguished theories from hypotheses, making the former acceptable and the latter unacceptable and purely conjectural.[68]

Beyond his frequent insistence that he constructs theories, in contrast to his rivals' systems and hypotheses (*HN*, i.66, 98-99, 129, 232-33, 324, 453), Buffon says little about what distinguishes theories and hypotheses in the *Histoire et théorie*. Some sixty pages into the text he comes as close as he ever will to defining his terms, saying that systems are based on possibilities, whereas theories come from facts (*HN*, i.129). While reminiscent of Newton's distinction, this is not very helpful. The passage from facts to non-conjectural theories remains notably mysterious. Another, more helpful passage associates hypothesisers

66. Haller to Charles Bonnet, 23 July 1759, in *Correspondence Haller-Bonnet*, p.173. See also Condillac, *Traité des animaux*, p.340-41; Malesherbes, *Observations sur l'histoire naturelle*, ii.137-41; Vicq d'Azyr, 'Eloge de Buffon', p.liv; *Mémoires et journal inédit du marquis d'Argenson*, 5 vols (Paris 1857-1858), v.117.

67. 'Théorie', in *Encyclopédie*, xvi.253. For a more usual view see Joseph-Jérôme Le François de Lalande, 'Hypothèse', in *Supplément à l'Encylopédie*, iii.487.

68. Paul K. Feyerabend, 'Classical empiricism', in *Methodological heritage of Newton*, p.159-60n.

with geological catastrophes and historians with the everyday (*HN*, i.98-99). Here, thirty pages into the text, the reflective reader may assimilate history to theories and understand the essence of Buffon's dichotomy. For though never linked explicitly with theories and systems, there are many remarks on method in the *Histoire et théorie* and the 'Premier discours' that clarify the opposition of the catastrophic and the everyday. Again and again Buffon returns to the idea that physical truth is founded on repetition and the ordinary, while conjectures depend on the rare and the singular:

> Ces causes dont l'effet est rare, violent, et subit, ne doivent pas nous toucher, elles ne se trouvent pas dans la marche ordinaire de la nature, mais des effets qui arrivent tous les jours, des mouvements qui se succèdent, et se renouvellent sans interruption, des opérations constantes et toujours réitérées, ce sont là nos causes et nos raisons.[69]

What Buffon seems to have meant but never made explicit was that theories must be premised on everyday facts, facts observable repeatedly in 'normal' circumstances, whereas systems can be premised on extraordinary events such as natural cataclysms. His examples of systems and hypotheses make sense in this light, since all invoke singular, catastrophic events. Buffon's geology, on the other hand, is generally actualist, like most French geologies of the early eighteenth century.[70] The geological prime movers he uncovers in the *Histoire et théorie* are erosion and sedimentation, the least cataclysmic and most humdrum of forces. His Earth acquires contours slowly and undramatically as water wears down coasts and mountains and deposits sediment elsewhere. Even mountains are traced not to volcanoes or earthquakes but to oceanic deposition over many centuries, the results of which are revealed by the sea's gradual drop to lower levels.

Natural cataclysms are not ignored in the *Histoire et théorie*. The devastating, sublime power of volcanoes, earthquakes, tempests, waterspouts, and hurricanes is evoked frequently and vividly, and the 'Preuves' following the main text include chapters devoted to such phenomena. Nor are these extraordinary events denied any role in shaping the Earth. A very few mountains, Buffon concedes, may have been raised by volcanoes and earthquakes (*HN*, i.80-81). But cataclysms are regularly derided as 'particular' or 'accidental causes'. Like the miraculous biblical deluge invoked by Buffon's British predecessors, they are too quirky and chaotic to support reliable science. Natural philosophers must therefore confine themselves to 'general causes', those constant enough to be studied with certainty. Significantly, the summary at the end of the main text neglects accidental causes completely, citing water alone as the motor of geology (*HN*, i.124).

69. *HN*, i.99. See also *HN*, i.55, 57, ii.77.

70. For more on the *Histoire et théorie*, its actualism, and the character of French geology in Buffon's time see Gabriel Gohau, 'La *Théorie de la terre* de 1749', in *Buffon 88*, p.343-52; Roger, *Buffon*, p.135-51; Ehrard, *Idée de nature*, p.199-210; François Ellenberger, 'Les sciences de la terre avant Buffon', in *Buffon 88*, p.334-37.

For the many post-eighteenth-century commentators who have discovered everyday recurrence as the basis of Buffon's distinction between theories and systems and thus of the methodology of the *Histoire et théorie*, further confirmation can be found in his analysis of induction in the 'Premier discours' and the 'Essai d'arithmétique morale'. In both texts Buffon confronts a problem familiar to philosophers since the seventeenth century – namely, how can universal, general truths be derived from finite sequences of individual instances?[71] His solution is to recognise that such inductive truths inevitably lack the certainty of mathematical 'evidence', but to suggest that enough instances can make them so probable that they can be considered 'physically' or 'morally' if not mathematically certain (*HN*, i.52-57; *HNS*, iv.47-79).

In non-quantitative form the same analysis had been offered by many previous thinkers, among them Claude Buffier and the Port-Royal theorists.[72] Inspired by Leibniz, a few philosophers had gone farther and attempted not just to legitimate 'uncertain certainties' but to calculate their value using the techniques of probability theory. In the late seventeenth century Jakob Bernoulli and Huyghens proposed that probabilities below certain fixed values such as 1/1000 or 1/100,000 should be considered equal to zero for human purposes, a supposition that would justify the inductive leap from instances to universals.[73] Probability theory was also applied to specific scientific problems in the years immediately preceding the *Histoire naturelle*, notably in proofs by *reductio ad absurdum* that observed regularities were not due to chance. In 1734 Daniel Bernoulli won a prize in the Académie des sciences for a paper that used probability theory to prove that chance alone could not be responsible for angular and directional regularities in the planets' orbits.[74] In 1743 Mairan offered a similar calculation regarding one of Winslow's twelve-toed and -fingered monsters to conclude that the probability of its forming fortui-

71. On the origins of the problem see Peter Dear, *Discipline and experience: the mathematical way in the scientific revolution* (Chicago 1995), p.11-21.

72. See for example Claude Buffier, *Traité des premières vérités et de la source de nos jugements*, in *Cours de sciences*, p.602-605 (pt 1, ch. 19); Antoine Arnauld and Pierre Nicole, *La Logique, ou l'art de penser*, ed. Emile Charles (Paris 1878), p.437-48 (pt 4, ch. 12). More generally see Lorraine Daston, *Classical probability in the Enlightenment* (Princeton, N.J. 1988), p.56-67; Roger, *Sciences de la vie*, p.533-34; Bernard Bru, 'Statistique et bonheur des hommes', *Revue de synthèse* 109 (1988), p.72-74.

73. Jacques Dutka, 'On the St Petersburg paradox', *Archive for history of exact sciences* 39 (1988), p.17, 33; Lenay, 'Hasard chez Buffon', p.623; Gérard Jorland, 'The Saint Petersburg paradox, 1713-1937', in *The Probabilistic revolution*, ed. Lorenz Krüger, Lorraine J. Daston and Michael Heidelberger, 2 vols (Cambridge, Mass. 1987), i.157-90, at p.159.

74. Daniel Bernoulli, 'Quelle est la cause physique de l'inclinaison des plans des orbites des planètes', in *Die Werke von Daniel Bernoulli*, ed. D. Speiser *et al.* (Basel 1987), iii.303-26. On John Arbuthnot's similar, even earlier probabilistic analysis of 'the regularity observed in the births of both sexes' see Stephen M. Stigler, *The History of statistics: the measurement of uncertainty before 1900* (Cambridge, Mass. 1986), p.225-26. Without using mathematics Newton too had argued that the solar system's regularities could not be due to chance. See Roger, *Sciences de la vie*, p.469-70. On the probabilistic *reductio ad absurdum* in general see Bru, 'Statistique et bonheur des hommes', p.75-77.

tously was less than $1/10^{80}$.[75] Buffon's friend Maupertuis used the same reasoning to show that heredity, not simple chance, was responsible for the recurrence of polydactylism in a certain Berlin family, a conclusion he published in 1752.[76]

Buffon, an adept of probability theory since the 1730s, was probably thinking of such examples when he proposed to mathematise induction in the 'Premier discours' and the 'Essai d'arithmétique morale'. In the 'Preuves' following the *Histoire et théorie de la terre* he seems to have adapted Bernoulli's argument regarding the planetary orbits to suit his own needs. First he calculated the odds that the six planets would turn in the same direction because of chance as a mere 1 to 2^6 or 64. The odds he was looking for were in fact 1 to 63, if not 1 to 31, but like Mairan, Bernoulli and other early probabilists, Buffon often translated probabilities of $1/n$ into odds of 1 to n rather than odds of 1 to $n-1$. Next, the planes of the planets' orbits being in a swath of just 7.5 degrees, that is, within a mere 1/24 of possible values, Buffon calculated the odds that such regularity was produced by chance as 1 to 24^5 or 7,692,624 (*HN*, i.134) – a mode of reasoning nearly identical with Bernoulli's earlier one. For Buffon as for others, such odds were negligible in moral if not mathematical terms, an assertion he would defend in detail in the 'Essai d'arithmétique morale'. Ergo, the solar system was not due to chance. But here, where Bernoulli had leapt to conclusions about the effects of a solar 'atmosphere', Buffon saw only evidence that the solar system had developed from a single determining cause, quite possibly a comet that had just grazed the sun (*HN*, i.134-35).

Later, in the 'Essai d'arithmétique morale', Buffon examined a classic if less practical question with even more bearing on the problem of induction. Given that human beings have individually and collectively experienced just a finite number of sunrises, on what basis do we believe that the sun will continue to rise or 'return'?[77] Here Buffon seems to have relied on the same line of reasoning he took from Bernoulli in the problem above, the same one used by Mairan, Maupertuis, and other philosophers of the first half of the century.[78]

75. *HARS* 1743 (1746), p.60-63. See also Roger, *Sciences de la vie*, p.418; Tort, *Ordre et les monstres*, p.196-99.

76. Pierre-Louis Moreau de Maupertuis, *Lettres*, in *Œuvres de Maupertuis*, ii.217-372, at p.307-10; Maupertuis, *Système de la nature*, in *Œuvres de Maupertuis*, ii.135-84, at p.159-61. On another early scientific use of the probabilistic *reductio ad absurdum* see Bernard Bru, 'Laplace et la critique probabiliste des mesures géodésiques', in *La Figure de la terre du XVIIIe siècle à l'ère spatiale*, ed. Henri Lacombe et Pierre Costabel (Paris 1988), p.223-27.

77. Non-quantitative analyses of our belief in continued sunrise can be found in Blaise Pascal, *Pensées*, ed. Dominique Descotes (Paris 1976), p.77 (no. 91 Brunschvicg); Gottfried Wilhelm Leibniz, *La Monadologie*, in *Opera philosophica*, p.707 (no. 28); Claude Buffier, *Eléments de métaphysique à la portée de tout le monde*, in *Cours de sciences*, p.935-36; Sandy L. Zabell, 'The rule of succession', *Erkenntnis* 31 (1989), p.293-97; Peter Dear, 'Miracles, experiments, and the ordinary course of nature', *Isis* 81 (1990), p.665-66; Brunet, *Physiciens hollandais*, p.65; Bru, 'Statistique et bonheur des hommes', p.73; Etienne Bonnot de Condillac, *Traité des sensations*, in *Œuvres philosophiques de Condillac*, i.219-319, at p.292-93.

78. Sandy Zabell has suggested that Buffon's calculation of the probability of sunrise was copied from a similar calculation published in 1764 by Richard Price as an application of his

In the 6000 years of recorded human history (2,190,000 days, leap days excepted) the sun has been observed to return 2,189,999 times – that is, every day, since it could only return from the second day onward. Assuming it had an equal probability of returning or not returning, the likelihood that it would have returned by chance for 2,189,999 days is $1/2^{2189999}$. This probability, a product of an argument by *reductio ad absurdum*, shows only one thing, namely that the sun's 'movements' are probably not due to pure chance. But Buffon appears to have translated this numerical by-product of his *reductio ad absurdum* into expectation about tomorrow's sunrise. Significantly, a nearly identical misinference can be found in his probabilistic treatment of the planets' orbits (*HN*, i.134). Overlooking his misstep, and giving his usual rounded odds, he comes up with odds of $2^{2189999}$ to 1 that the sun will rise tomorrow (*HNS*, iv.52-53).

Buffon's recourse to probabilistic analysis to prove causal connections and justify induction suggests what he had in mind when he championed the use of probability in natural history. The example above also sheds light on his distinction between theories and systems. Like sunrises, the recurrent, everyday events backing Buffon's theories have been observed so consistently that probabilistic calculation should be able to demonstrate not just their continuation but also their occurrence in the distant past, before anyone could observe them. In contrast the singular, rare occurrences underpinning systems and hypotheses constitute too short a series to push the probability of their past or future continuance into the realm of moral or physical certainty. The boundary between frequent and rare occurrences may be ill-defined in colloquial language, but Buffon's mathematisation of regularity, combined with his assumption of cut-off probabilities for moral and physical certainty, allowed him, in principle, to make these distinctions in a precise way.

Unfortunately for Buffon, contemporary critics paid little attention to his opposition of theories to systems or to the whole series of parallel oppositions on which he based it – the frequent versus the rare, the everyday versus the miraculous, the gradual versus the cataclysmic, the general versus the accidental. From the moment the *Histoire et théorie* appeared, Buffon's image as a systematiser was essentially fixed in public perception. Not until after his death would anyone examine the fine points of his criteria for distinguishing theories and systems.[79] Even Malesherbes, nowadays considered one of Buffon's most careful contemporary readers, failed to appreciate Buffon's distinction. Like

friend Thomas Bayes's work on probability. This is a possibility, but Buffon's solution to the 'problem' of sunrise is more likely to have developed from a simple, flawed argument by *reductio ad absurdum*. See below and Sandy L. Zabell, 'Buffon, Price, and Laplace: scientific attribution in the 18th century', *Archive for history of exact sciences* 39 (1988), p.173-81. For a revised view see Sandy L. Zabell, 'The continuum of inductive methods revisited', in *The Cosmos of science: essays of exploration*, ed. John Earman and John D. Norton (Pittsburgh, Penn. 1997), p.367-69. An account textually similar to Buffon's and Price's can be found in Condillac, *Traité des sensations*, p.292-93. I am grateful to Bernard Bru for his comments on this argument.

79. Sloan, 'Buffon, German biology', p.122. An early attempt to appreciate the distinction can be found in Flourens, *Histoire*, p.56-57.

many thinkers of the mid-eighteenth century, Malesherbes judged theories to be acceptable and systems less so, but he relegated the *Histoire et théorie* to the latter category without hesitation, completely ignoring Buffon's thoughts on the subject.[80] Ironically, he even reproached Buffon for not paying attention to recurrent events, apparently not realising that this was Buffon's methodology![81]

In large part, indifference toward Buffon's methodology stemmed from his failure to establish practical boundaries between theories and systems. Probability theory offered a means of doing so in principle, but despite Buffon's promises of mathematical rigour in the 'Premier discours' and elsewhere, the only two cases in which he actually calculated the likelihood of his theoretical assertions are those discussed above. In hundreds of other instances his judgements regarding the frequent and the rare remained vague and intuitive, a state of affairs he sometimes acknowledged: 'Chacun a sa mesure, surtout pour estimer des probabilités de cette nature' (*HN*, i.153). Buffon would continue to claim probabilistic rigour for many of his affirmations, suggesting that they could be mathematised if anyone wished to, but his knowing references to 'les règles des probabilités' (*HN*, i.196, 258) were little more, all in all, than attempts to intimidate through mathematical expertise. Ultimately his proposal for mathematising induction may have been in advance of his time, as John Lyon suggests, but it was not a proposal Buffon was able to implement.[82]

Going beyond the *Histoire et théorie*, where the scientific importance of recurrence is most prominently stressed, some scholars have argued that attention to repeated events marked Buffon's natural history in other domains.[83] Even without its probabilistic basis, Buffon's proposed method of generalising from the frequent and neglecting the rare has a certain distinctiveness. As Phillip Sloan has noted, it excludes 'crucial experiments', unique experiments set up to decide in themselves whether theories are true.[84] Still, in the watered-down form it assumes without probability, Buffon's method of cumulative induction is not hard to find in the works of many eighteenth-century thinkers. Nor did Buffon always propound it as the only way of arriving at physical truth. In the 'Essai d'arithmétique morale', for example, he also allows for immediate intuition from a single observation.[85] Perhaps for this reason there is great

80. See for example Malesherbes, *Observations sur l'histoire naturelle*, i.219-20, ii.1-3, 6-9, 137, 149-50.

81. Malesherbes, *Observations sur l'histoire naturelle*, i.226-27, 234. For other examples of Buffon's 'method' being used against him see Antonio Di Meo, 'Macquer e Buffon: una controversia scientifica del Settecento', *Cultura e scuola* 29 (1990), p.256; [Lignac], *Lettres à un Américain*, pt 2, letter IV, p.39-42.

82. Lyon, '"Initial discourse"', p.143. For other problems with Buffon's method see Sloan, 'Hypothétisme de Buffon', p.216; Duchesneau, *Physiologie des Lumières*, p.263-64.

83. Sloan, 'Hypothétisme de Buffon', p.213-19; Wohl, 'Buffon and his project', p.189-91; Hanks, *Buffon avant l'"Histoire naturelle"*, p.226-28; Ibrahim, 'Pensée de Buffon', p.181; Hodge, 'Two cosmogonies', p.245-51; Di Meo, 'Macquer e Buffon', p.248-29.

84. Sloan, 'Hypothétisme de Buffon', p.219.

85. *HNS*, iv.62; Sloan, 'From logical universals to historical individuals', p.117n.

variation in Buffon's practical requirements for the number of recurrences needed for truth. In his 'Essai' (*HNS*, iv.51-59) he rather arbitrarily calculates the bounds of moral and physical certainty, making the former achievable through just fourteen repetitions of the same event (2^{14} being greater than his cut-off probability of 1/10000) and the latter achievable through 2,189,999 repetitions (sunrise being the epitome of physical certainty). In practice, however, Buffon made no effort to match either criterion. As Hanks has noted, he regularly generalised from very small numbers of instances, often five or fewer.[86] As a sceptic, on the other hand, he once suggested that more than forty-six observations would be required to persuade him that birds will only incubate eggs of their own species or those of the cuckoo (*HNO*, vi.322-23). This is not to question the value of Buffon's probabilistic methodology as a theoretical ideal or the general habit he had of studying series, but it does help to understand why his contemporaries remained unconvinced of his methodology's relevance.

Other factors as well undermine Buffon's dichotomy between systems and theories. The very breadth of his geology weakens the opposition between the usual and the unusual. A contemporary French reader just opening the *Histoire et théorie* would probably have accepted Buffon's assumption that natural cataclysms are rare and thus marginal geological forces – forces more at home in fanciful systems than scientific theories. To such a reader, sedimentation and erosion would probably have seemed more believable forces, given their presence in everyday reality. But as pages are turned, Buffon's perspective stretches to take in the whole planet and several thousand years of history. Hundreds of natural cataclysms are briefly alluded to, and one is led to wonder if the geological normalcy of eighteenth-century France is an exception rather than the rule in the big scheme of things. But if cataclysms are normal on the scale of the universe, actualism and catastrophism merge at their limits, and theories and systems are hardly stark opposites.

Furthermore, natural cataclysms do play a role in Buffon's theory of the Earth, and despite their suppression from the final synopsis, their role is not negligible. Volcanoes, earthquakes, and other 'local' events are treated as accidents – a typical category in rationalistic eighteenth-century geologies – and like other accidents elsewhere in the *Histoire naturelle*, they are generally dismissed as superficial disturbances to an underlying order. But neither in the *Histoire et théorie* nor in the rest of his natural history was Buffon able to deny accidents any role in creating general and lasting effects.[87] The Mediterra-

86. Hanks, *Buffon avant l'*'*Histoire naturelle*', p.160-61, 180-81, 210, 222. See also *HN*, xii.145-46; *HNS*, iii.15.

87. See for example *HN*, i.109-10, xi.228-31. For more on accidents see Rhoda Rappaport, 'Borrowed words: problems of vocabulary in eighteenth-century geology', *British journal for the history of science* 15 (1982), p.38-44; Bremner, 'Buffon and the casting out of fear', p.80-81; Taylor, '*Epoques de la nature*', p.379-80. Elsewhere in the *Histoire naturelle* the category of the accidental came to include monsters (*HNS*, iv.578), fur colour among animals (*HN*, ix.205-206), the musk-rat's internal anatomy (*HN*, x.7), crystal structure (*HNM*, i.242), pearls (*HNM*, iv.128), and even

nean Sea, he suggests in the *Histoire et théorie*, was produced by an accident, quite possibly an earthquake that opened the Strait of Gibraltar to an influx of ocean (*HN*, i.99-100). More generally, the subsidence of the oceans that led to the uncovering of present-day continents was itself the result of the accidental collapse of dike-like natural walls, underwater caverns, or even of an entire Atlantis-like continent (*HN*, i.95-98). And as all Paris knew in 1749, Buffon's Earth developed from the 'splash' of a most accidental near-collision between the sun and a comet (*HN*, i.127-67).

These last two suggestions Buffon tagged as hypotheses rather than theories, as he did a number of less sweeping conjectures scattered throughout the *Histoire et théorie*. Like many eighteenth-century natural philosophers, he obviously wanted to forestall accusations of systematising by labelling his hypotheses as such in the text. Newton's 'Queries' were a model for the epoch's 'separation' of conjectures from certainties. In France scientists made it a point of honour to declare their conjectures throughout the eighteenth century, a practice encouraged by Fontenelle's 1699 'charter' for the Académie. As Alan Gross points out, such declarations have an important rhetorical function, namely to assure the reader of the non-hypothetical character of everything that is not identified as hypothetical.[88] In the *Histoire et théorie*, introducing his hypothesis about the Earth's origin, Buffon is explicit regarding this inference (*HN*, i.129):

> Nous nous refusons d'autant moins à publier ce que nous avons pensé sur cette matière, que nous espérons par là mettre le lecteur plus en état de prononcer sur la grande différence qu'il y a entre une hypothèse où il n'entre que des possibilités, et une théorie fondée sur des faits; entre un système tel que nous allons en donner un dans cet article sur la formation et le premier état de la terre, et une histoire physique de son état actuel, telle que nous venons de la donner dans le discours précédent.

But whatever their purpose, and however forthrightly he marks them off from his 'physical history', Buffon's conjectures and hypotheses are hard to write off as insignificant and disconnected exercises in the midst of an independent theory. First, of all his ideas in the *Histoire et théorie*, his hypothesis about the Earth's origin was the one that aroused the most popular interest.[89] The prominence of this hypothesis in public consciousness undoubtedly went a long way toward establishing his reputation for Cartesian system-building. But even beyond this popular reading, or misreading of the text, the *Histoire et théorie* is hardly a theory without its hypotheses, for its theories and hypotheses are interdependent. On at least one occasion Buffon's theory of the Earth explicitly refers to his hypothesis of the comet as a supporting piece of evidence.[90] Similarly, his geology requires sea level to fall irreversibly in

platinum (*HNM*, iii.318-19). Opposed to the real and the essential, it was especially important in defining species and varieties, as in *HN*, xi.292.

88. Gross, *Rhetoric of science*, p.122-26.

89. See Malesherbes, *Observations sur l'histoire naturelle*, ii.19.

90. *HN*, i.258-65. Buffon's comet is read as an integral part of his theory in [Lignac], *Lettres à un Américain*, pt I, letter III, p.2-5; Malesherbes, *Observations sur l'histoire naturelle*, ii.150. In contrast see Fréron, *Lettres sur quelques écrits de ce temps* (4 January 1750), p.7-14.

order to expose mountains that developed as underwater ridges. Yet the only explanation Buffon can imagine for this subsidence involves collapsing barriers, continents, and caverns, and he explicitly designates this a conjecture.[91] Faced with this intermingling of conjecture and theory, Buffon himself seems uncertain at times which he is writing (*HN*, i.263). The fact that his 'tags' on conjectures are often retrospective and referentially vague only adds to the difficulty of fixing precise boundaries between theory and hypothesis.

As we have seen, Buffon opposes theories to systems and hypotheses on numerous occasions in the *Histoire et théorie*, playing up his methodological 'differences' with respect to his rivals. But the text also contains terminological slips – a hint that his notions of theories and systems were not as well-defined as his highlighted slogans made them. Just after his opening condemnation of systems, Buffon refers to Newton's 'system' (*HN*, i.66). Twice, describing ideas that are no threat to his own, Buffon uses 'theory' and 'system' as synonyms (*HN*, i.193, 195-96) – an understandable fault, given the period's still ambiguous usage, but also a strike against the absoluteness of his earlier distinction. Similarly, Buffon's hypothesis about the Earth's origin is later called a theory (*HN*, i.150, 152). Here the 'slip' appears tactical, an example of the age-old strategy of sneaking ideas in as conjectures and then tacitly upgrading them to truths and certainties.[92] As Sloan has noted, Buffon effected a similar 'slip' with his supposedly hypothetical interior mould in another text of 1749, the *Histoire générale des animaux*.[93] In both cases the loose semantics of ordinary language allow local polemical interests to undermine a proposed philosophical distinction – as well as a key element of Buffon's wider-reaching scientific rhetoric, namely his distinction between systems and theories. Paradoxically, Buffon's rhetoric sometimes got in its own way.

Why Buffon's distinction of systems and theories was so little-heeded by his contemporaries should now be clearer. First, he never defined the terms 'theory', 'system', or 'hypothesis' explicitly. To pick up the methodology behind the oft-repeated opposition, the *Histoire et théorie* must be read and carefully synthesised in light of the critical last section of the 'Premier discours'. Theories, in particular, have to be assimilated to history and physical truths, an equivalence the text itself does little to publicise. Second, the methodology backing the opposition, while not wholly general or devoid of content, is vague and intuitive, lacking a promised foundation in probability theory. Third, the *Histoire et théorie* is replete with conjectures and hypotheses. In principle, they are all labelled and remain independent of Buffon's 'theory', but the text creates ambiguities and interdependence. Beyond his grand catch-phrases Buffon himself was inconsistent in his usage of 'system' and 'theory'.

91. *HN*, i.95-98. See also Gohau, '*Théorie de la terre*', p.347-48.

92. The technique is described in Pera, *Discourses of science*, p.85-88; Royou, *Monde de verre*, p.93.

93. See Sloan, 'History of the concept of the biological species', p.321-22; *HN*, ii.33-34, 56; *HNM*, i.11-12.

All these 'faults' were widespread in the quarrel over systems in eighteenth-century France. To begin with, few methodologists bothered defining terms such as 'system'. Nor was it relevant for the most part, since participants in the quarrel tended to use habitual assertions in habitual ways. In contrast, Buffon's distinction between theories and hypotheses was generally novel, even if based on a standard methodology involving recurrence, yet Buffon presented it piecemeal all through his text, as if to imply that it was no more than the usual empiricist's babble. Second, methods regularly promised more than they could offer, whether probabilistic certainty, as in Buffon's case, or a science of pure facts, as in the case of many an empiricist. Stripped of their often fantastic pretensions, methods would have lost much of their rhetorical appeal. Third, the prevalent practice of allowing marked conjectures or hypotheses gave much-needed flexibility to the supposedly empirical sciences of the early eighteenth century even as it reinforced and valorised the simplistic dichotomy of facts to speculation. But narratives often blended the conjectural and factual, and authors often converted the first to the second midway through papers. The ladder of near synonyms stretching from 'fact' and 'observation' to 'hypothesis' and 'system' facilitated strategic renaming as well as counterproductive terminological slips.

But if Buffon's failure to communicate a persuasive methodology was due to factors typical of the quarrel over systems, it does not mean that the eighteenth century's appeals to method were ineffective in general. Buffon's task as a methodologist was exceptionally difficult. Not content just to exalt facts and experiments or decry *esprit de système*, he sought to legitimate extrapolation of the distant past in light of the present. Historical science had earned little favour in the Académie des sciences in the first half of the century, partly because of the hostility of Newtonians such as John Keill.[94] The same year Buffon published his *Histoire et théorie*, Condillac denounced 'les physiciens [qui] entreprennent de découvrir par quelles voies s'est formé et se conserve l'univers'.[95] Resistance and doubt were inevitable. In contrast, most would-be methodologists had lesser goals, and their rhetoric probably helped them in lesser ways. Furthermore, Buffon insisted on accomplishing his already difficult task within the vocabulary of Newtonian empiricism. In their equally ambitious apologies for a new scientific rationalism, Mairan and Du Châtelet made peace with the terms 'system' and 'hypothesis' respectively, but Buffon, early on, refused to commit either sacrilege, a decision that was galling to his enemies and confusing to readers unused to historical 'theory'.

It may be, in any case, that Buffon's methodological rhetoric in the *Histoire et théorie* did not fail as completely as available contemporary reactions might lead us to believe. Jacques Roger has asserted that the rhetoric of experience was taken more seriously by journalists and 'half-scientists' than by the

94. Phillip R. Sloan and John Lyon (eds), intro. to *From natural history to the history of nature*, p.14-15; Sloan, 'Buffon, German biology', p.110-11; Sloan, 'Natural history', forthcoming.

95. Condillac, *Traité des systèmes*, p.198.

scientific elite, and methodological statements in general may have been more persuasive at popular levels.[96] Certainly vulgarisers have long considered methodology important in defining science for the public.[97] All through the century the *Histoire de l'Académie* outdid the *Mémoires* in its sententiousness regarding proper method, even though the authors of papers for the *Mémoires* had more stake in legitimating their ideas and methods than the mostly merely summarising author of the *Histoire*. In their stints as secretary around 1740, Mairan and Grandjean de Fouchy were especially good at finding methodological 'meaning' in the research of their colleagues – and little did it matter if some of their meanings contradicted each other. Likewise, reviews of the *Histoire et théorie* in eighteenth-century periodicals often respectfully paraphrased the text's remarks on methodology. After so doing, however, the *Journal des savants* also referred casually – and seemingly without malice – to Buffon's 'system'.[98] Such superficiality suggests that considerable negligence accompanied any popular credulity regarding methodology.

In conclusion, the Enlightenment's quarrel over systems cannot be understood, nor Buffon's engagement in it, without recognition of its rhetorical aspects. Scientific knowledge is more than mere rhetoric, at least in the sense in which 'rhetoric' is used here, and scientific methods are almost never just arbitrary propaganda. What has been said here is not meant to challenge Laudan's or Roger's views on the general evolution of eighteenth-century methods. Nor is it meant to refute Sloan's reconstruction of the ideal outlines of Buffon's methodology. Overall, hypotheses did enjoy more favour after 1740 than in the early part of the century. In this context Buffon imagined a novel means of giving certainty to truths of induction, especially the extrapolatory truths of historical natural history. But pronouncements on method regularly advertised rigour, consistency, and simplicity beyond what their authors could possibly offer, and Buffon was no exception. Though attention to recurrence marked aspects of his work, his proposal to calculate causes as probabilities was mostly impracticable. His and others' grandiose promises regarding methodology were meant at least partly to bolster their results through the prestige of abstract method. Representative though it may have been of practice at times, methodology was also a means of legitimation defined against a context of previous such attempts.

96. Roger, *Sciences de la vie*, p.251.
97. Schuster and Yeo, intro. to *Politics and rhetoric of scientific method*, p.xiii-xiv.
98. *Journal des savants* (October 1749), p.653.

5. Mathematics

MATHEMATICS, like systems, furnished matter for a great deal of seventeenth- and eighteenth-century scientific rhetoric, even in unmathematised natural history. 'Mathematical intimidation', one form of such rhetoric, remains familiar today. In addition, eighteenth-century authors used a contrary rhetoric, one denouncing mathematics and abstraction in general in the interest of weakening ostensibly mathematical and abstract rival theories. But the rhetorical aspect of such anti-mathematical discourse has long been neglected because of its attribution to a mistrust of mathematics on the part of specific authors and the eighteenth century in general. Here it is necessary to re-examine elements of the French Enlightenment's alleged 'crisis of mathematics'.[1]

Historians from Jean-Etienne Montucla onward have seen the eighteenth century as one of mathematical quiescence compared to its predecessor. Not only did eighteenth-century mathematics lack spectacular and clear-cut accomplishments, it was pronounced dead and tautological by many philosophers. Already at the turn of the century Fontenelle and Jean-Paul Bignon imagined mathematics to be all but wrapped up, an attitude that would echo all through the century. Among the most famous of the obituaries of mathematics was Diderot's. In his *Pensées sur l'interprétation de la nature* (1753) Diderot suggested among other things that experimental philosophy would be better off with no help from 'geometry' and that within a hundred years Europe would count fewer than three living geometers.[2] On a certain literal level his prediction was mistaken, but scholars have acknowledged its accuracy in other senses. Thomas Hankins, for example, notes that Diderot was right in predicting the end of mathematics as he knew it, since classic analysis was nearly exhausted.[3] Similarly, for Peter Gay, Diderot's anti-mathematical pronouncements are indicative of growing interest in new, non-mathematical scientific disciplines.[4]

1. For more on the 'crisis of mathematics' in eighteenth-century France see Yvon Belaval, 'La crise de la géométrisation de l'univers dans la philosophie des Lumières', *Revue internationale de philosophie* 6 (1952), p.337-55; Ehrard, *Idée de nature*, p.181-85; Henk J. M. Bos, 'Mathematics and rational mechanics', in *The Ferment of knowledge*, ed. G. S. Rousseau and Roy Porter (Cambridge 1980), p.327-33; Vartanian, *Diderot and Descartes*, p.178-89; Gay, *Science of freedom*, p.150-66; Paul Hazard, *European thought in the eighteenth century*, translated by J. Lewis May (Cleveland, Ohio 1963), p.130-31.

2. In *DPV*, ix.1-111, at p.28-31 (no. 2-4). For a similar obituary see *HNS*, iv.128-29. On Fontenelle and Bignon see Hankins, *Science and the Enlightenment*, p.17; Torlais, *Réaumur*, p.208.

3. *Science and the Enlightenment*, p.17-20.

4. *Science of freedom*, p.150-51. See also Thomas S. Kuhn, *The Essential tension* (Chicago 1977), p.35-59.

Buffon has always loomed large in discussions of the eighteenth century's devaluation of mathematics. Not only were his ideas the probable inspiration for Diderot's stance against mathematics in the *Interprétation* and other works,[5] Buffon's career seemed emblematic of the end to mathematics that Diderot predicted. As we have seen, Buffon achieved early mathematical success with a 1733 paper inaugurating the study of geometric probability. In the Académie des sciences he took his place in a mathematical section. As late as 1736 he expressed the hope that he would eventually get back to mathematics as a full-time occupation, but his focus had been elsewhere for some time already. In 1749 the 'Premier discours' formalised his change of interest, suggesting that mathematics had few applications and no inherent meaning outside its own symbols. While not as incendiary as Diderot's, the anti-mathematical passages of the 'Premier discours' are often cited as evidence of a retreat from mathematics in the mid-eighteenth century.

Arguments can be offered for doubting the profundity of this retreat from mathematics. Throughout the period mathematicians made discoveries and benefited from new positions and institutions, while intellectual culture in general became increasingly quantified.[6] Still, 'revolutions' are hard to identify in the theoretical advances of eighteenth-century mathematics. Perhaps more importantly, conflicts over infinitesimals, *vis viva*, and other abstractions in the early eighteenth century intensified claims about the unreality of mathematics. But if this was a crisis of mathematics, it is important to be clear about how exactly Diderot and Buffon embodied it. As Jean Mayer has argued, the anti-mathematical outlook of Diderot's *Interprétation* was fuelled by his attachment to old-fashioned analysis, his enthusiasm for experimental philosophy, and perhaps by some bitterness about his unrealised mathematical dreams. Diderot remained curious and confident about mathematics long after his statements about its futility.[7] In practice he registered far less scepticism about it than he did in his manifestos, a state of affairs J. L. Heilbron suggests may have been common.[8] As we will see, Buffon's case is similar. His aggressive claims notwithstanding, he did not uniformly or fundamentally mistrust mathematics. Furthermore, his anti-mathematical statements are not placed at random in the *Histoire naturelle*, but rather in contexts where they are polemically favourable to his own arguments. They constitute a key element of his scientific rhetoric.

5. Roger, 'Diderot et Buffon en 1749', p.230-31; Jean Varloot, 'Buffon et Diderot', in *Présence de Diderot*, ed. Siegfried Jüttner (Frankfurt am Main 1990), p.300-301; Hankins, *Jean d'Alembert*, p.74-76, 89-90.

6. See Jean Dhombres, 'Quelques rencontres de Diderot avec les mathématiques', in *Colloque international Diderot (1713-1784), Paris-Sèvres-Reims-Langres, 4-11 juillet 1984*, ed. Anne-Marie Chouillet (Paris 1985), p.270; J. L. Heilbron, intro. to *Quantifying spirit*, p.1-23.

7. Jean Mayer, *Diderot, homme de science* (Rennes 1959), p.72-74, 99-104. See also Dhombres, 'Quelques rencontres de Diderot', p.269-80.

8. 'Experimental natural philosophy', in *Ferment of knowledge*, p.363-64.

Mathematical rhetoric

Mathematics developed rapidly in the seventeenth century. Impressed with its achievements in the hands of Descartes, Pierre de Fermat, Blaise Pascal, and others, many French philosophers assumed that it would continue to flourish and bring a new kind of certainty to knowledge as a whole. So bolstered, its philosophical prestige was such as to guarantee that it would give rise to a rhetoric. Mathematicians enjoyed great success in applying their techniques to old and new problems, but they and their imitators claimed even greater successes. Texts of all kinds assumed the forms of mathematics, while thinkers of all kinds assured readers that their ideas enjoyed mathematical certainty. This was the epoch of Spinoza's *Ethica ordine geometrico demonstrata* (1677), which made use of numbered definitions, axioms, propositions, and proofs to establish God's properties, among other things. Even naturalists were not immune to the fashion for Euclidian presentation, particularly after the publication of Michel Adanson's purportedly mathematical *Familles des plantes* (1763).[9] In 1750 Maupertuis warned sarcastically that 'les mots de *lemme*, de *théorème* et de *corollaire* ne portent pas partout la certitude mathématique',[10] but eighteenth-century writers went on exploiting the rhetoric of mathematics, though usually more subtly than Spinoza or Christian Wolff. 'Proof by intimidation', known since antiquity, had acquired a new and more powerful form.

Seventeenth- and eighteenth-century thinkers who presented their conclusions as mathematically certain were not in most cases deliberately deceiving readers. Especially after Descartes, philosophers often equated mathematics with logic or rigorous thinking. In his *Discours de la méthode* (1637) Descartes charged sceptics of his theory of the heart with failing to appreciate 'la force des démonstrations mathématiques'. Above all his implication was not that his physiology could be mathematised but rather that it was as rigorous as his mathematics.[11] In this light the hackneyed eighteenth-century compliment of calling someone a geometer without geometry makes perfect sense, as does the liberality with which writers of the period applied the expression 'esprit géométrique'. Still, 'mathematical' philosophers were surely aware of the rhetorical advantage they stood to gain by generalising the notion of mathematical method. Other ways of describing rigorous thinking would have been found if not for the prestige attached to such expressions as 'geometry', 'theorem', and 'quod erat demonstrandum'. Non-mathematical mathematics was a rhetorical if also a conceptual innovation.

9. For an extreme case – a paper on botany organised around 'propositions', 'scholia', 'problems', and 'theorems' – see Narcisse Clavillart, 'Description botanico-mathématique de la Clavi-lakea', BMHN, Ms. 1063. On Adanson's 'mathematical' taxonomy see chapter 6, p.162-63.

10. *Essai de cosmologie*, p.22.

11. Dear, *Discipline and experience*, p.212; Vartanian, *Diderot and Descartes*, p.179-81. For another example of Descartes's mathematical rhetoric see Roger, *Sciences de la vie*, p.144.

Like Descartes and many other seventeenth- and eighteenth-century philosophers, Buffon exploited the rhetoric of mathematical authority. His 1740 study of pyrotechnics is organised around 'demonstrations', even though mathematics is never invoked.[12] As we saw in the last chapter, he suggests periodically that probabilistic analysis underlies his assertions in the *Histoire naturelle*, but he manages to supply it in only two cases. Another unrealised mathematical reverie, this one for calculating the shape of matter's constituents, adds pseudo-credibility to his assertion that chemical affinity is just a special case of gravity.[13] In the *Epoques de la nature* he judges his theory of terrestrial cooling to be confirmed 'to the point of demonstration' (*HNS*, v.26). Used several times in the *Epoques*, this mathematical bullying drew ironic outcries from Thomas-Marie Royou.[14] And in the 'Essai d'arithmétique morale' Buffon prefaces a dubious proof of the badness of gambling with the following warning: 'Ceci n'est point un discours de morale vague, ce sont des vérités précises de métaphysique que je soumets au calcul ou plutôt à la force de la raison; des vérités que je prétends démontrer mathématiquement à tous ceux qui ont l'esprit assez net et l'imagination assez forte pour combiner sans géométrie et calculer sans algèbre' (*HNS*, iv.68).

Echoing complaints from Buffon's lifetime, Gaston Bachelard has argued that the *Histoire naturelle* makes use of a rhetoric of accuracy. As Bachelard points out, Buffon is often extremely precise with the figures he calculates. Having discovered a formula for the time iron spheres take to cool from their liquid state, Buffon applies it to the Earth and concludes that our once molten planet took 96,670 years and 132 days to cool to its present temperature (*HNS*, i.157-58). Later he combines his analysis of cooling spheres with his belief that life develops as a function of temperature to calculate the fate of life on other planets (*HNS*, ii.513). On Saturn, for example, he concludes that life appeared 62,906 years after the formation of the planets and that it will continue for another 187,188 years before dying out because of the cold. In other contexts Buffon rounds his figures arbitrarily, or even adjusts them to bring them into line with intuition. In all of this, concludes Bachelard, Buffon manifests a 'prescientific' attitude toward mathematics.[15]

Buffon's hyper-precision is easy to interpret as another effort to intimidate through mathematics, but caution is in order. As Roger observes, Buffon's most exact figures are those given immediately after calculations.[16] Thereafter he typically rounds them. And if his rounding seems arbitrary, it is important to remember that studies of mathematical error were still in their infancy in

12. Hanks, 'Buffon et les fusées volantes', p.145.

13. *HN*, xiii.xii-xv. See also Marco Beretta, *The Enlightenment of matter* (Canton, Mass. 1993), p.142, 151.

14. Royou, *Monde de verre*, p.93, 146-50. Royou indulged in his own form of mathematical intimidation. See for example Royou, *Monde de verre*, p.108n.

15. *Formation de l'esprit scientifique*, p.212-16. See also Hanks, *Buffon avant l''Histoire naturelle'*, p.206; Wilda Anderson, 'Error in Buffon', *MLN* 114 (1999), p.695-701.

16. Roger, *Buffon*, p.537-38.

the late eighteenth century. Nor is it hard to find 'over-precision' in the works of other thinkers. In a footnote to his computation of the 'life spans' of the various planets Buffon minimises the importance of numerical accuracy: 'Je ne me suis pas piqué d'une exactitude rigoureuse, parce qu'elle n'aurait produit que de légères différences' (*HNS*, ii.515n.). But his arbitrary adjustments to certain calculated figures make his exactness elsewhere appear somewhat misleading. At the very least, his hyper-precision highlights the mathematical character of his natural history – an act of considerable rhetorical value. As Antoine de La Salle noted, mocking Buffon, 'On ne saurait croire combien un livre plein de chiffres, ou de signes algébriques, en impose au vulgaire des gens de lettres.'[17]

To readers familiar with Buffon's uncharitable view of mathematics in the 'Premier discours', his appeals to its authority may come as a surprise. But in simultaneously aping and denouncing it, Buffon was only conforming to his century's usage. As Ernst Cassirer has noted, eighteenth-century thinkers at the same time idolised and vilified mathematics.[18] The coexistence of these two attitudes dates from the early seventeenth century. Once mathematics was recognised as a promising form of knowledge, philosophers borrowed its trappings to frame non-mathematical claims. Conversely, once it began overflowing with potentially disruptive knowledge, strategies were perfected for contesting its primacy. Arguments against it were not hard to find. Ever since antiquity mathematics and abstraction had been opposed in many contexts to physical reality, an opposition that invited denunciations of mathematics as arcane and unreal. But only as greater authority accrued to experience and experiment in early modern Europe did it become rhetorically devastating to oppose mathematics and reality, and only after the mathematical revolutions of the seventeenth century did mathematics acquire enough prominence to make its denunciation a rhetorical commonplace.

Jean Ehrard has argued that early reactions to Newtonian physics opened the way for the eighteenth century's denigration of mathematics.[19] The very notion of attraction made Newtonians question earlier, 'geometrical' views of matter. Anti-Newtonians further aggravated the split between mathematics and physics by insisting that Newton's laws were just mathematical artifacts, not explanations of how things really were. The tumult over Newtonianism was certainly important in refining charges against mathematics that would flare up with new vigour around 1750. But the rift that the eighteenth century discovered between mathematics and reality had been opening for a long time. Mathematics and physics had been placed in opposition by both Plato

17. *Désordre régulier*, p.200-203. See also Royou, *Monde de verre*, p.65, 70. For more on the rhetoric of precision in the eighteenth-century sciences see Sven Widmalm, 'Accuracy, rhetoric, and technology: the Paris–Greenwich triangulation, 1784-88', in *Quantifying spirit*, p.195-98; Bensaude-Vincent, *Lavoisier*, p.204-207.

18. *The Philosophy of the Enlightenment*, translated by Fritz C. A. Koelln and James P. Pettegrove (Princeton, N.J. 1979), p.15-16.

19. *Idée de nature*, p.151-57, 181-84.

and Aristotle, with the latter anticipating later critiques of mathematics by claiming that it was too abstract for application to physics. The medieval curriculum maintained the opposition, and efforts to mathematise physics and astronomy in the sixteenth and seventeenth centuries were forced to confront and to some extent neutralise it. Critics of these efforts were quick, for their part, to seize and enhance traditional rhetorical strategies for highlighting the difference between mathematics and physics. Kepler, for example, was forced to address widespread scepticism about mathematical astronomy, scepticism which had only deepened after Copernicus. Doubters such as Pierre Bayle had an easy time undermining Kepler's discoveries by suggesting that he was a great mathematician but an incompetent physicist.[20] Jesuit thinkers such as Christopher Clavius, eager to bring mathematics into natural philosophy, had to contend with objections that 'physico-mathematics' was a contradiction in terms.[21] Descartes was the target of similar complaints. Though modelled on mathematics, his physics was generally non-mathematical. Still, he took pains to associate it with mathematics and thus invited the same attacks that detractors had wielded against Kepler and Galileo. François Bernier, for example, linked Cartesian philosophy with abstract mathematics and thus with the unreal in his *Abrégé de la philosophie de Gassendi* (1678). As late as 1715 Descartes's detractors were accusing him of having illegitimately mathematised physics.[22] Just or not, the denunciation had force as long as Descartes was seen as the epitome of a mathematical philosopher.

By the early eighteenth century a new, manifestly more mathematical physics was taking hold in France. With the rise of Newtonian physics, Cartesians changed their image. All at once their philosophy seemed less mathematical than physical, and rhetoric for and against it was adjusted accordingly. Having been attacked as abstract, mathematical, and out of touch with reality since the mid-seventeenth century, Cartesians now turned the accusations around. Among Newton's enemies it became a cliché to acknowledge the mathematical correctness of his results even while denying their physical significance. The law of gravity might work as a calculating device, but for Cartesians and others it said next to nothing about physical process and did not mean that bodies actually attracted each other. Newtonians mostly recognised that their law of gravity was not a traditional physical explanation. Citing Locke or the Bible, many doubted that such an explanation was within human grasp. For Cartesians, however, this 'positivism' reflected the weakness of Newtonian physics.

Another opportunity for denouncing Newtonian philosophy as abstract and unreal came in the debates over absolute time and space. For Leibniz and his

20. See Nicholas Jardine, 'Epistemology of the sciences', in *The Cambridge history of Renaissance philosophy*, ed. Charles B. Schmitt *et al.* (Cambridge 1988), p.693-702; Paul Burrell, 'Pierre Bayle's *Dictionnaire historique et critique*', *SVEC* 194 (1981), p.93-94.

21. Dear, *Discipline and experience*, p.32-37. See also Galileo's resolution of the problem in Pera, *Discourses of science*, p.62-63.

22. Ehrard, *Idée de nature*, p.103-105, 113-14.

followers as for Descartes, space and time were relational, not absolute entities. Space only existed as relations among objects, while time was no more than a succession of events. For Samuel Clarke and the Newtonians, space and time existed independently of objects and events. Commonsensical as the Newtonian position might seem, Leibnizians managed to cast themselves as defenders of reality while relegating their opponents to the inherently stigmatised side of abstraction. Though not focused on mathematics, the campaign was fought with the dichotomy 'abstraction versus reality', among other weapons, and it may have helped to popularise this moralistic opposition.

In other fields Newtonians deployed the same charges against their antagonists. Voltaire especially had no trouble identifying and ridiculing 'unreal' abstractions in Leibnizian philosophy, among them *vis viva*. *Vis viva*, or living force, was a measure promoted by Leibnizian physics as a replacement for Cartesian momentum. Calculated as mass times velocity squared (mv^2), *vis viva* was said to remain constant in the universe. Needless to say, its proponents saw *vis viva* as entirely real. Along with Cartesians, Newtonians agreed that the quantity mv^2 was conserved in elastic collisions, but with the notable exceptions of Willem Jacob 's Gravesande and Du Châtelet, they mostly considered it a mathematical artifact. The dispute over *vis viva* and other kinds of force would continue throughout the century, despite syncretic 'resolutions' by d'Alembert and others. As Yvon Belaval has argued, one effect of the lingering quarrel was to exacerbate the divide between mathematics and reality.[23] Another was to sharpen and popularise anti-mathematical rhetoric.

Also important in eliciting scepticism about mathematics were eighteenth-century controversies over calculus and infinity. As a technique, it was widely agreed that calculus worked. Despite much scepticism about mathematics in general, few expressed scepticism about the achievements of calculus or its utility. But the fact that it appeared to depend on the existence of infinitesimals embarrassed practitioners. Dissatisfaction with the foundations of 'fluxions' probably contributed to Newton's reticence regarding his discoveries. Leibniz, for his part, declared that infinitesimals were just useful heuristics with no real existence – a credo that separated mathematics from reality in order to justify the continued application of the former to the latter! In France Fontenelle took up the problem of the infinite. His *Eléments de la géométrie de l'infini* (1727) was read, if little appreciated, by mathematicians throughout Europe. In the *Eléments* he argued that infinite quantities were as real as finite ones, both being intellectual realities if nothing else.[24] His goal was to legitimate the infinite, not to sever intellectual from concrete reality. Geometry, he insisted, was intimately linked with physics as well as metaphysics. But

23. 'Crise de la géométrisation', p.345-46, 352. For more on the quarrel see Hankins, *Science and the Enlightenment*, p.30-33; Hankins, *Jean d'Alembert*, p.204-207; David Beeson, *Maupertuis: an intellectual biography*, *SVEC* 299 (1992), p.18-24.

24. p.50. For more on the *Eléments* see Fontenelle, *Eléments de la géométrie*, introduction, p.7-32; Andrea A. Rusnock (ed.), *The Correspondence of James Jurin (1684-1750): physician and secretary to the Royal Society* (Amsterdam 1996), introduction, p.38.

his efforts to legitimate the infinite led him to treat mathematics as purely conventional: 'Il n'y a dans la géométrie, pour ainsi dire, que ce que nous y avons mis.'[25] Similar to Leibniz in his formalism, Fontenelle divided mathematics and reality in the process of trying to legitimate their union.

Discomfort with infinitesimals grew more acute with the publication of George Berkeley's *Analyst* in 1734 and his *Defence of free-thinking in mathematics* in 1735. Neither work challenged the practical validity of calculus, but both highlighted inconsistencies in its purported foundations and urged mathematicians to find clearer methods. The problem with calculus, Berkeley suggested, was that mathematicians had ignored his strictures against Locke's 'abstract ideas'. Infinitely small quantities were abstractions lacking a grounding in the realm of the 'concrete'. As such, they risked disconnecting mathematics from reality. Similar reasoning led Berkeley to repudiate absolute space and time. In both cases his presentation put mathematics and abstraction at odds with reality. Though scores of philosophers leapt forth to vindicate calculus, Berkeley's arguments were strong enough to remain troubling throughout the century.

Early eighteenth-century controversies pitting mathematics against reality were bound to have an effect on Buffon's generation. Buffon was engaged in nearly all of them. As an early French Newtonian he was undoubtedly familiar with the Cartesian refrain that Newton's law of gravity was true mathematically but physically meaningless. In 1729 he read and admired Fontenelle's *Eléments*. His letters to Gabriel Cramer over the next several years show his debts to the *Eléments* in their carefree manipulation of infinite quantities.[26] In 1740 he published *La Méthode des fluxions et des suites infinies*, a translation of John Colson's 1736 English translation of Newton's dated classic. Though his preface borrowed considerably from Fontenelle's *Eléments*, Buffon now repudiated Fontenelle's view of infinity, insisting that infinite numbers were less real than finite ones since the latter derived from real objects while the former were merely privative ideas.[27] Buffon also used his preface to dismiss Berkeley's *Analyst*, perhaps out of personal loyalty to Berkeley's enemy James Jurin.[28] Ironically, Buffon's new scepticism about mathematical abstractions

25. Fontenelle, *Eléments de la géométrie*, p.59. On mathematics' links with physics and metaphysics see Fontenelle, *Eléments de la géométrie*, p.59-60.

26. Weil, 'Correspondance Buffon–Cramer', p.101-109.

27. Georges-Louis Leclerc de Buffon, preface to Isaac Newton, *La Méthode des fluxions et des suites infinies*, translated by Buffon (Paris, Debure 1740), p.vii-x. On the preface's debts to Fontenelle see Hanks, *Buffon avant l'"Histoire naturelle'*, p.112-14. Fontenelle too saw infinity as a privative idea, but with different consequences. See Fontenelle, *Eléments de la géométrie*, introduction, p.13-15. For other examples of the same view see Hanks, *Buffon avant l'"Histoire naturelle'*, p.117-18; Roger, *Buffon*, p.62-64.

28. See Buffon to James Jurin, 17 January 1741, in Rusnock (ed.), *Correspondence of James Jurin*, p.429. On Jurin's involvement with Berkeley see Rusnock (ed.), *Correspondence of James Jurin*, introduction, p.37-40.

with no concrete basis was similar to Berkeley's – but Buffon may not have read Berkeley carefully.[29]

Between the early 1730s and 1740, then, Buffon was transformed from a carefree Fontenellian formalist to a critic of mathematics beyond the intuitive. This would be his posture for the rest of his life. Fontenelle himself may have invited the turn-around with his suggestion that mathematics was human: 'Il n'y a dans la géométrie, pour ainsi dire, que ce que nous y avons mis.' Dropping the 'pour ainsi dire' and devalorising abstraction, we arrive at something not far from the 'Premier discours'. If Buffon took it seriously, Berkeley's critique of abstraction might also have pushed him toward a new scepticism.[30] On a practical level his doubts about mathematics were probably encouraged by his encounter with the Saint Petersburg paradox in the early 1730s.[31] In this notorious paradox, known since 1713 but unnamed when Buffon considered it, player A tosses a coin until heads appears. If heads appears for the first time on toss number n, B must pay A 2^{n-1} units of money, after which the game ends. Following established rules, eighteenth-century mathematicians calculated A's expectation as the sum of the products of each outcome's probability and value:

$$(1/2)(1) + (1/4)(2) + (1/8)(4) + \cdots + (1/2^n)(2^{n-1}) + \cdots$$

But this sum is infinite – '$\infty/2$', as Buffon put it in a letter to Cramer around 1730.[32] Common sense, on the other hand, told Buffon and his contemporaries that A stood to gain little from playing the game, since heads would almost certainly turn up in just a few tosses. Buffon went so far as to verify this experimentally by having a child play the game 2048 times. Consistent with common sense, A's average return was just 5.5 units of money (*HNS*, iv.84-87). Mathematics, in short, seemed out of touch with reality. As a result, Fontenelle's moderate formalism, with its cavalier refusal to enter into dialogue with commonsense notions, may have been tarnished in Buffon's perception. Buffon's letter to Cramer discussing the paradox bears no trace of his later anti-mathematical rhetoric. Significantly, his resolution of the Saint Petersburg paradox was not to reject mathematics, but rather to mathematise more realistically. Still, occurring as it did immediately after his 'conversion' to Fontenelle's mathematics, the episode could have set the stage for Buffon's later tirades against empty formalism.

29. See Roger, *Buffon*, p.64-65; Sloan, 'From logical universals to historical individuals', p.110-12; Milliken, 'Buffon and the British', p.187-91; Hanks, *Buffon avant l'‘Histoire naturelle’*, p.120-22.

30. Sloan, 'Hypothétisme de Buffon', p.209; Sloan, 'From logical universals to historical individuals', p.110-12.

31. Hanks, *Buffon avant l'‘Histoire naturelle’*, p.30-33; Roger, *Buffon*, 30-32. For more on the Saint Petersburg paradox see Jorland, 'Saint Petersburg paradox', p.157-90; Dutka, 'On the St Petersburg paradox', p.13-39.

32. For Buffon's calculations and a comment on their dating see Weil, 'Correspondance Buffon–Cramer', p.115-18.

In 1740 or 1741, shortly after finishing the *Méthode des fluxions*, Buffon read Du Châtelet's *Institutions de physique* and praised it to their mutual friend Claude-Adrien Helvétius. Phillip Sloan has argued that his reading of the *Institutions* was important in defining Buffon's view of abstraction.[33] In the *Institutions* Buffon encountered a detailed presentation of the case for *vis viva*, which he rejected.[34] He probably also encountered Du Châtelet's arguments against absolute space and time, which highlighted the gap between abstract ideas and 'concrete' reality. As for mathematics, Du Châtelet associated it with abstraction, and in terms reminiscent of Buffon's *Méthode des fluxions* she warned of the danger of assuming its reality: 'Les notions imaginaires, qui nous aident infiniment dans nos recherches, viennent très dangereuses, lorsque nous les prenons pour des réalités.'[35] Though Du Châtelet, unlike Buffon, did not dramatically moralise the gap she evoked between abstraction and reality, the *Institutions* may have deepened Buffon's suspicions about 'imaginary' objects without concrete referents.

In sum, Buffon's experiences from the 1720s to the early 1740s guaranteed his awareness of mathematics' potential for betraying reality. Again and again issues of perceived conflict between mathematics and reality impinged on his life when he was in his twenties and thirties. Whatever his dispositions, this context goes a long way toward explaining his anti-mathematical tendencies from 1740 onward. After the publication of the *Méthode des fluxions* Buffon regularly assumed the role of the pragmatic realist demolishing abstractions. In the late 1740s, for example, he refuted Descartes's claim that burning mirrors were impossible by constructing and testing several: 'Il fallait quelque chose de plus fort que des raisons.'[36] The cause of scientific reality had found a new champion, or so it would seem from Buffon's presentation – for the rhetorical side of his position must not be forgotten. Thanks to his self-promoting commentary, Buffon's crusade for reality is conspicuous in the *Histoire naturelle*, but the series also features abstractions such as interior moulds. Announcing one's alliance with reality against mathematical abstraction was too useful polemically to be just self-disclosure. Philosophers of all sorts flaunted their allegiance to concrete reality and their rivals' attachment to mathematical abstraction whenever they thought they could make a good case. Expressed attitudes toward mathematics were rarely outright lies, but neither did they

33. 'Hypothétisme de Buffon', p.211-13, 216-17; Sloan, 'From logical universals to historical individuals', p.138-39. For Buffon's praise of the *Institutions* see Helvétius to Du Châtelet, [early 1741], in Alan Dainard *et al.* (eds), *Correspondance générale d'Helvétius* (Toronto, Oxford 1981-1991), i.51.

34. Buffon to Jurin, 17 January 1741, 20 June 1742, 13 March 1743, in Rusnock (ed.), *Correspondence of James Jurin*, p.429, 440, 448.

35. *Institutions physiques*, p.111. On absolute space and time see Du Châtelet, *Institutions physiques*, p.94-134.

36. *HNS*, i.400. For more on the mirrors see David Carramolino del Valle, 'El Espejo ustorio de Buffon: entre la leyenda de Siracusa y la tradición del análisis químico', *Asclepio* 45 (1993), p.269-89; Findlen, *Possessing nature*, p.208, 329-31; Roger, *Buffon*, p.83-85; Hanks, *Buffon avant l'"Histoire naturelle"*, p.219.

attempt to capture rhetorically counterproductive nuances in their authors' ways of reasoning. The 'crisis of mathematics' of the early eighteenth century may have checked Buffon's confidence in Fontenellian symbol-shuffling, but just as importantly it taught him a powerful rhetoric.

Buffon's denigrations of mathematics reappear constantly in the *Histoire naturelle*, almost always in contexts where rival theories can be stigmatised as mathematical. The discussions of mathematics in the 'Premier discours' and the 'Essai d'arithmétique morale' are at the same time the most detailed and the least obviously subservient to polemical goals. In the last several pages of the 'Premier discours' Buffon turns to explication of the concept of truth, arguing that it comes in different forms. At the extremes are mathematical truths and physical truths, which he presents as absolute opposites: 'En mathématique on suppose, en physique on pose et on établit; là ce sont des définitions, ici ce sont des faits; on va de définitions en définitions dans les sciences abstraites, on marche d'observations en observations dans les sciences réelles; dans les premières on arrive à l'évidence, dans les dernières à la certitude' (*HN*, i.55).

So different and distant are mathematics and physics that Buffon doubts that the sciences will ever be mathematised beyond astronomy, optics, and rational mechanics (*HN*, i.59-60). In 1774, on the other hand, he would urge chemists to reduce their science to rational mechanics and thus to mathematics.[37] Perhaps his views changed, but it is worth noting that in the 'Premier discours', just a few pages after pronouncing mathematics largely irrelevant, Buffon proposes mathematising all natural history through the use of probability theory as a gauge for induction (*HN*, i.62). Neither here nor in optics, astronomy, or rational mechanics is it obvious how he imagines mathematics to be relevant to physics, since he reduces the former to a formal, empty exercise: 'Il n'y a donc rien dans cette science que ce que nous y avons mis, et les vérités qu'on en tire ne peuvent être que des expressions différentes sous lesquelles se présentent les suppositions que nous avons employées.'[38]

Buffon was not the first to reduce mathematics to a meaningless 'art'. Hobbes and other philosophers of the seventeenth century considered mathematics a human creation.[39] The Port-Royal *Logique* (1662), following Pascal, emphasised the potentially formal character of mathematical truth. Even arbitrary *definitio nominis* could produce valid reasoning, the *Logique* suggested, as long as these definitions were consistently used. But the *Logique* assumed the existence of non-arbitrary axioms and *definitio rei*, and promoted intuitive, Cartesian inference alongside its formalism.[40] A more radical formalism was

37. *HNS*, i.75-58. Buffon's vision of a reformed, mathematical chemistry was already established by 1765 (*HN*, xiii.xii-xv), but his attack on 'non-physical' chemistry would not be explicit for another ten years. For more on contemporary debates over autonomous and 'mathematical' chemistry see Bensaude-Vincent, *Lavoisier*, p.62-66; Di Meo, 'Macquer e Buffon', p.248-49.

38. *HN*, i.53-54. See also *HNS*, iv.47-48.

39. Dear, *Discipline and experience*, p.213.

40. See Arnauld and Nicole, *Logique*, p.111-21 (pt 1, ch. 12-13); p.404-37 (pt 4, ch. 3-11).

inaugurated by Leibniz. By putting predicates in their subjects and by making rational truths virtually identical, Leibniz reinforced the idea that mathematical inference was a matter of renaming, not construction or discovery. Significantly he referred to mathematical proof as a 'catena definitionum' ('chain of definitions').[41] Closer to Buffon, Fontenelle reacted to the paradoxes of calculus by transcending common sense with mathematical formalism. As strict analytic sensationism gained ground among philosophers, and as algebraic manipulation supplanted geometric construction as the standard for mathematical proof, formal views of mathematics became increasingly common, even fashionable according to Malesherbes, by the mid-eighteenth century.[42] Condillac, d'Alembert, and Condorcet came to speculate that all knowledge, not just mathematics, was a matter of translation. Under these circumstances few directly disputed Buffon's characterisation of mathematics as mechanical translation.[43]

If Buffon had company in making mathematics a conventional art, his rhetoric was nonetheless unusually strong. Not only did he stress the uncreative nature of mathematical inference, he also judged mathematics to be fundamentally easy. As a corollary to its abstractness, the simplicity of mathematics relative to other forms of knowledge had long been asserted. Pascal, for example, argued in his *Pensées* that once we perceive geometrical principles, 'il faudrait avoir l'esprit tout à fait faux pour mal raisonner sur des principes si gros'.[44] D'Alembert too associated simplicity with abstraction and distance from reality.[45] Buffon went further, pushing simplicity into facility and utter transparency: 'Comme nous sommes les créateurs de cette science, et qu'elle ne comprend absolument rien que ce que nous avons nous-mêmes imaginé, il ne peut y avoir ni obscurités ni paradoxes qui soient réels ou impossibles, et on trouvera toujours la solution en examinant avec soin les principes supposés' (*HN*, i.56). In the 'Essai d'arithmétique morale' Buffon would be even more categorical: 'Il ne s'agit, pour résoudre le problème le plus difficile, que de le bien entendre' (*HNS*, iv.132-33).

The 'Premier discours' thus disparages mathematics on three major scores, all familiar to contemporaries in one form or another. First, Buffon opposes mathematics to physics in the starkest of terms, placing the former so far from

41. Belaval, 'Leibniz et la chaîne', p.67.

42. See Henk J. M. Bos, *Lectures in the history of mathematics* (Providence, R.I. 1993), p.28-31; Malesherbes, *Observations sur l'histoire naturelle*, i.176-77n.

43. On Condillac, d'Alembert, and Condorcet see O'Neal, *Authority of experience*, p.51-57; Hankins, *Jean d'Alembert*, p.111-14; Keith Michael Baker, *Condorcet: from natural philosophy to social mathematics* (Chicago 1975), p.125-26; Keith Michael Baker, 'An unpublished essay of Condorcet on technical methods of classification', *Annals of science* 18 (1962), p.108. Contemporary rebuttals of Buffon's characterisation of mathematics can be found in [Lignac], *Lettres à un Américain*, pt 3, letter VIII, p.1-37; Malesherbes, *Observations sur l'histoire naturelle*, i.187-206; *Nouvelles ecclésiastiques* (6 February 1750), p.21-23.

44. p.49 (no. 1 Brunschvicg). See also Etienne Bonnot de Condillac, *La Langue des calculs*, in *Œuvres philosophiques de Condillac*, ii.417-529, at p.429, 470.

45. D'Alembert, 'Discours préliminaire', p.viii-ix; Jean Le Rond d'Alembert, *Traité de dynamique* (Paris, David 1758), p.i-iii.

reality that it ends up appearing useless. Second, he suggests that mathematics is a formal art, a matter of combining definitions in mechanical ways. Lastly, he argues that mathematics is easy once its principles have been mastered. All three of these claims are repeated many times in the *Histoire naturelle*, and scholars have made them the foundation for Buffon's 'philosophy of mathematics' – whence his reputation as an anti-mathematical thinker and his emblematic status in the retreat from mathematics around 1750. Still, there are grounds for wondering how deeply Buffon believed in this philosophy of mathematics. As we have seen, he was not averse to using mathematics to bolster his authority in natural history. Likewise, his eagerness to mathematise induction suggests tacit acceptance of mathematics' bearing on reality.[46] Furthermore, his sermons on the sterility of mathematics and abstraction in general almost always occur in contexts where his ideas can be cast as factual and those of his enemies as abstract. When his own claims depend on mathematics and abstraction, on the other hand, his critique of abstraction is either forgotten or else mentioned and then neutralised with the assurance that his abstractions are close enough to reality to be very useful.[47]

So comfortable did Buffon become in denouncing mathematics that he used it as a scapegoat even when his arguments involved other issues. In the *Histoire des minéraux*, for example, he condemns the nascent science of crystallography for imposing geometry on nature: 'C'est par abstraction que nous avons formé les figures géométriques et régulières, et par conséquent nous ne devons pas les appliquer comme des propriétés réelles aux productions de la nature' (*HNM*, iii.433). Yet as his other arguments reveal, his objection to crystallography has less to do with the ideal nature of the shapes it imposes than with their unreliability in defining mineral species. Buffon himself employs geometrical terms when evoking the variety in a given mineral's crystals.[48] The mathematical associations of crystallography merely make geometry a convenient pretext for adverse comment.

This is not meant to suggest that Buffon's critique of mathematics was no more than rhetoric. As we will see, his confidence in it was limited, and in interesting ways. But his mistrust never lived up to the heady declarations of the 'Premier discours', and it coexisted with faith in certain forms of mathematics. In light of the dual role of mathematics in seventeenth- and eighteenth-century scientific rhetoric, Buffon's insistence on its absolute disjunction from physical reality is no more surprising than his appeals to its authority. Both rhetorics miscast his approach to natural history in the interest

46. See Hanks, *Buffon avant l'"Histoire naturelle"*, p.60n.; Lenay, 'Hasard chez Buffon', p.613-16. Notice also Buffon's insistence on the 'applications' of his research in geometrical probability in Hanks, *Buffon avant l'"Histoire naturelle"*, p.60-61.

47. For an example of an argument that is mercifully spared contact with Buffon's critique of abstractions see *HNS*, i.119-20. For an example of an argument that excuses abstractions see *HNS*, i.289. See also chapter 6, p.168-70.

48. For other denunciations of crystallography as geometry, and for examples of Buffon's recourse to geometric vocabulary see *HNM*, i.241-42, 432-33, 567.

of adding weight to particular arguments. Even in the 'Premier discours', where the polemical goal of denouncing mathematics is less obvious than it usually is in the *Histoire naturelle*, the dim view of mathematics caps off and confirms what amounts to a chapter-long attack on Linnaeus's formal systematics. Just as importantly, Buffon's critique of mathematics sets his chosen discipline of natural history at the centre of the sciences by marginalising the whole class of mathematical sciences. A more moderate view of mathematics might have come closer to his practical attitudes, but it would hardly have been as rhetorically strong.

Mathematical practice

Consciously, Buffon probably believed in his characterisation of mathematics in the 'Premier discours', but his use of mathematics points to other assumptions. Consider the three qualities ascribed to mathematics in the 'Premier discours' – easiness, formality, and inutility. How easy did he find mathematics when actually doing it? The absence of mathematics beyond basic algebra in all but the *Supplément* of the *Histoire naturelle* suggests that he avoided difficult mathematics. His mathematical papers of the 1730s, reprinted in the *Supplément*, prove that he was capable of working with probability and calculus, but as Lesley Hanks has shown, even here in his most sophisticated mathematical work, Buffon had a tendency to underestimate problems. At times he had clever, even brilliant intuitions, but overconfidence in his abilities led him to bungle calculations and miss any number of subtle complexities.[49] In the case of his greatest mathematical achievement, his foundational study of probabilistic geometry, his penchant for 'philosophical mathematics' was undoubtedly an asset. It took a broad vision to imagine the synthesis of two independent mathematical fields. But his discovery of probabilistic geometry, like all his other mathematical achievements, was less mathematical than meta-mathematical. In mathematics proper, in his geometry and calculus of the 1730s, Buffon was sloppy, thanks to his apathy toward formal detail, and he erroneously simplified difficult problems.

The most mathematical text of Buffon's mature years, the 'Essai d'arithmétique morale', provides further evidence that his approach to mathematics made it easier than it should have been. The 'Essai' forms a composite of mathematical work he did before 1749, probably reworked and supplemented in the 1760s.[50] Here, among other things, Buffon claims to resolve the venerable

49. Hanks, *Buffon avant l'"Histoire naturelle'*, p.24-26, 45-61. See also Weil, 'Correspondance Buffon–Cramer', p.112. On Buffon's intuitions see Roger, *Buffon*, p.30-33, 36-41; Philip Holgate, 'Studies in the history of probability and statistics XXXIX: Buffon's cycloid', *Biometrika* 68 (1981), p.712-16.

50. On the genesis of the 'Essai' see Hanks, *Buffon avant l'"Histoire naturelle'*, p.42n.; Milliken, 'Buffon and the British', p.180-81n.; Roger, *Buffon*, p.504-505; Sloan, 'From logical universals to historical individuals', p.132-33; Jacques-Louis Binet and Jacques Roger (eds), *Un autre Buffon* (Paris 1977), p.25-26.

problem of squaring the circle. Made famous by Euclid, the problem runs as follows. Given the standard Euclidean tools – an unmarkable straight edge and a compass that collapses when removed from the paper – can a square be drawn that is equal in area to some given circle? In 1736 the Académie des sciences appointed Buffon to examine a proposed 'quadrature', and this may have initiated his interest in the problem.[51] In any case the problem was widely discussed, as evidenced by the publication of Montucla's *Histoire des recherches sur la quadrature du cercle* (1754). In 1775, overwhelmed with proposed methods for squaring the circle, the Académie des sciences decided to stop even reading them. Summarising the decision in the *Histoire*, Condorcet came close to pronouncing the construction impossible, but he refrained from elevating his judgement to the status of proof.[52] Buffon, in contrast, claimed in his 'Essai' to have proved the impossibility of squaring the circle. The essence of his argument was simply to assert that curves and straight lines are necessarily incommensurable: 'On voit clairement que la ligne droite ne peut pas plus mesurer la ligne courbe, que celle-ci peut mesurer la ligne droite' (*HNS*, iv.136). As Maupertuis pointed out in 1752, not necessarily in response to Buffon, Archimedes had long ago refuted this argument by squaring the parabola.[53] A valid proof of the impossibility of squaring the circle would not be discovered till 1882.[54]

Mathematics may have been easy the way Buffon did it, but it cannot be described as formal. On the contrary, his approach was highly intuitive and relied more on pictorial than symbolic evidence.[55] At one point, revealingly, he condemned mathematicians 'qui n'auront appris de la géométrie que l'usage des signes et des symboles, lesquels sont la langue et non pas l'esprit de la science' (*HNS*, iv.134). The mathematics of Buffon's rhetoric, this mythical entity that moves in strict progression from definition to definition, was plainly not the mathematics with which he worked, especially after his dalliance with Fontenellian infinity. Nor was it the mathematics with which contemporary mathematicians generally worked. Alexis-Claude Clairault's popular mathematical textbooks may have been criticised for substituting

51. Hanks, *Buffon avant l'"Histoire naturelle'*, p.104.

52. *HARS* 1775 (1778), p.61-66. See also Hahn, *Anatomy of a scientific institution*, p.140-47; Mary Terrall, 'Metaphysics, mathematics, and the gendering of science in eighteenth-century France', in *The Sciences in Enlightened Europe*, ed. Willam Clark, Jan Golinski, and Simon Schaffer (Chicago 1999), p.254-55. Maupertuis too refused to pronounce the squaring impossible in *Lettres*, p.372.

53. Maupertuis, *Lettres*, p.366-67. For a bizarrely flattering analysis of Buffon's 'proof' see Maurice Fréchet, 'Buffon comme philosophe des mathématiques', in *Œuvres philosophiques de Buffon*, p.438. For a more sober judgement see I. Todhunter, *A History of the mathematical theory of probability from the time of Pascal to that of Laplace* (New York 1949), p.349. Notice Diderot's similar naïvety regarding unsolved constructions in Dhombres, 'Quelques rencontres de Diderot', p.275-79.

54. On the history and resolution of the problem of squaring the circle see Morris Kline, *Mathematical thought from ancient to modern times* (New York 1972), p.38-42, 763-64, 981-82.

55. See Holgate, 'Studies in the history of probability', p.714-15; Hanks, *Buffon avant l'"Histoire naturelle'*, p.32, 55. Notice also Buffon's mention of mathematical truths seized by 'les yeux de l'esprit avec la dernière évidence' (*HN*, i.60). Milliken's view of Buffon as an 'uncompromising formalist' stems from his nearly exclusive attention to Buffon's pronouncements. See Milliken, 'Buffon and the British', p.193.

intuition for rigorous proof,[56] but mathematicians into the late eighteenth century continued to distrust algebra for its failure to offer intelligible proofs. Even the period's more formalistic mathematicians – Fontenelle, for example, or Joseph-Louis Lagrange – seem loose in their reasoning when compared with Buffon's ideal of mechanical deduction. Though probably based on his disappointing experience with Fontenelle's *Eléments*, Buffon's depiction of mathematics as perfectly formal is fancifully overstated. Exaggeration, naturally, makes it all the more suitable for rhetorical engagement. The more formal and mechanical he makes mathematics, the easier it is to ridicule 'mathematical' enemies.

Next comes the matter of the relevance of mathematics to natural history. According to Roger, Buffon's realism 'l'a prévenu contre l'usage des mathématiques dans ses théories physiques'.[57] But if Buffon found calculus useless in natural history, the same cannot be said of arithmetic and algebra, on which he drew regularly throughout his career, perhaps modelling his style on that of Stephen Hales.[58] Buffon's use of mathematics in his early work on arboriculture and his later work on metallurgy has long been remarked on. Most famously, he calculates the Earth's age in the *Epoques de la nature* by extrapolating from his experiments with cooling spheres. Similar calculations regarding other planets rebuffed the *Journal encyclopédique* in 1776: 'Ces recherches sont trop remplies de calculs pour pouvoir en donner l'analyse.'[59] Already in 1749, though, in the same volume of the *Histoire naturelle* in which he doubts mathematics' pertinence of to natural history, Buffon uses mathematics to prove that the solar system was caused by a single event (*HN*, i.134-35), to correlate the planets' densities and orbital speeds (*HN*, i.143-47), to frame rules about the speed of rivers (*HN*, i.343-47), to prove that earthquakes could theoretically have formed the highest mountains (*HN*, i.523-24), and to calculate the amount of water flowing into the ocean each day (*HN*, i.355-56). Unhappy with this last case of mathematisation, Malesherbes turned Buffon's rhetoric around, reproaching him for applying mathematics where it did not yet belong.[60]

In his studies of the living world Buffon's recourse to mathematics is less remarkable than in geology, but even outside arboriculture he relies on it periodically. Early in the series he observes that the age at which animals' growth ends varies proportionally with life expectancy (*HN*, ii.569). This 'law' is cited and updated, often numerically, through the rest of the series. In his article on the deer Buffon suggests typically that age at full development can be multiplied by seven to calculate average life span (*HN*, vi.93). With regard to humans he goes into detail, computing life expectancy for every age. Significantly, he considers his results 'vérités physiques' despite their manifestly

56. Pierre Brunet, *La Vie et l'œuvre de Clairault (1713-1765)* (Paris 1952), p.62-75.
57. *Buffon*, p.563.
58. Hanks, *Buffon avant l'"Histoire naturelle"*, p.27, 90.
59. 4.1 (15 May 1776), p.43.
60. *Observations sur l'histoire naturelle*, ii.226-27.

mathematical form (*HN*, ii.602). Elsewhere, inspired by Leibniz, he proposes a new geometry for studying malleable figures in order to conceptualise foetal development, an idea perhaps foreshadowing recent applications of topology to morphogenesis.[61] In his study of the senses Buffon draws on geometry and an analogy from mechanics to show why two eyes see only 'one-thirteenth' better than one eye alone (*HN*, iii.327-28), and he traces aesthetics to the preference humans have for ideal shapes and Pythagorean ratios among pitches (*HN*, iii.341-42). In the 'Discours sur la nature des oiseaux' alone he offers a mathematical proof that birds can fly above clouds (*HNO*, i.10-11n.), a calculation of the strength of birds' 'cries' relative to those of quadrupeds (*HNO*, i.18-21), and a numerical comparison of animals' speeds (*HNO*, i.31-33).

Despite Buffon's quickness to denounce mathematics as sterile, the only mathematics he did not use after the 1730s was 'higher mathematics'. The irrelevance of the latter to his work in natural history was heightened by the flexibility he gave to arithmetic by making it intuitive rather than formal. Consider how he theorised numerical data. On at least three occasions he was faced with large amounts of such data – first in his research on arboriculture in the 1730s and 1740s, then with his study of human life expectancy in the late 1740s, and finally with his experiments on cooling metal balls in the late 1760s. In all these endeavours Buffon organised and synthesised numbers with minimal use of mathematical symbols. In his reports he arrayed data in vast, redundant tables, not only to facilitate synthesis but also to make his findings accessible and to flaunt his work's 'utility'.[62] Then he charted tendencies, sometimes wholly verbally, and sometimes with a mixture of verbal and symbolic algebra. Like Fontenelle, he was perhaps most at home doing verbal mathematics.[63] Ordering his data, Buffon sought relations expressible in ordinary language, notably inverse and direct variation. When his data were at variance with such relations he reacted in one of two ways. Sometimes he gave up, declaring the problem too involved for mathematics. At other times he proposed 'adjustments'.[64] In one study of wood strength he concluded as follows (*HNS*, ii.177-78):

> Dans les pièces de même grosseur, la règle de la résistance de la raison inverse de la longueur, s'observe d'autant moins que les pièces sont plus courtes [...]. Cependant à tout prendre, on peut se servir de la règle générale [...] car en jetant les yeux sur cette même table, on voit un grand accord entre la règle et les expériences [...] et il règne un ordre assez constant dans les différences, par rapport aux longueurs et aux grosseurs, pour juger de la modification qu'on doit faire à cette règle.

In 1776 the naval engineer Louis-Edmé-Gabriel Du Maitz de Goimpy claimed to discern a more precise law in Buffon's data on wood strength: 'Le

61. *HN*, ii.372-74. See also Roger, *Buffon*, p.205; Ferrari, 'Kant, lecteur de Buffon', p.159.

62. Notice the use of Buffon's tables in Nicolas Lenglet Du Fresnoy, 'Pont militaire', in *Encyclopédie*, xiii.53, 59. See also Licoppe, *Formation de la pratique scientifique*, p.227-32, 264-65.

63. See Hanks, *Buffon avant l'"Histoire naturelle'*, p.29-30, 82.

64. See for example Roger, *Buffon*, p.69-70; Hanks, *Buffon avant l'"Histoire naturelle'*, p.49-50, 202-13, 222; *HN*, i.341; *HNS*, ii.511-12.

moment de l'effet qui travaille à les rompre, dépend de la longueur du bras de levier élévé à une puissance dont 4/3 ou 5/4 est l'exposant.'[65] But Buffon preferred adjusting his more commonsense variations. Throughout his life he made similar adjustments to calculations, rules, and even raw data. His hunches about the failings of empirical and mathematical particulars were frequently accurate. Studying human life span, for example, he 'smoothed' demographic data to compensate for rounding on the part of informants, a procedure Maurice Fréchet has praised as modern and sophisticated, though paradoxical in light of Buffon's 'distaste' for abstraction.[66] Rightly suspecting that the 75,000-odd years he had calculated for the Earth's age in the *Epoques de la nature* understated the truth, Buffon 'fudged' his calculations to get higher figures and penned in an age of about three million years in an unpublished manuscript.[67] Without the confirmation of hindsight his flexibility with figures and formal mathematics was bound to appear arbitrary. In the *Histoire et théorie de la terre*, for example, he sought to bolster his hypothesis about the Earth's origin by proving that the planets' densities had originally varied inversely with their orbital speeds. Clearly the proportion no longer held, but a whimsical, numerically sloppy adjustment for 'condensation' allowed him to calculate that the Earth's original density was only half what it now is – almost exactly the value his theory required![68] This self-serving tampering provoked grumbling from Malesherbes: 'Avec la faculté de se tirer d'affaire par de pareilles suppositions, on ne doit pas être embarrassé de faire des systèmes.'[69] The same could be said of Buffon's rationalisation of exceptions to his 'law' of cooling spheres (*HNS*, i.159-60n.).

In sum, Buffon not only used basic mathematics but also the para-mathematical technique of fudging. While hardly unusual or cause for condemnation, the latter fact is interesting in light of his famous quarrel with Clairaut over the precession of the moon's apsides, the extreme points of its orbit. For its 1747 prize the Académie des sciences asked mathematicians why the moon's apsides precessed over nine years instead of eighteen years, the value that seemed to follow from Newton's law of gravity. Unable to reconcile a nine-year precession with Newton's law, Clairaut shocked the scientific community by suggesting that the law of gravity needed revising. One possibility he offered, none too original, was a law using two terms rather than one. Where Newton made the force of gravity proportional to m_1m_2/d^2, with m_1 and m_2 masses and d the distance between them, Clairaut proposed that the force

65. Quoted in Licoppe, *Formation de la pratique scientifique*, p.241.

66. 'Buffon comme philosophe des mathématiques', p.435-36.

67. Roger, *Buffon*, p.539-43; Roger, intro. to Buffon, *Epoques*, p.lxiii-lxvii. On Buffon's casual attitude toward all such figures see Anderson, 'Error in Buffon', p.695-701.

68. *HN*, i.146-47. In the first edition of the *Histoire naturelle* Buffon's adjustment for condensation puts the Earth's original relative density at 400⁷/₈, but subsequent editions replaced this with the more plausible 440⁷/₈. Even this last figure only approximately follows from Buffon's suppositions. See Seidengart, 'Traitement du problème cosmologique', p.319-20.

69. *Observations sur l'histoire naturelle*, ii.15. See also [Lignac], *Lettres à un Américain*, pt I, letter II, p.100-103.

might be $G_1m_1m_2/d^2 + G_2m_1m_2/d^4$, with G_1 and G_2 undetermined constants. Such a law would converge to Newton's as *d* became infinite, but it could be different enough at the moon's distance to resolve the discrepancy in the apsides' movement. Presented as no more than an illustrative possibility, Clairaut's 'law' would be treated as a definitive proposal by Buffon and others.[70]

Early the next year Buffon denounced Clairaut's 'law' in the Académie. Later, unrepentant, he revised his 'Réflexions sur la loi de l'attraction' for republication in the *Supplément*. By this time Clairaut had succeeding in reconciling the moon's movements with Newton's law, thus allowing Buffon to claim an unearned victory. The reasons Buffon gave for dismissing Clairaut's law were diverse and numerous, especially early on, but he chose not to publish a handful of the weaker ones.[71] Clairaut's law, he claimed at one point, was mathematically nonsensical, an assertion he supported with a misguided algebraic analysis.[72] Less miserably rebutted was his analogy between gravity and odour or light. As Buffon knew, particles radiating from a centre disperse in proportion to the square of their distance, but he had no answer for Clairaut's riposte that gravity was not particulate.[73] In the end metaphysics proved to be Buffon's only enduring source of arguments. Intuitive as always in his approach to mathematics, he contended that the discrepancy in the apsides' movements was 'almost imperceptible' and that the old law's successes were too numerous to let it fall to a single exception. It would be better, he argued, to explain the discrepancy through an additional cause, perhaps a magnetic field or a hidden asymmetry in the moon's mass.[74] At the same time he maintained that every term and constant in a mathematical law of nature must represent something. Any effect not describable with a single mathematical term was a composite of several more basic effects. Finally he argued that mathematically adjusted laws are not really laws:

Exprimer la loi d'attraction par deux ou plusieurs termes [...] me paraît n'être autre chose que d'ajuster une expression de telle façon qu'elle corresponde à tous les cas; ce n'est plus une loi physique que cette expression représente, car en se permettant une

70. Philip Chandler, 'Clairaut's critique of Newtonian attraction: some insights into his philosophy of science', *Annals of science* 32 (1975), p.373. For more on the controversy see Chandler, 'Clairaut's critique of Newtonian attraction', p.369-78; Roger, *Buffon*, p.85-91; Phillip R. Sloan, 'Buffon on Newton's law of attraction (1749)', in *From natural history to the history of nature*, p.77-78.

71. For a glimpse of Buffon's earliest arguments see Alexis-Claude Clairaut, 'Réponse aux réflexions de M. de Buffon sur la loi de l'attraction et sur le mouvement des apsides', *MARS* 1745 (1749), p.532-48.

72. See Georges-Louis Leclerc de Buffon, 'Addition au mémoire qui a pour titre: Réflexions sur la loi de l'attraction', *MARS* 1745 (1749), p.551-52; *HNS*, i.136-38; Alexis-Claude Clairaut, 'Réponse à la réplique de M. de Buffon', *MARS* 1745 (1749), p.578-80; Buffon, 'Seconde addition au mémoire qui a pour titre: Réflexions sur la loi de l'attraction', *MARS* 1745 (1749), p.580-81; *HNS*, i.138-40.

73. Buffon, 'Réflexions sur la loi de l'attraction', *MARS* 1745 (1749), p.493; *HNS*, i.126-27; Clairaut, 'Réponse aux réflexions de M. de Buffon', p.538-39.

74. See Clairaut, 'Réponse aux réflexions de M. de Buffon', p.543-46; Buffon, 'Réflexions sur la loi de l'attraction', p.495-97; *HNS*, i.129-31.

fois de mettre un second, un troisième, un quatrième terme, etc. on pourrait trouver une expression qui dans toutes les lois d'attraction représenterait les cas dont il s'agit [...] et par conséquent cette supposition [...] donnerait entrée à toutes les lois possibles et imaginables.[75]

Newtonians had long imagined analogues to the law of gravity to account for other phenomena. Newton himself concluded that explaining chemical attraction would require a force varying as $1/d^n$, with d the distance between particles, and n a higher power than the canonical 2. The threat of multiple laws of attraction inspired accusations that Newtonians were coming up with new laws for every phenomenon: 'Ils en prescrivent d'abord des lois générales [...] tout prêts à en imaginer de nouvelles encore, à mesure qu'il se présentera des phénomènes.'[76] Buffon himself proposed a new kind of attraction in his theory of nutrition and reproduction of the late 1740s, namely a *force pénétrante* that organised living matter, but in 1765 he insisted that all attraction, chemical affinity included, was based on Newton's standard inverse-square law of gravity. Newton's error, he claimed, was to overlook the possibility that oddly shaped particles could make ordinary attraction appear to vary as $1/d^3$ or otherwise (*HN*, xiii.xii-xv). Perhaps he imagined his *force pénétrante* along these lines in 1749. In any case, it was unfair of him to invoke a banal criticism of 'proliferating' laws of attraction against Clairaut's 'solution' to the problem of the apsides. For contrary to Buffon's suggestion, Clairaut made it clear that he was seeking a unique law of attraction, not a law for each phenomenon.[77] On close inspection Buffon's 'worries' about multiple laws appear mostly rhetorical.

Buffon's stand against Clairaut is often seen as reflecting the anti-mathematical philosophy of the 'Premier discours'. On this reading Clairaut's law epitomises mathematical abstraction without a real basis, while Buffon's objections represent his hostility toward such abstraction. To Clairaut's law, assert Sloan and Lyon, Buffon 'opposed the "concrete" simplicity of the [old] law'.[78] This is certainly the interpretation Buffon promoted, for he repeatedly called Clairaut's law abstract and unreal. But there are reasons for wondering if Buffon was not just applying a favourite rhetoric to a situation where its pertinence was far from determined. Even Jacques Roger, who traces Buffon's position to his alleged separation of mathematics and reality, acknowledges that the explanation is paradoxical. If Buffon believed mathematics was cut

75. 'Réflexions sur la loi de l'attraction', p.497; *HNS*, i.131-32. For his claim that terms and constants have to represent realities see Buffon, 'Réflexions sur la loi de l'attraction', p.498-99; Buffon, 'Seconde addition au mémoire', p.581-82; *HNS*, i.132-34, 140-41.

76. Falconet, intro. to Fontenelle, *Théorie des tourbillons cartésiens*, p.xv-xvi.

77. See Alexis-Claude Clairaut, 'Du système du monde dans les principes de la gravitation universelle', *MARS* 1745 (1749), p.359-61n.; Roger, *Sciences de la vie*, p.483n. See also Du Châtelet, *Institutions physiques*, p.338-39.

78. Sloan and Lyon, intro. to *From natural history to the history of nature*, p.22. See also Sloan, 'Buffon on Newton's law of attraction', p.77-78; Sloan, 'From logical universals to historical individuals', p.112, 130; Roger, *Buffon*, p.89.

off from reality, why was it he, not Clairaut, who insisted on a correspondence between individual forces and mathematical symbols?[79]

And the paradox gets bigger. As we have seen, fudging was a key part of Buffon's mathematics. Even while accusing Clairaut of adjusting the law of gravity to make it fit data, Buffon was indulging in algebraic hocus-pocus to 'demonstrate' the impossibility of Clairaut's law. His efforts to find money's subjective value in the 1730s led him to adjust formulae in the same way as Clairaut. Just as Clairaut proposed replacing Newton's law with a more complex one designed to match experience, Buffon replaced money's face value v with the 'moral' value v^k, where k was the constant log(9/5)/log 2, so calculated in order to corroborate his experimental 'determination' that expectation in the Saint Petersburg game was 5 units of money.[80] Buffon's complaints about Clairaut's fudging concerned a physical law, not moral arithmetic, but as we have seen, Buffon adjusted his rules for wood strength and cooling balls too, if only verbally. Nor with his own laws did he require terms and coefficients to represent real entities. If someone had asked him why there were two terms in his formulae relating a ball's diameter to the time required to cool it, he might have answered enigmatically that the second term represented what Newton called 'hidden causes'.[81] But had someone asked him what the seven meant in his rule giving animals' life spans as seven times their adult age, he would have found the question ridiculous. And he never answered Clairaut's ironic question about the meaning of the exponent 2 in the standard law of gravity.[82] Yet he insisted that the terms and constants of Clairaut's law needed meaning. As Lesley Hanks has noted, Buffon had little understanding of mathematical functions, but the fact that he wanted Clairaut's constants as well as his terms to be representative shows that his resistance involved more than an inability to see unity in a two-term expression.[83] For Clairaut, Buffon's demands about symbolic representation were tantamount to a kind of Pythagorean mysticism: 'Demander qu'on trouve des coefficients ou d'autres quantités de même espèce, existants par eux-mêmes, me paraît une prétention aussi bien fondée que si on voulait

79. Roger, *Buffon*, p.88-90.

80. Fréchet, 'Buffon comme philosophe des mathématiques', p.444-45. As noted above, Buffon's experiments put the game's value at 5.5 units of money, but other considerations, typically, induced him to lower this figure. See *HNS*, iv.84-87.

81. Buffon never formulated a general law relating temperature, time, and diameter for cooling iron balls, but he gave formulae relating time and diameter for different falls in temperature and assumed that the time of cooling would equal $df(t) + g(t)$, where d is diameter and f and g are functions of cooling. See *HNS*, i.148-58; Roger, *Buffon*, p.516-18; Lucien Leclaire, 'L'*Histoire naturelle des minéraux* ou Buffon géologue universaliste', in *Buffon 88*, p.355-62.

82. Alexis-Claude Clairaut, 'Réponse au nouveau mémoire de M. de Buffon', *MARS* 1745 (1749), p.583-84. Fontenelle asks the same question, albeit seriously, in *Théorie des tourbillons cartésiens*, p.196.

83. See *Buffon avant l'"Histoire naturelle'*, p.211; Fréchet, 'Buffon comme philosophe des mathématiques', p.437-38.

trouver dans la nature l'existence des lettres et des mots qu'on emploie à définir des choses.'[84]

The laws Buffon propounded were meant as heuristics, as indifferent instruments for approximating facts and 'saving phenomena'. He held Newton's law to higher standards, notably to the standard of term-by-term intelligibility. As Hanks has argued, his dispute with Clairaut suggests that Buffon saw the law of gravity as an absolute truth, no longer subject to critical scrutiny. He also saw the controversy as a personal matter – a chance to 'defend Newton and his commentators', as he described it – not as a simple philosophical disagreement.[85] The sheer number of his arguments against Clairaut's proposal indicates a willingness to conquer by any means, including abstract algebra. Buffon's stance in the controversy was above all not motivated by contempt for abstractions or mathematical fudging, both of which he exploited in his own arguments. Here if not elsewhere his 'scorn' for mathematicians and their arbitrary adjustments seems to mask a naive confidence in mathematics' ability to represent reality number by number and symbol by symbol.[86] Beyond this the philosophy of mathematics he used against Clairaut was largely rhetorical.

As always Buffon's rhetoric was susceptible to being turned against him. All through the controversy Clairaut portrayed himself as the ally of facts and reality, and Buffon as the abstract, out-of-touch party. As noted above, Clairaut called attention to Buffon's Pythagorean understanding of mathematical language, thereby associating him with useless abstraction. Likewise Clairaut accused Buffon of fudging, not only with his algebra but also by imagining unverifiable causes to explain away the discrepancy in the apsides' movements.[87] Accusations of fudging were a staple of scientific rhetoric throughout the Enlightenment. Like rhetorical contrasts between mathematics and reality, they pointed to subjectivity and thus to unreliability in others' abstractions. Buffon made such accusations many times against his enemies, and some of his critics made them right back.[88] Underneath the rhetoric everyone

84. 'Réponse au nouveau mémoire de M. de Buffon', p.585.

85. See Hanks, *Buffon avant l'"Histoire naturelle'*, p.90-91, 220-23; Buffon to Jean-Bernard Le Blanc, 23 June 1750, in Nadault (ed.), *Correspondance générale [de] Buffon*, i.71. See also Paolo Casini, 'Buffon et Newton', in *Buffon 88*, p.302-304. On the heuristic status of the rules Buffon formulated see Licoppe, *Formation de la pratique scientifique*, p.228-32, 264-65.

86. See Hanks, *Buffon avant l'"Histoire naturelle'*, p.114-17, 225-26. Though based on the pronouncements of the 'Premier discours', Baker's claim that Buffon was mathematically 'non-metaphysical' may be broadly accurate, but it does not account for his position vis-à-vis Clairaut. See Baker, *Condorcet*, p.113. On the other hand the fact that he made 'non-empirical assumptions' about Newton's law need not mean that Buffon had a coherent 'metaphysics of mathematics'. For a critique of efforts to find such metaphysics throughout the sciences see Gary Hatfield, 'Metaphysics and the new science', in *Reappraisals of the scientific revolution*, p.93-102.

87. Clairaut, 'Réponse aux réflexions de M. de Buffon', p.530.

88. See for example *HNO*, vi.322-32; Malesherbes, *Observations sur l'histoire naturelle*, ii.17, 212-13, 217, 296. See also Kuhn, *Essential tension*, p.25.

realised that mathematics and reality were unlikely to harmonise without being adjusted to each other to a degree.

This is not to say that Buffon's dealings with mathematics were entirely generic. Though his practical attitudes lacked the simplicity of his famous declarations, they were not incoherent. Among his more obvious mathematical regularities were reliance on arithmetic and verbal mathematics and avoidance of calculus after the 1730s. Beyond this Lesley Hanks has made the most detailed reconstruction of the epistemology implicit in Buffon's use of mathematics. According to Hanks, Buffon was convinced that simple ideas, including mathematical ones, were only valid when they reflected simple sensations. Once ideas became complex Buffon often rejected them as abstract and unreal, though his tolerance could be greater when it suited his purposes.[89] Hanks's reconstruction is to some extent oversimplified. Only against Clairaut did Buffon insist that mathematical terms and constants represent something real. Nor is it apparent that he was a conscientious sensationist. Still, Hanks's analysis is useful in emphasising that Buffon's attitudes toward mathematics were more mixed than his rhetoric often suggests. A more plausible if murkier key to Buffon's use of mathematics is what Jacques Roger has called his realism. More commonsense than philosophical in Roger's analysis, Buffon's realism was that of rural practicality and everyday life.[90] His mathematics, characterised by petty reckoning, verbal fudging, and intelligible relationships, was real in the same way. Common sense as much as anything was what defined Buffon's use of mathematics in the *Histoire naturelle*.

Looking back at the anti-mathematical rhetoric of the 'Premier discours', we can see that it both represents and misrepresents Buffon's assumptions. Mathematics may have been easy the way Buffon did it – deceptively easy on many occasions – but it was not very formal. His image of mathematics as formal probably grew out of his disappointment with Fontenellian infinity and survived for its value in demeaning 'mathematical' theories. Likewise, despite the glum views of the 'Premier discours', he did not see mathematics as irrelevant to natural history. On the contrary, he was an enthusiast of commonsense algebra and quantification. What he did see as irrelevant was sophisticated mathematics – probability excepted – and mathematics lacking intelligibility. Borrowing from his preface to the *Méthode des fluxions*, the 'Premier discours' turned his arguments against infinity into a versatile weapon for attacking mathematics and abstractions in general. Rival theories plausibly identifiable as abstract or mathematical could now be discredited as arbitrary tautologies out of touch with reality. In so attacking mathematics, Buffon was building on a longstanding tradition, but he, along with Diderot, raised the rhetoric to a new level.

89. Hanks, *Buffon avant l'"Histoire naturelle'*, p.114-17, 225-26. On Buffon's tolerance for his own abstractions see below, chapter 6, p.168-70, and Hanks, *Buffon avant l'"Histoire naturelle'*, p.52n.

90. Roger, *Buffon*, p.560-63; Sloan, 'Buffon studies today', p.472-73. Sloan's reconstruction of Buffon's epistemology, mathematical and otherwise, is discussed in chapters 4 and 6, p.116-26, 168, 175-76. See also Sloan, 'Hypothétisme de Buffon', p.213-17.

His polemical innovations aside, the material of this chapter gives us reason for questioning Buffon's identification with any fundamental new scepticism about mathematics. He did abandon a career in mathematics to study natural history, but so too did others – Réaumur, for example. And if the *Histoire naturelle* neglects mathematics beyond simple algebra and arithmetic, how surprising is this? No one used calculus in natural history.[91] Though now lauded as a pioneer in 'quantitative research on insects' for his census of the beehive,[92] Réaumur used mathematics no more than Buffon in his *Mémoires pour servir à l'histoire des insectes.* In some domains, notably arboriculture, Buffon was more mathematical than contemporaries studying the same phenomena.[93] Reputations might be different if Réaumur, not Buffon, had aggressively rationalised his general non-use of mathematics. In any case, as his arguments against Clairaut indicate, Buffon was not always sceptical about the ability of mathematics to represent reality. On the contrary he seems to have believed in the law of gravity as a mirror of nature. His rhetoric, however, is hard to ignore. Its forceful generalisations have constantly 'identified' him as a sceptic of mathematics.

Buffon's reputation as an enemy of mathematics has long been reinforced by his unfriendly relationship with d'Alembert. Some scholars see d'Alembert as participating in his century's retreat from mathematics,[94] but he often remains cast as a foil to the anti-mathematical rebels Buffon and Diderot. Many discussions of the Enlightenment's crisis of mathematics quote a letter from d'Alembert to Cramer, Buffon's sometime correspondent, in which d'Alembert opposes mathematicians, including himself, to detractors of mathematics, among them Buffon: 'A propos de calculs, et de géométrie, vous nous trouverez bien maltraités dans le nouvel ouvrage de Mr de Buffon. Il est vrai qu'avec du calcul et de la géométrie, il n'eût peut-être pas tant hasardé de choses sur la formation de la terre, et qu'il en aurait même rayé plusieurs.'[95]

What d'Alembert found offensive in Buffon's opening volumes is not immediately obvious. His second sentence suggests vexation at the 'absence' of calculation in the *Histoire et théorie de la terre.* All through his life he expressed scorn for loose hypotheses and non-mathematical physics.[96] But if he considered the *Histoire et théorie* too little mathematical and too hypothetical to be of any value, why did he draw on it for his articles on geology, geography, and several related topics in the *Encyclopédie*? In part, perhaps, he lacked more acceptable authorities, but d'Holbach managed to ignore the *Histoire et théorie*

91. On the relatively few eighteenth-century 'applications' for higher mathematics see Bos, *Lectures*, p.113-22.

92. J. B. Gough, 'Réaumur, René-Antoine Ferchault de', in *Dictionary of scientific biography*, xi.332.

93. On arboriculture see Licoppe, *Formation de la pratique scientifique*, p.210, 225, 230.

94. See for example Hanks, *Buffon avant l'"Histoire naturelle'*, p.26-27; Hankins, *Jean d'Alembert*, p.70-71.

95. Buffon to Gabriel Cramer, 21 September 1749, quoted in Hanks, *Buffon avant l'"Histoire naturelle'*, p.27.

96. See Hankins, *Jean d'Alembert*, p.75-96.

almost entirely in his own articles on geology for the *Encyclopédie*.[97] Furthermore, d'Alembert's articles praise the *Histoire et théorie* and even register agreement with Buffon's cautious approach to mathematising geology. In 'Fleuve', for example, d'Alembert accepts Buffon's calculation for the amount of water flowing into the ocean each day, and ends up concurring that 'geometric' theories of rivers are not very accurate.[98] In none of his articles does he criticise the *Histoire et théorie* for being non-mathematical.

There is, of course, one episode of the *Histoire et théorie* that d'Alembert conspicuously neglects in the *Encyclopédie*, namely Buffon's hypothesis about the Earth's origin. This was undoubtedly a major contributor to d'Alembert's unhappiness with Buffon's geology. Not only did Buffon decline to use mathematics in his discussion of the origin of the planets (*HN*, i.136), he did so in the eminently mathematical context of theoretical astronomy. Worse still, Buffon's hypothesis appeared to contradict the rules of mechanics. If a comet grazing the sun had dislodged planet-sized globs of matter, it stood to reason that these proto-planets would return to their point of origin with each revolution – but in fact, none of the planets comes close to the sun. Aware of the problem, Buffon attempted to explain it away verbally in three different ways, but he refused to attempt a mathematical analysis.[99] Here, then, was a section of the *Histoire et théorie* that almost certainly struck d'Alembert as contemptibly non-mathematical. His letter to Cramer evokes a plurality of non-mathematical conjectures in the *Histoire et théorie*, but perhaps he was referring to the details of Buffon's hypothesis.

The first sentence of the passage cited above suggests d'Alembert's displeasure with the rhetoric of the 'Premier discours': 'A propos de calculs, et de géométrie, vous nous trouverez bien maltraités dans le nouvel ouvrage de Mr de Buffon.' Accordingly many scholars have focused on the 'Premier discours' as the source of d'Alembert's annoyance with the *Histoire naturelle*, proposing that he was offended or even 'shocked' by the limitations Buffon placed on mathematics.[100] But in his 'Discours préliminaire' d'Alembert adopted Buffon's characterisation of mathematics as mechanical translation, only adding that it still took genius to discover correct principles. Throughout his career he expressed reservations about applying mathematics to other fields. Electricity, for example, he did not consider suitable for mathematical analysis, and he was pessimistic about the very possibility of probability theory, let alone about making it the foundation for the sciences in the way Buffon wanted.[101] From

97. Jean Ehrard, 'Diderot, l'*Encyclopédie*, et l'*Histoire et théorie de la terre*', in *Buffon 88*, p.137-38. In addition to the geological articles mentioned by Ehrard, d'Alembert cites the *Histoire et théorie* in 'Cataracte', 'Courant', 'Détroit', 'Fentes perpendiculaires', 'Flux et reflux', 'Glace', and 'Lune'.

98. Jean Le Rond d'Alembert, 'Fleuve', in *Encyclopédie*, vi.868, 872-73. Compare *HN*, i.341.

99. *HN*, i.139-43. See Seidengart, 'Traitement du problème cosmologique', p.316-19.

100. See for example Hankins, *Jean d'Alembert*, p.76-77; Roger, *Buffon*, p.263.

101. See d'Alembert, 'Discours préliminaire', p.vii-ix; Hankins, *Jean d'Alembert*, p.81-82, 88, 146-49; Heilbron, 'Experimental natural philosophy', p.382-83. See also Michel Paty, 'D'Alembert et la théorie physique', in *Jean d'Alembert, savant et philosophe*, p.238-39, 254-55; Pietro Redondi, 'D'Alembert et la technologie. L'affaire du canal de Picardie', in *Jean d'Alembert, savant*

the standpoint of ideas it is hard to imagine d'Alembert being shocked by the view of mathematics in the 'Premier discours'. Much of what Buffon wrote had been or would be written by d'Alembert.

But d'Alembert was a mathematician criticising his field from the inside, while Buffon had become an outsider by 1749, a 'Newtonien non géomètre' as Clairaut put it.[102] D'Alembert's complaints to Cramer about the *Histoire naturelle* probably reflected on Buffon as much as on his arguments. Among other things, d'Alembert's comments sought to solidify a disciplinary boundary that would isolate Buffon and unite himself and Cramer. Like a certain number of his colleagues in the Académie des sciences, d'Alembert may have disliked Buffon before 1749. Certainly he was displeased with Buffon's hypothesis about the Earth's origin as well as perhaps other aspects of the *Histoire naturelle*. What better way to demean Buffon in the eyes of Cramer, a mathematician, than to dwell on and play the innocent with the anti-mathematical rhetoric of the 'Premier discours'? Little did it matter that Buffon's use of mathematics in the *Histoire et théorie* hardly bore out the extremism of the 'Premier discours'. Nor was d'Alembert about to waste time and rhetorical advantage by observing that he too had questions about mathematics' applicability to physics. The point of his remarks was to denigrate Buffon, not to analyse his positions impartially or carefully.

To make Buffon an emblem of a retreat from mathematics is to some degree to repeat d'Alembert's sleight of hand. Buffon used arithmetic and algebra throughout his career, he proposed mathematising induction, and he saw the law of gravity as a representation of reality, not as a mere vehicle for making predictions. Many contemporary naturalists used less mathematics, but few directly challenged its place in the hierarchy of the sciences. It was here, in his theoretical, rhetorical views, that Buffon distinguished himself as an enemy of mathematics. Like mathematical intimidation, anti-mathematical rhetoric had been commonplace since the early seventeenth century, but it became particularly virulent in the early eighteenth century, thanks in part to the period's mathematical quarrels. Buffon adopted this newly fortified rhetoric, even as he adopted the technique of mathematical intimidation, but his denunciations overstated his mistrust of mathematics. Like other enthusiasts of anti-mathematical rhetoric, he invoked it when his rivals were prone to being seen as mathematical, but never seriously applied it to his own mathematics, although certain critics did. Helpful as it was in discrediting rivals and calling attention to natural history as an autonomous science, his anti-mathematical rhetoric paved the way for criticism like d'Alembert's. Even now it distracts from his more moderate and ambiguous working assumptions.

et philosophe, p.437; Marie-Elisabeth Duchez, 'D'Alembert diffuseur de la théorie harmonique de Rameau. Déduction scientifique et simplification musicale', in *Jean d'Alembert, savant et philosophe*, p.491-92.

102. Roger, *Buffon*, p.90-91.

6. Nominalism

In the Enlightenment, a period much invested in classification, there developed a profusion of scientific rhetorics for contesting and legitimating taxonomic projects. A major form of such rhetoric turned on the doctrine of nominalism, according to which universals have no real existence beyond their status as names. As an ontology and an epistemology, nominalism is often considered a distinctive philosophical tenet of the early modern period and especially of the Enlightenment. In a lecture of 1903 Charles Peirce went so far as to lament the 'tidal wave of nominalism' that swept over philosophy after the Renaissance.[1] While Peirce and others saw medieval critiques of universals as a unified, full-scale precedent for later nominalism, intellectual historians have increasingly emphasised the diversity and sometime realism of the 'movement' once known as medieval nominalism.[2] No doubt there were links between ancient and medieval nominalists and their subsequent namesakes, especially in the Renaissance, but the 'genuine', more coherent nominalism of the seventeenth and eighteenth centuries was in large measure a product of novel developments. Above all, the mechanical philosophy of Descartes, Boyle, and others promoted nominalism through its reductionism. No longer shored up by Aristotelian essences, traditional natural kinds were artifacts to mechanists – linguistic conveniences or epiphenomena to be explained away atomistically. More generally, nominalism was favoured by the method of analysis, which many in the Enlightenment vaguely professed. Applied to epistemology, analysis made knowledge dependent on atomised sensations rather than on innate ideas or intuited forms. Such sensationism made universals potentially deceptive constructs. Together, these factors pushed the Enlightenment in the direction of nominalism. Significantly, theories of language came to emphasise the relation between language and thought, not the relation between language and reality – the usual focus of grammarians before the seventeenth century.[3]

1. Charles Hartshorne and Paul Weiss (eds), *Collected papers of Charles Sanders Peirce*, 6 vols (Cambridge, Mass. 1931-1935), i.4-6. See also John Wild, *Introduction to realistic philosophy* (New York 1948), p.29-31.

2. Alister McGrath, *The Intellectual origins of the European Reformation* (Oxford 1987), p.69-75. See also R. I. Aaron, *The Theory of universals* (Oxford 1952), p.18-21. The word 'nominalism' will be applied here to conceptualism as well as to 'genuine' nominalism. See Michael J. Loux, 'Nominalism', in *Routledge encyclopedia of philosophy*, ed. Edward Craig, 10 vols (London 1998), vii.18.

3. G. A. Padley, *Grammatical theory in western Europe, 1500-1700*, 2 vols (Cambridge 1985), i.295-99. On the relationship between nominalism and eighteenth-century theories of language see Sylvain Auroux, *La Sémiotique des encyclopédistes. Essai d'épistémologie historique des sciences du langage* (Paris 1979), p.83-84, 102-13.

But if eighteenth-century nominalism was a philosophical tendency, it was also a rhetoric, as we will see in the case of Buffon. Buffon has frequently and with some justice been identified as a nominalist. In the 'Premier discours' he denounces taxonomy as merely conventional, insisting that only individuals exist. Forced to organise animals somehow in the *Histoire des quadrupèdes*, he first chose a plainly artificial arrangement, treating the quadrupeds in order of their 'relevance' to human affairs. In subsequent volumes of the *Histoire naturelle* he adopted and recognised various 'real' ways of ordering nature, including one based on historical filiation. Many critics have concluded that he abandoned the nominalism of the 'Premier discours' once he gained experience as a natural historian. More recently Phillip Sloan has argued that the nominalism of the 'Premier discours' is superficial and masks a vague preview of the 'real' means of classifying that Buffon soon adopted. Lacking supporting documents to clarify the 'Premier discours', we may never know precisely how much evolution or constancy to assign to Buffon's early attitudes toward taxonomy. But contemporary discussions of classification shed light on the logic of his offensive against it. Like previous 'nominalists', notably Locke, Buffon formulated his critique of universals less as a positive, general doctrine than as a means of impugning a specific class of universals. His nominalism, in other words, was eminently polemical.[4] Furthermore his 'misunderstanding' of the goals of nomenclature in the 'Premier discours' hinged on the ambiguities of taxonomical discourse. For eighteenth-century classifiers no less than nominalists had philosophical weaknesses and rhetorics for masking them. Placed against the backdrop of contemporary polemics, the 'Premier discours' was neither unique nor wholly unjustified in its contradictory stance against classification. To a degree it merely turned the tables on a contrary rhetoric exaggerating the rationality and certainty of authors' taxonomies.

Natural systems

Classification is a necessity of human survival. In particular, all humans classify living beings. As Scott Atran has argued, these pre-scientific 'folk taxonomies' are in many ways similar, not in their content, since humans inhabit different ecosystems, but rather in their structure. All humans appear to have an idea of species, to classify living beings by overall morphology, and to use additive naming, as in the progression from 'oak' to 'shingle oak' and 'spotted shingle oak'. Folk taxonomies everywhere recognise the same 'life-forms', notably birds, fish, land animals, bugs, grasses, shrubs, and trees. Except for the bugs – a residual category including small organisms and 'leftovers' – these life-forms differ little from modern zoological classes, but they run counter to modern botanical taxa. On the basis of such regularities, Atran makes a strong case for an innate common sense directing if not limiting our

4. For a similar argument about Darwin see David N. Stamos, 'Was Darwin really a species nominalist?', *JHB* 29 (1996), p.127-44.

classification of living beings. In the 'scientific breakaway' from folk taxonomy, which Atran sees as culminating in the late eighteenth century, taxonomic common sense was superseded in its scientific role by less specialised mental faculties.[5]

Most early philosophical orderings of nature stayed close to common sense, though alphabetic, medicinal, and other 'artificial' arrangements dominated practical texts such as encyclopedias and herbals. In the Renaissance, for the first time, naturalists faced the challenge of organising living beings not from a single ecosystem but from all over the Earth, a challenge that raised questions about the status and rationale of folk taxonomy. Not only did the observed species of temperate Europe require reconciliation with the species reported in Greek and Latin antiquity, there were also, increasingly, new plants and animals being discovered in the Americas, Africa, and the Pacific. The wealth of new data did not in itself entail what would become a 'veritable mania' for classing by the eighteenth century,[6] but it was clearly a crucial factor. The spread of the printing press made this 'overload' of species all the more pressing by reifying it repeatedly in catalogues and lists.[7] Adapted for ordering local environments, where species are mostly alone in their genus, commonsense taxonomy could not be extended without a conscious reappraisal of its theoretical foundations.[8]

A leader in this enterprise was the sixteenth-century physician and philosopher Andrea Cesalpino. Where Aristotle saw species as tendencies to be realised under normal conditions, Cesalpino made them into eternal, neo-Platonic forms. With regard to plants, which he focused on, he considered fructification the point of contact between divine rationality and the material world. The organs of reproduction were thus his window of access to God's eternal order. Within this window – the basis for his system – Cesalpino felt free to choose taxonomic *differentiae*, or distinguishing characters, that were not obviously functional – the number of seeds in each fruit, for example.[9] In fact his choice of fructification as a basis for classifying was motivated by practical as well as philosophical considerations. Like other taxonomists, Cesalpino recognised that plants' reproductive systems have more distinctive organs than their leaves, roots, or other parts. This informational richness

5. Atran, *Cognitive foundations of natural history*. Atran's book has been faulted for discounting cultural and symbolic influences. While reasonable, such criticism does not invalidate his main argument as invoked in this chapter. See Paula Findlen, review of Scott Atran, *Cognitive foundations of natural history*, *JHB* 24 (1991), p.537-40.

6. Vernon Pratt, 'System-building in the eighteenth century', in *The Light of nature*, ed. J. D. North and J. J. Roche (Dordrecht 1985), p.422-23; Phillip R. Sloan, review of Wolf Lepenies, *Das Ende der Naturgeschichte*, *Isis* 72 (1981), p.123. On eighteenth-century systematics and the 'quantifying spirit' see John E. Lesch, 'Systematics and the geometrical spirit', in *Quantifying spirit*, p.83-88.

7. Dirk Stemerding, *Plants, animals, and formulae: natural history in the light of Latour's 'Science in action' and Foucault's 'The Order of things'* (Entschede, Netherlands 1991), p.91-95.

8. See Atran, *Cognitive foundations of natural history*, p.27-29.

9. Larson, *Reason and experience*, p.147-48; Atran, *Cognitive foundations of natural history*, p.138-40, 151-52.

makes fructification an ideal basis for classification. But in presenting his system Cesalpino emphasised logic and *a priori* reasoning over utility, leading Julius von Sachs and other scholars to contrast him with the less metaphysical and supposedly more empirical herbalists of his day. Examined in their own right, Cesalpino's taxa often coincide with the herbalists' intuitive ones. In practice he worked backward from commonsense groupings, seeking *differentiae* that would make them follow logically.[10]

Toward 1800, after a half-century of ambiguous near-synonymy, the terms 'system' and 'method' came to be distinguished in their taxonomic application, the former referring to an 'artificial' classification relying on a single locus of differentiation, usually the fructification, the latter referring to a 'natural' classification based on multiple characters.[11] Cesalpino, the first systematist in Linnaeus's view, did not identify his taxonomy as a system or a method, but his *De plantis* was a model for subsequent systems in demonstrating the possibility of rationalising taxonomy through attention to the structures of fructification. Following Cesalpino, botanical systems were typically billed as derived via logic from a few select characters, but they were almost inevitably retrospective reconstructions of groupings perceived because of gross similarity. Retrospective or not, logical divisions based on limited characters theoretically made taxonomies easier to use. For this reason and others, exaggerating the part of rationality in one's taxonomy became standard practice. Like the mathematical rhetoric of philosophers such as Spinoza and Christian Wolff, the rationalistic rhetoric of botanical systematists commanded attention in a period in which clarity and reason were seen as ideals. Trumpeting the rationality of one's approach to taxonomy was especially important in that evaluating actual taxa was a complicated task, and not one conducive to clear-cut conclusions. The fact that intuition guided taxonomical judgements meant that proposed taxa, for all their diversity, were regularly less different than their official rationales. Early modern botanists had similar conceptions of a number of genera and even a few families such as *Leguminosae* and *Umbelliferae*. By the mid-eighteenth century agreement between 'opposed' taxonomists was often considerable.[12] Rationalistic methodologies heightened the

10. See Larson, *Reason and experience*, p.6-30; Atran, *Cognitive foundations of natural history*, p.153-58; A. G. Morton, *History of botanical science: an account of the development of botany from ancient times to the present day* (London 1981), p.140.

11. See Emile Callot, 'Système et méthode dans l'histoire de la botanique', *RHS* 18 (1965), p.47-48; Foucault, *Les Mots et les choses*, p.150-58. According to Stevens, arranging nature, not classifying it, was the main goal of Antoine-Laurent de Jussieu and other 'methodists' believing in natural continuity. See *Development of biological systematics*, p.11-13, 89-90, 256.

12. On common ground among taxonomists see Callot, 'Système et méthode', p.52; Edward Lee Greene, *Landmarks of botanical history*, ed. Frank N. Egerton, 2 vols (Stanford, Calif. 1983), i.113; Jean-Etienne Guettard, *Observations sur les plantes*, 2 vols (Paris, Durand 1747), i.xxvii; Frans Stafleu, 'Adanson and his *Familles des plantes*', in *Adanson: the bicentennial of Michel Adanson's 'Familles des plantes'*, ed. George H. M. Lawrence, 2 vols (Pittsburgh, Penn. 1963), i.215-18; Stevens, *Development of biological systematics*, p.264-66. But see also Phillip R. Sloan, 'John Locke, John Ray, and the problem of the natural system', *JHB* 5 (1972), p.49-50; A. J. Cain, 'Logic and memory in Linnaeus's system of taxonomy', *Proceedings of the Linnaean Society of London* 169 (1958),

contrast between rivals' offerings. Useful as it was in fostering distinction, respect, and an impression of convenience, the rhetoric of reasonableness of botanical classifiers would also end up being turned against them.

Cesalpino's system did not extend to animals. Others did, with some even encompassing minerals, diseases, mathematical curves, and other phenomena, but systems of plants and animals evolved very differently. By the mid-eighteenth century botanical taxonomy had attained considerable stability in taxa up to genus and beyond, whereas the only stable taxa in zoology before the nineteenth century were the species and the life-form. In part this was due to practical impediments. Plant specimens were easier to preserve, breed, and transport. In consequence European menageries were consistently less cosmopolitan than botanical gardens, and botanists had more data than those studying animals. Classification also concentrated on exterior, visual character before the nineteenth century, and whereas plants have salient external organs with discrete parts and forms, animals do not, except perhaps the insects.[13] Botany, for these reasons, was the principal field for taxonomical systems. Buffon, significantly, seems to have preferred studying animals, to judge by the contents of the *Histoire naturelle*. This probably contributed to his mistrust of taxonomy.

Seventeenth-century nominalism, nursed on analysis and the mechanical philosophy, constituted an explicit challenge to the claims of taxonomists. Already in Hobbes' *Leviathan* (1651), which proclaims that there is 'nothing in the world universal but names', nominalism is associated with an attack on 'the vain philosophy of Aristotle', notably its essences – the basis, in an idealised form, of Cesalpino's taxonomy.[14] Both accidents and essences were widely interpreted as subjective 'secondary qualities' on the principles of the new sciences. This equation led Boyle, among others, to see 'genders and species' as artificial or at best probable linguistic conveniences.[15] The threat of nominalism to taxonomy was spelled out most clearly in Locke's *Essay concerning human understanding* (1690). In the *Essay*, after announcing that 'all things that exist are only particulars', Locke goes on to apply this doctrine to the taxa of natural history and other sciences. Why, he enquires with typical scepticism, are a poodle and a hound 'not as distinct species as a spaniel and an elephant'?[16] His answer, of course, is that species are determined by

p.156; Ritvo, *Platypus and the mermaid*, p.19-22. On the rhetorical abuse of 'principles' in systematics see Greene, *Landmarks of botanical history*, i.113-14.

13. Atran, *Cognitive foundations of natural history*, p.190-93; Larson, *Interpreting nature*, p.44. On the exteriority of eighteenth-century taxonomy see Foucault, *Les Mots et les choses*, p.144-50.

14. Thomas Hobbes, *Leviathan, or the matter, form, and power of a commonwealth ecclesiastical and civil*, ed. J. C. A. Gaskin (Oxford 1996), pt I, ch. 4, no. 6; pt IV, ch. 46, no. 13-19. On Aristotle, Cesalpino, and essentialism see Atran, *Cognitive foundations of natural history*, p.83-85, 96-97, 138-40.

15. Sloan, 'John Locke', p.14-20; Margaret J. Osler, 'John Locke and the changing ideal of scientific knowledge', *Journal of the history of ideas* 31 (1970), p.6-10.

16. Locke, *Essay concerning human understanding*, bk III, ch. 3, no. 6; bk III, ch. 6, no. 38. Chapter 6 of book III deals extensively with the kinds of natural history.

conventions, not nature. Natural kinds, in Locke's view, are determined by real as well as nominal essence, but both, paradoxically, reflect human usage. Nominal essence is the essence we choose for establishing taxa, while real essence is a postulated physical explanation for the phenomenal resemblance informing our choice. Real essence, in this sense, is merely a real explanation for a nominal category.[17]

Locke's critique of universals was ostensibly directed toward universals in general, but he applied it most relentlessly to the universals of common parlance and everyday experience, including those of natural history. His repeated assertion that only particulars exist points to unbounded nominalism, but all through the *Essay* he assumes that phenomenal kinds can at least be explained by matter's inner constituents. Real essence, he insists, is unknown and perhaps unknowable, but he outlines it on the suppositions of the corpuscular philosophy, making it a product of the primary qualities of matter's 'minute constituents'. It is not obvious why real essence, understood with such pragmatism, must remain forever unknowable. More seriously, how can real essence depend on the 'figure, size, texture, and motion' of matter's 'insensible parts' if none of these qualities exists more than nominally?[18] Nowhere in the *Essay* does Locke allow his nominalism to jeopardise the universals implicit in his vision of micro-physics in its finished state. There is, as Rom Harré puts it, a tension between Locke's sensationist epistemology and his 'realistic conception of what a completed science would be like'.[19] While sensationism and nominalism dominate the *Essay*, invalidating taxonomy among other sciences, corpuscular micro-mechanism survives as a possibility in a haven of casual realism.

Locke's *Essay* was not meant to rule out probable knowledge in taxonomy or any other science. His doubts about knowledge concern absolute, certain knowledge, not knowledge in the broader sense. Regarding living kinds – a major target of his nominalism – Locke admits that we can know and indeed already know enough for 'common use and discourse': 'By the colour, figure, taste, and smell, and other sensible qualities, we have as clear and distinct ideas of sage and hemlock, as we have of a circle and a triangle.'[20] Writing as a physician and herbalist around the time of the first, already nominalistic draft of the *Essay*, Locke gives every sign of trusting in the reality of standard plant species.[21] Even the *Essay* sometimes hints that natural taxa may exist, and in approximate correspondence with human ideas: 'I cannot see how it can be properly said, that nature sets the boundaries of the species of things:

17. See Locke, *Essay concerning human understanding*, bk III, ch. 6, no. 6; Michael Ayers, *Locke: epistemology and ontology*, 2 vols (London 1991), ii.68-74.

18. Locke, *Essay concerning human understanding*, bk III, ch. 3, no. 18; bk IV, ch. 3, no. 25. On primary and secondary qualities see Locke, *Essay concerning human understanding*, bk II, ch. 8. See also Ayers, *Locke*, p.66-67; E. J. Lowe, *Locke on human understanding* (New York 1995), p.162-63.

19. 'Knowledge', in *Ferment of knowledge*, p.23.

20. *Essay concerning human understanding*, bk IV, ch. 3, no. 26.

21. See the entries from his journals in Dewhurst, *John Locke*. See also Sloan, 'John Locke', p.21-22.

or, if it be so, our boundaries of species are not exactly conformable to those in nature.'[22] Faced with concrete examples, Locke often softened his nominalism, at least on the surface. To an enquiry from William Molyneux he replied, 'I hope I have nowhere said, "there is no such sort of creatures in *nature* as birds".' But having made this concession to commonsense realism, he then proceeded to interpret the 'real' existence of birds in a manner compatible with his atomistic nominalism.[23] Locke's nominalism, in sum, was susceptible to being reconciled both rhetorically and philosophically with intuitions about natural kinds. On the other hand there remains an asymmetry in his treatment of the kinds of natural history and the kinds of microphysics in that only the former are directly confronted with his nominalistic scepticism. Like most contemporary nominalists, Locke brandished his nominalism against secondary qualities and Aristotelian natural kinds, but conspicuously not against primary qualities.

In an entry in his journal for 1676 Locke endorsed a view of species that was seemingly well-protected from the carping of nominalism: 'Nature in things that have life keeps them in the distinct classes by the order of generation wherein those of the same species produce their like.'[24] Already Cesalpino and Ray had proposed the same criterion for demarcating species – the biological criterion, or the criterion of interbreeding, as it is called. But in the *Essay* Locke rejects this conception of species for reasons unrelated to his theoretical nominalism. At the level of facts he cites the well-known ambiguities of cross-breeding and monsters as well as first-hand experience: 'I once saw a creature that was the issue of a cat and a rat.' In addition Locke highlights the impracticality of determining species by reproduction: 'Must I go to the Indies to see the sire and dam [...] to know whether this be a tiger?'[25] The criterion of interbreeding and these same arguments against it would return again and again in the next hundred years.

Seventeenth-century philosophical nominalism first entered the world of Europe's major taxonomists through Ray's later works. In his *Methodus plantarum nova* (1682) Ray adopted much of Cesalpino's methodology, but he also based taxa on characteristics unrelated to fructification – notably on seed leaves in his seminal division of monocotyledons from dicotyledons. In a later dispute with the German botanist Augustus Rivinus, whose system was based exclusively on plants' 'essential' organs – that is, their fructification – Ray invoked the new sciences' truism that essences are unknown to us to

22. Locke, *Essay concerning human understanding*, bk III, ch. 6, no. 30. For other suggestions that taxa exist in nature see Locke, *Essay concerning human understanding*, bk III, ch. 3, no. 13; bk III, ch. 6, no. 36-37; Sloan, 'John Locke', p.23n. Elsewhere Locke doubts that nominal essence can have much to do with reality, unless tautologically. See for example Locke, *Essay concerning human understanding*, bk III, ch. 6, no. 19-20; bk III, ch. 10, no. 20.

23. Locke to William Molyneux, 20 January 1693, in *Essay concerning human understanding*, ii.87n.

24. Dewhurst, *John Locke*, p.74.

25. *Essay concerning human understanding*, bk III, ch. 6, no. 23. See also Sloan, 'John Locke', p.24-25.

conclude that 'characteristic accidents' are acceptable, even inevitable *differentiae* in botanical systems. At this point, in his 1695 'Epistola ad Rivinum', Ray offered little justification for his new stance against essence. That same year, however, he obtained a copy of the just-published *Eléments de botanique* by the formidable 'essentialist' Joseph Pitton de Tournefort, where he found himself attacked for his eclectic *differentiae.* In response Ray took up 'philosophical studies', clearly of Locke's *Essay.* His reply to Tournefort shows his assimilation of the main themes of the *Essay*, although he refrained from adopting Locke's scepticism regarding the existence of natural kinds.[26]

Nominalism and taxonomy had finally met, but 'essentialism' in botany continued to flourish. Though Ray's taxonomy was popular in England, Rivinus and Tournefort dominated the Continent, with Tournefort the clear leader outside Germany. Tournefort's survival of the challenge of nominalism hinged on his system's ambiguity and simplicity. Like other 'essentialists', he identified fructification as the 'essential' character of differentiation for its structural richness as well as for abstract, philosophical reasons. Furthermore, when his 'essence' proved indiscriminate in establishing certain genera, he turned to secondary characters to pare down his taxa. But these points were often lost in his rationalistic rhetoric. Deductive as he often made botany seem, it was fitting that his *Eléments de botanique* took its title from Euclid's *Elements.* Not only did he repeatedly describe fructification as the 'essential', 'true' character of plants, he suggested that this essence was a God-given quality: 'L'auteur de la nature [...] a imprimé un caractère commun à chacune de leurs espèces, qui doit nous servir de guide pour les ranger à leur place naturelle.'[27] In this sense Ray's complaints about essence were relevant to Tournefort's presentation if not to his assumptions.[28] Tournefort's success, due above all to his system's simplicity, was facilitated by another misunderstanding. Many of his nominalistic contemporaries assumed that his system had been proposed as artificial, since his arguments in favour of it hinged on 'practical' necessities such as preserving usual groupings and creating manageable taxa. Tournefort considered it natural though incomplete, but his arguments via practicality opened the way for the 'charitable' misinterpretation of many of his readers.[29]

Here as elsewhere in the Enlightenment, the word 'natural' and its cognates were a source of endless confusion and of considerable possibilities for rhetorical claims. The 'natural' was perhaps best defined by what it was not – unna-

26. Sloan, 'John Locke', p.26-53. See also Atran, *Cognitive foundations of natural history*, p.161-65.

27. Joseph Pitton de Tournefort, *Eléments de botanique, ou méthode pour connaître les plantes*, 2 vols (Paris, Imprimerie royale 1694), i.20. See also Atran, *Cognitive foundations of natural history*, p.165-69; Stroup, *Company of scientists*, p.222.

28. For Ray's complaints see Sloan, 'John Locke', p.43-44.

29. Atran, *Cognitive foundations of natural history*, p.169-70, 173; Sloan, 'John Locke', p.51. For language that made his system appear artificial see for example Tournefort's statements in Daudin, *De Linné à Jussieu*, p.33; Sloan, 'John Locke', p.48. See also Ehrard, *Idée de nature*, p.188-89.

tural, arbitrary, or artificial. 'Jusqu'à présent', complained Adanson, 'personne n'a pu [...] donner une définition juste du genre naturel, mais seulement de l'artificiel'.[30] In taxonomy an artificial system was one not intended to mirror reality but merely serving as a convenient way of ordering, usually on the basis of limited *differentiae*. Though not necessarily unnatural or arbitrary – terms largely reserved for one's enemies' means of ordering – alphabetical and medicinal orders offer clear-cut examples of artificial systems. Calling a system 'natural', on the other hand, could mean a number of things. Above all a natural system was one reflecting nature. On the one hand natural systems were often restricted to those preserving intuitive groupings, even 'non-scientific' ones such as 'trees', 'shrubs', and 'grass'. Consensus regarding taxa was thus a typical criterion for calling them natural. Natural systems could also be systems based on numerous *differentiae* or on comparison of 'functional', 'essential', or 'natural' organs, the latter term a rhetorically invaluable synonym for the reproductive parts. Finally, natural systems could be systems that came to the mind naturally in view of their size and layout. Tournefort, it appears, considered his system natural in this sense. For by shifting attention from the species to the genus, he managed to cut the number of taxa down to the 'natural' size of folk taxonomy even while retaining a commonsense basis, his genus being the global analogue of the species of folk taxonomy.[31]

Toward the middle of the eighteenth century partisans of Tournefort, Ray, and Rivinus were confronted with a powerful alternative in the system of Linnaeus. Linnaeus, the inventor of modern binomial nomenclature, was the first to classify nature comprehensively and the first to apply the discovery of plant sexuality throughout his taxonomy. The title of his magnum opus, the *Systema naturae* (1735), suggests that Linnaeus saw his goal as a system of nature, a natural system.[32] In effect he is often portrayed as a misguided essentialist ignorant of his century's philosophical modernity and convinced of his special power to perceive God's designs. There is some truth to this view. But Linnaeus also conceded that his system was not natural but merely 'orthodox' and closer to naturalness than anything previous. His genera were natural in being reasonably commonsensical, in replacing folk species with a cognitively manageable substitute, and in being defined by 'natural' parts only, but

30. Michel Adanson, *Familles des plantes*, 2 vols (Paris, Vincent 1763), i.cv.

31. Atran, *Cognitive foundations of natural history*, p.166-67, 170; Scott Atran, 'The common-sense basis of Buffon's *méthode naturelle*', in *Buffon 88*, p.229-30. Antoine-Laurent de Jussieu also pared down his taxa to cognitively 'natural' sizes. See Stevens, *Development of biological systematics*, p.45-49. On the ambiguities of natural systems see Stevens, *Development of biological systematics*, p.176-77; Sloan, 'John Locke', p.1-8; Atran, *Cognitive foundations of natural history*, p.30; Stafleu, 'Adanson', p.131-33; Ritvo, *Platypus and the mermaid*, p.26-27. According to Callot, all classifying botanists considered their systems natural in some sense of the word. See Callot, 'Système et méthode', p.48. See also Stevens, *Development of biological systematics*, p.183.

32. See Daubenton, *HN*, iv.162; Daubenton, 'Histoire naturelle', in *Séances des écoles normales. Leçons*, i.437, iv.130; Daudin, *De Linné à Jussieu*, p.102n.

Linnaeus failed to find similarly natural orders and classes.[33] Sometimes he doubted that a natural system was possible, but in his *Fragmenta methodi naturalis* (1738) he sketched out intuitively and holistically 'natural' orders and then tried to justify them by fructification. A complete natural method, he argued, would require more data and the discovery of new organs of fructification.[34]

Thanks to its convenience, accessibility, and scientific air, the Linnaean system was popular with amateurs into the nineteenth century.[35] Linnaeus's nomenclature, if not his system, was widely accepted in the scientific community within a decade of the publication of *Species plantarum* (1753).[36] To the delight of Linnaeus, even Buffon's stronghold, the Jardin du roi, switched to the new nomenclature in 1774 on the initiative of André Thouin and Antoine-Laurent de Jussieu.[37] But in Paris at least, botanists never fully accepted the Linnaean system. In the second half of the eighteenth century they turned increasingly to natural 'methods', classifications based on overall morphology.[38] Antoine-Laurent de Jussieu's method, which relied on a hierarchy of *differentiae* to array plants in a continuum, proved the most enduring. Adanson's method, more short-lived, remains noteworthy for the rhetoric by which it was justified. Unlike Jussieu, Adanson wanted to classify plants and animals on the basis of their total habitus, or external aspect. Habitus, of course, is theoretically limitless, and Adanson's comparisons were inevitably limited to familiar characteristics. In this sense previous botanists had used his method implicitly, but their pretensions to the contrary perhaps justify Adanson's claim to have discovered a new method. The rhetoric of rationalistic systematists certainly enhanced the impression of his originality.[39]

A second 'novelty' of Adanson's method was its mathematical aspect. If habitus were formalised, Adanson reasoned, taxa could be calculated as correlations of characters with the certainty of geometry. Not surprisingly this remained little more than a brilliant pipe dream until the twentieth century. Adanson, for his part, correlated characters intuitively, while Condorcet

33. Atran, *Cognitive foundations of natural history*, p.170-81; Eriksson, 'Linnaeus the botanist', p.88-94; Larson, *Reason and experience*, p.61-62, 73; Frans A. Stafleu, *Linnaeus and the Linnaeans: the spreading of their ideas in systematic botany, 1735-1789* (Utrecht 1971), p.28, 63-67.

34. Larson, *Reason and experience*, p.62-72.

35. Anne Secord, 'Artisan botany', in *Cultures of natural history*, p.391; Duris, *Linné et la France*, p.225-26. In France, Duris shows, Linnaeus's popularity was connected with Rousseau and the politics of the French Revolution. On the attractiveness of Linnaeus's appearance of infallibility see Allen, *Naturalist in Britain*, p.41; *HN*, i.18-19, ix.123.

36. Stafleu, *Linnaeus and the Linnaeans*, p.109-10, 338.

37. Yves Laissus, 'Le Jardin du roi', in *Enseignement et diffusion des sciences*, p.310; Larson, *Interpreting nature*, p.17; Roger, *Buffon*, p.93. See also Duris, *Linné et la France*, p.65, 104; Spary, 'Political, natural, and bodily economies', p.188.

38. On Linnaeus's fortunes in France see Stafleu, *Linnaeus and the Linnaeans*, p.267-336; Duris, *Linné et la France*, p.39; Hahn, *Anatomy of a scientific institution*, p.113-14. Alongside French aspirants toward natural methods, the nominalistic Lamarck sought a merely convenient system. See Burkhardt, *Spirit of system*, p.23-26, 202-203.

39. Atran, *Cognitive foundations of natural history*, p.196-99.

treated the mathematics of the problem in abstract generality.[40] Adanson's mathematism, in other words, was largely rhetorical. The method he practiced in his *Familles des plantes* was no different from the one motivating Linnaeus's *Fragmenta.* Nor were Adanson's results so different from those of Linnaeus. Ironically both botanists denounced the other's work as 'unnatural'.[41] Both were probably right in one sense or another, but the very ambiguity of the word 'natural' in botanical discourse is symptomatic of a desire to take rhetorical advantage of its colloquial associations. Linnaeus, who was as guilty as anyone of expropriating culturally charged terms for specialised uses, was well-aware of the process: 'Quid magis frequens apud botanicos quam unius methodi summus honos, alterius vero summus neglectus; quod illa magis naturalis sit, haec autem minus; licet res contrario modo saepe se habeat, et a fama methodi et idiotismo botanici solum multoties pependit pretium' ('what is more frequent among botanists than to give the greatest honour to one method and completely dismiss another because the former is more natural and the latter less so; often the opposite might just as well be maintained, and often a method's reputation and the botanist's words alone determine its value').[42]

As James Larson has shown, a 'professional contradiction' in early modern botany produced hesitation between different modes of classifying. To facilitate identification in practical situations, botanists defined taxa by a small number of salient structures, almost always organs of fructification. But their very success in this project made them dream of alternatives, since they could not help but notice more 'natural' relations among the plants they were studying.[43] Paralleling and disguising this 'contradiction' were two rhetorical commonplaces of botanical discourse. First, to make their systems appear convenient and rational, taxonomists regularly claimed to consider just a few essential *differentiae*, but to make their systems credible to common sense, they almost always took stock of overall habitus, often surreptitiously.[44] Second, while acknowledging for the most part that their systems fell short of being perfectly 'natural', taxonomists tended to apply the word 'unnatural' only to their rivals' work. Conversely, they were diligent in making their own systems seem at least mostly natural, true, and objective. It is worth noting, for example, that Linnaeus's 'artificial' system was entitled *Systema naturae*, not

40. See Adanson, *Familles des plantes*, i.cc-cci; Atran, *Cognitive foundations of natural history*, p.200-201; Peter H. A. Sneath, 'Mathematics and classification, from Adanson to the present', in *Adanson*, ii.471-98, at p.483; Baker, *Condorcet*, p.119-24; Baker, 'Unpublished essay of Condorcet', p.99-123. On the 'neo-Adansonian' movement see Stafleu, *Linnaeus and the Linnaeans*, p.313; Clive A. Stace, *Plant taxonomy and biosystematics*, 2nd edn (London 1989), p.44.

41. Atran, *Cognitive foundations of natural history*, p.197-99; Stafleu, *Linnaeus and the Linnaeans*, p.125-26; Stafleu, 'Adanson', i.50-51.

42. Quoted in Stafleu, *Linnaeus and the Linnaeans*, p.127.

43. Larson, *Reason and experience*, p.47-49.

44. Larson, *Reason and experience*, p.38-40, 145-46; Stevens, 'Metaphors and typology', p.172-73.

Systema naturae artificiale – a title that would have sounded a good deal less impressive.

According to Larson the most important question to be asked about taxonomies is 'how have practical dispositions and natural affinities been reconciled in theory and in practice'.[45] This formulation of the dimensions of taxonomy clarifies the role of rhetoric in classification. Even beyond the clear-cut cases of alphabetical and medicinal ordering, systematists' results differed in the early modern period, enough so, for example, to make Tournefort's system preferable to Ray's or Rivinus's in the eyes of the sceptical, 'utilitarian' Boerhaave.[46] But thanks to the ascendancy of certain natural groupings, taxonomical results differed less than the theories behind them – a classic situation in the history of methodology. Independent of practice to a degree, theories of classifying functioned in two ways as rhetoric and as scientific propaganda. First, to distinguish themselves and their taxonomies, botanists had every reason to play up their differences with respect to their peers. Second, they used their theories to accentuate and idealise traits of their systems, notably simplicity and naturalness. Botanists' balance between 'practical dispositions' and 'natural affinities' was thus partly a matter of polemical positioning. Rhetoric could be blatant in the world of botany, as in the vainglorious self-promotions of Linnaeus or Robert Morison, but it was more often a matter of subtle exaggeration and prudent attention to the connotations of words.[47]

It is important to recognise the rhetoric of botanical systematists in that Buffon is often accused of inadvertently or deliberately misunderstanding taxonomy in his 'Premier discours'. But to some extent the 'Premier discours' merely capitalised on contradictions inherent in taxonomical discourse. This is not to imply that Buffon's critique of taxonomy was neutral or even-handed. In fact it made use of a whole contrary rhetoric. As we have seen, seventeenth-century philosophical nominalism had its own internal tensions and means of downplaying them. Pre-eminent among them was the impossibility of being a 'pure' nominalist, a problem noted by philosophers from Kant to our own days. Significantly, scholars regularly speak of Hobbes', Locke's, or Buffon's tendency to nominalism rather than their nominalism, as if to acknowledge that they were involved in a discursive movement, not poised in a tenable epistemological position. In its polemical application seventeenth- and eighteenth-century nominalism was inevitably focused on certain types of universals – above all the kinds of the phenomenal world – not the universals presumed by micro-mechanists. Within natural history nominalism could be

45. *Reason and experience*, p.49.

46. Sloan, 'John Locke', p.51. Notice, in contrast, Philibert Collet's claim that Tournefort had done little more than copy Ray's work in Sturdy, *Science and social status*, p.234. On Boerhaave and classifying see also Stemerding, *Plants, animals and formulae*, p.32; Brunet, *Physiciens hollandais*, p.78.

47. On the megalomaniacal rhetoric of Linnaeus and Morison see Lindroth, 'Two faces of Linnaeus', p.55-59; Adanson, *Familles des plantes*, i.xix; Tournefort, *Eléments de botanique*, i.19.

selectively directed toward species, higher taxa, or any number of universals.[48] But the slogans of nominalism often went farther than their authors' assumptions, in part because such extremism followed from the modern faith in analysis and sensationism, in part because it promised sharper presentation. The resultant discord between expression and application would characterise nominalism throughout the Enlightenment.

Buffon the nominalist

In 1749 Buffon's 'Premier discours' appeared in volume i of the *Histoire naturelle*. As noted above, the 'Premier discours' includes arguments about truth, mathematics, and other subjects, but its overwhelming focus is on the ills of taxonomy, especially Linnaeus's. Seemingly a nominalist, Buffon asserts in the 'Premier discours' that only individuals exist (*HN*, i.38) and that nature is a perfect mathematical continuum. He excoriates Linnaeus for lumping together dissimilar plants and animals (*HN*, i.18, 37-41), for neglecting habitus in favour of fructification (*HN*, i.20-22), for creating a dogmatic jargon (*HN*, i.18-19), and for forcing botanists to carry a microscope to identify plants (*HN*, i.19).

In place of 'arbitrary' systems such as Linnaeus's, Buffon proposes three forms of 'real' or 'natural' taxonomy. First he suggests that living beings be sorted by the similarity and dissimilarity of all of their parts (*HN*, i.21-22). This was a truism in the discourse of taxonomy – and an unworkable truism in most people's view – but Adanson came to see it as foreshadowing his more limited natural method.[49] Like all 'modest proposals', Buffon's has an obvious rhetorical function, namely highlighting the excessive focus of Linnaeus's system, and this may be its main function, for it soon fades from the 'Premier discours' as a specific proposal. Instead Buffon turns to a second, not obviously compatible way of classifying. To see which taxa come 'naturally' to human beings he indulges in the first of several thought experiments concerning an adult man awakened to life. His conclusion is that certain basic taxa cannot be avoided from our human perspective, among others the three kingdoms (animal, vegetable, and mineral) and the zoological life-forms of folk taxonomy.[50] This second proposal shades into a third one as Buffon's natural man goes beyond 'necessary' intuition and begins to sort plants and animals on the basis of their utility and proximity to humans. 'Cet ordre, le plus naturel de

48. For a typology of nominalisms in natural history see Stevens, *Development of biological systematics*, p.176-77.

49. See Adanson, *Familles des plantes*, i.clv-clvi; Atran, *Cognitive foundations of natural history*, p.24-25, 196-200; Pratt, 'System-building', p.429; Stemerding, *Plants, animals and formulae*, p.28; *Herrn von Buffons Allgemeine Naturgeschichte*, 7 vols (Troppau, Joseph Georg Trassler 1785), i.40-42n. See also *HN*, i.38; *HNO*, i.41.

50. *HN*, i.31-32. See also Atran, *Cognitive foundations of natural history*, p.193n. For a reply see Malesherbes, *Observations sur l'histoire naturelle*, i.95-107. Notice Buffon's denial of the reality of the three kingdoms in *HN*, ii.26.

tous, est celui que nous avons cru devoir suivre', Buffon announces (*HN*, i.32-33). Accordingly the *Histoire des quadrupèdes* begins with domestic animals, leaving their wild 'cognates' for later, separate treatment.[51]

As with a number of Buffon's ideas, it is hard to know where exactly his 'nominalism' came from, but the 'Premier discours' itself can be traced to a presentation he made in a semi-annual public meeting of the Académie des sciences. The text of his presentation has never been found, but Buffon described it in a letter to Jean Jalabert dated 2 August 1745:

> J'ai lu, l'année passée, à la séance publique de l'Académie des sciences, un discours à ce sujet, dans lequel je crois avoir démontré les défauts et l'insuffisance des méthodes, et l'impossibilité de les rendre bonnes et générales en ne se servant que de quelques parties pour caractères; et je conclus par faire voir que la méthode de Linnaeus est de toutes la moins sensée et la plus monstrueuse, puisqu'il met sous la même classe, et souvent sous le même genre, des plantes absolument différentes.[52]

Citing this letter, Giulio Barsanti has argued that Buffon's original attack on Linnaeus contained only one element, his rejection of taxonomies based on just a few characters. According to Barsanti, the other arguments against Linnaean taxonomy in the 'Premier discours' – arguments about abstractions, the chain of being, human knowledge, and so on – are later additions intended to turn a 'technical' question into a philosophical one and thus to 'faire éclater une "affaire" de retentissement européen'.[53]

Two documents not cited in Barsanti's analysis shed further light on the 'original' 'Premier discours'. First, a notice in the *Mercure de France* gives the date of Buffon's presentation – 15 April 1744 – and a one-sentence summary: 'M. *de Buffon* termina la séance par une préface historique pour le catalogue raisonné du cabinet d'histoire naturelle du Jardin du roi.'[54] This suggests that Buffon's discourse, like certain sections of the printed 'Premier discours', focused on the historical development of systems and methods, undoubtedly from a predominantly critical perspective. Significantly, the *Mercure* does not call Buffon's discourse philosophical or methodological – two adjectives that describe his 1749 version at least as well as 'historical'. A second piece of evidence appears in a letter from Abraham Bäck to Linnaeus the month after Buffon's discourse. Here Bäck announces that a certain Buffon has denounced systems and methods in the Académie des sciences. According to Bäck, Buffon reserved special contempt for classifications of birds based on their feet and beaks, and held up Martin Lister's study of shellfish as his taxonomical ideal.[55] Lister, a

51. On the precedent for Buffon's anthropocentric order see Sloan, 'Natural history', forthcoming; Roger, *Buffon*, p.127-29. On the precedent for his separation of domestic and wild animals see Ritvo, *Platypus and the mermaid*, p.38-41.

52. In Eugène Ritter (ed.), 'Lettres de Buffon et de Maupertuis adressées à Jalabert', *Revue d'histoire littéraire de la France* 8 (1901), p.653.

53. 'Linné et Buffon', p.100-102.

54. *Mercure de France* (May 1744), p.970.

55. Abraham Bäck to Linnaeus, 1 May 1744, in Th. M. Fries (ed.), *Bref till och från Abraham Bäck*, in *Bref och Skrifvelser af och till Carol von Linné* (Stockholm 1910), ser. 1, iv.20-21.

friend of Ray's, classified shellfish using the biological criterion for species as well as multiple *differentiae* drawn from their shells and habitats.[56] The *Histoire naturelle* would endorse both approaches to classifying, but the published 'Premier discours' does not mention Lister. Perhaps Daubenton, who made his academic debut with a paper on the classification of shellfish (1743) and who chose to arrange the shells in the Jardin du roi in accordance with Lister's 'method',[57] was responsible for Buffon's early attention to Lister and shellfish. In any case Bäck's letter, like the notice in the *Mercure*, makes no reference to Buffon's later arguments about abstraction and continuity, and thus supports Barsanti's thesis that the 'Premier discours' became broadly philosophical only after 1744.

Buffon's contemporaries ventured two explanations for his nominalistic 'Premier discours' and its attack on Linnaeus. A few blamed inexperience with taxonomic natural history.[58] Buffon himself claimed to have learned botany three times. In his translation of the *Vegetable statics* he cited Tournefort's *Eléments de botanique* (1694) and other continental texts to identify Hales's plants. With Duhamel he designed a primitive but practical way of classifying oaks by their acorns in the late 1730s. Just how much taxonomy Buffon knew in the 1740s is unclear, but he certainly knew some.[59] Other contemporaries sensed Locke's presence in the 'Premier discours'.[60] Nothing proves that Buffon read the *Essay concerning human understanding*. Nor does he mention Locke or the *Essay* in any of his surviving letters or texts. Still, experts on Buffon have traditionally accepted Locke as a major contributor to Buffon's epistemology and to the nominalism of the 'Premier discours'. Certainly many of Buffon's ideas seem generally indebted to Lockean sensationism.[61] Several of his intimates were admirers of Locke, and scepticism about taxonomy was common among Lockeans from Ray to Condillac, if also among sensationists with non-Lockean pedigrees.[62]

56. J. Carr, 'The biological work of Martin Lister (1639-1712)', Doctoral dissertation, University of Leeds, 1974, p.199, 210-12.

57. Camille Limoges, 'Daubenton, Louis-Jean-Marie', in *Dictionary of scientific biography*, xv.111; Louis-Jean-Marie Daubenton, 'Coquille', in *Encyclopédie*, iv.185.

58. See for example Malesherbes, *Observations sur l'histoire naturelle*, i.4-5, 34, 129, 216-18; Atran, *Cognitive foundations of natural history*, p.194n.

59. Hanks, *Buffon avant l''Histoire naturelle'*, p.79, 137, 153, 159. For a suggestion that Buffon took Bernard de Jussieu's courses on botany see Stafleu, *Linnaeus and the Linnaeans*, p.275.

60. Sloan, 'Buffon–Linnaeus controversy', p.361.

61. Sloan, 'Buffon–Linnaeus controversy', p.361n.; Roger, *Buffon*, p.55, 62-64, 118-22, 213-18, 335-36, 410-12; Duchesneau, *Physiologie des Lumières*, p.261-62; Claude Blanckaert, 'La valeur de l'homme. L'idée de nature humaine chez Buffon', in *Buffon 88*, p.589-91, 594; Hanks, *Buffon avant l''Histoire naturelle'*, p.118n., 224-34; Hodge, 'Two cosmogonies', p.246.

62. On Buffon's Lockean acquaintances see Hanks, *Buffon avant l''Histoire naturelle'*, p.15n.; Roger, *Buffon*, p.24-25, 54-55. Locke had many enthusiasts in early eighteenth-century Dijon. See Marcel Bouchard, *De l'humanisme à l'Encyclopédie. L'esprit public en Bourgogne sous l'Ancien Régime* (Paris 1930), p.430. On the connection between nominalism and sensationism see O'Neal, *Authority of experience*, p.35n. One non-Lockean sensationist with nominalistic tendencies was Fontenelle. See Marsak, *Bernard de Fontenelle*, p.29, 31-33; Roger, *Sciences de la vie*, p.447; Sloan, 'John Locke', p.51n.; Sloan, 'History of the concept of the biological species', p.241-42.

While acknowledging formal similarities between Locke's and Buffon's statements regarding taxonomy, Phillip Sloan has argued that Buffon's mistrust of traditional classification originated elsewhere. Specifically, Sloan proposes that Buffon's critique of taxonomy grew out of his more general critique of abstraction, shaped not by Locke but by Leibniz, Wolff, Berkeley, and others.[63] Whether or not we accept Sloan's list of influences, it makes sense to study Buffon's 'nominalism' alongside his professed hostility toward abstraction and mathematics. For in the 'Premier discours' and elsewhere in the *Histoire naturelle*, taxonomy is associated with abstraction and opposed to reality in the same way mathematics is. Furthermore, taxonomy is characterised as 'une convention, une langue arbitraire, un moyen de s'entendre, mais dont il ne peut résulter aucune connaissance réelle' (*HN*, i.16). This is the same language Buffon uses to describe mathematics. Likewise his insistence that natural history's nomenclature is purely conventional parallels his assertion that mathematical notation and arithmetic scales can be chosen at whim. Significantly, non-representational accounts of mathematics of the kind Buffon offers in the 'Premier discours' are sometimes described as 'mathematical nominalism'.[64]

As shown in chapter 5, Buffon's denunciations of mathematics as formal and sterile misrepresent his assumptions. With his confidence in arithmetic and his faith in Newton's law as a mirror of reality, he was hardly the 'mathematical nominalist' he pretended to be. With regard to abstractions in general, the same case can be made. No doubt Buffon rejected many abstractions, often righteously citing his 'metaphysical' attachment to concrete reality. But he was also famous for inventing what others did not hesitate to label abstractions. Again and again his belief in the 'sublime simplicity' of natural law led him to reject data that contradicted his theories. Typically he doubted Charles-Marie de La Condamine's finding that there were no marine fossils on top of the Andes because his theory required them on top of all mountains.[65] Ironically he reproached Linnaeus for having the same faith: 'On pèche en physique en attribuant à la nature trop d'uniformité.'[66] The *Histoire naturelle* even champions abstraction. Already in the 'Premier discours' Buffon defends 'generalisation' and *grandes vues*, both in appearance more abstract than real (*HN*, i.4, 43, 50-51). In the *Histoire des minéraux* he accuses those not

Another non-Lockean critic of taxonomy was Philibert Collet, a Burgundian humanist whose 1697 attack on Tournefort may have inspired Buffon. See Bouchard, *De l'humanisme à l'Encyclopédie*, p.206-208, 434; Sturdy, *Science and social status*, p.234; Lepenies, *Ende der Naturgeschichte*, p.58.

63. Sloan, 'Buffon–Linnaeus controversy', p.62; Sloan, 'From logical universals to historical individuals', p.101-40.

64. For associations of 'regular' and 'mathematical' nominalism see Buffon, preface to Newton, *Méthode des fluxions*, p.xxii-xiv; Hanks, *Buffon avant l''Histoire naturelle'*, p.122-26; Condillac, *Langue des calculs*, p.429-30. In the case of calculus, Buffon's scorn for notation is borrowed directly from Newton's English supporters.

65. *HN*, i.294-95. For statements of his belief in nature's simplicity see for example *HN*, iv.104, 107; *HNS*, i.7.

66. Buffon to Jalabert, 2 August 1745, in Ritter (ed.), 'Lettres de Buffon', p.653.

accepting his interior mould of being too limited to deal with abstract ideas (*HNM*, i.11). But this does not herald an end to his rhetoric, for at the end of the same volume he dismisses pure earth as an 'abstraction': 'Cette terre idéale n'existe nulle part' (*HNM*, i.386).

This is not to suggest that Buffon's critique of the abstract was entirely hypocritical. As Jacques Roger has argued, there were certainly senses in which Buffon was attached to real, commonsensical, palpable things.[67] But alongside his 'realism', which never amounted to a consistent philosophical position, Buffon no less than Linnaeus cultivated abstractions.[68] Naturally Buffon claimed that his abstractions were different. So did almost everyone in a period keenly conscious of the ills of abstraction. Like *esprit de système*, 'empty abstraction' was inevitably only cultivated by one's rivals and enemies in the Enlightenment. Buffon, for his part, occasionally defended his abstractions as grounded on repeated events and physical truth. But as we saw in chapter 4, his criteria for physical truth were highly indeterminate. This is obvious, for example, in his justification of the interior mould as 'un fait constant' (*HNM*, i.12). If a once 'hypothetical', unobservable entity was susceptible to being interpreted as a physical truth and a 'fact' of induction, what abstraction could not be? Buffon's rejection of abstraction was neither uniform nor epistemological, though it claimed to be both. In this it resembled his related rejection of 'useless' natural history. As Christian Licoppe has noted, Buffon vaunted the utility of his scientific work with exceptional vigour. Despite the ingratiating rhetoric, many of his 'useful' ideas proved highly impractical – his anthropocentric taxonomy and his proposals for exploiting burning mirrors, for example.[69] Like the pragmatic realism in which he took pride, Buffon's 'utilitarianism' was not a philosophy but a rhetoric born of real but sporadic intellectual preference.

As we have seen, valorising reality at the expense of abstraction was a widespread rhetorical tactic in the seventeenth and eighteenth centuries. Critics inevitably applied it to the *Histoire naturelle* despite Buffon's claim to represent concrete reality. After pointing out that abstractions are ubiquitous and thus inescapable, Malesherbes, for example, criticised Buffon for imputing undue significance to the two lines he 'imagined' across the New World and the Old World in the *Histoire et théorie de la terre*: 'Toutes ces observations seraient curieuses si elles avaient de la réalité.'[70] Kant, though appreciative of Buffon's avoidance of metaphysics, reproached him for placing ideas above reality in his contention against experience that there could be no 'double currents' in

67. *Buffon*, p.560-61; Sloan, 'Buffon studies today', p.472-73.

68. Stemerding, *Plants, animals and formulae*, p.75-77; Hayes, *Reading the French Enlightenment*, p.38.

69. On Buffon's rhetoric of utility see Licoppe, *Formation de la pratique scientifique*, p.225-26. The examples referred to can be found in *HN*, i.31-34; *HNS*, i.452-63. Buffon's mirrors were never applied to five uses he proposed for them. See Carramolino del Valle, 'Espejo ustorio de Buffon', p.285-86.

70. *Observations sur l'histoire naturelle*, i.190-91, ii.20-36. Buffon's discussion of his lines is in *HN*, i.204-11.

the Strait of Gibraltar. Citing unnamed 'laws of hydrostatics', Buffon dismissed the currents in the following terms: 'Il y a grande apparence que les expériences ont été mal faites, puisque la chose est impossible et qu'elle répugne à toutes notions que l'on a sur le mouvement des eaux.'[71] The same attachment to theory led Buffon to quarrel with his Dutch correspondent, the naturalist Arnout Vosmaer. Vosmaer believed a sloth had been discovered in Ceylon. Long since convinced that the Old and New Worlds shared no tropical animals, Buffon denied it, thus allowing Vosmaer to accuse him of relying on *propositions idéales* rather than facts. Buffon retorted that his 'rule' of South America's faunal singularity was merely a 'fact', but it was clearly intertwined with his theories in the *Epoques de la nature*.[72]

As these examples suggest, criticism of abstraction in the *Histoire naturelle* often verged on criticism of Buffon's hypotheses. Buffon himself made the connection, drifting imperceptibly from denunciations of abstraction to denunciations of hypotheses as well as vice versa.[73] The link between hypothesising and 'abstract' taxonomy was especially close in that the word 'system' served to designate both theories and taxonomies. Before the eighteenth century naturalists used a range of words to refer to their arrangements of natural objects – 'method, 'catalogue', 'synopsis', and so on. Tournefort used 'method' for the most part but also sometimes 'system', most often when focusing on taxa above the genus, such taxa being 'systematic' – that is, hypothetical – in the usage of the time.[74] Linnaeus, for his part, did not clearly distinguish systems and methods. His use of 'system' for artificial taxonomies stemmed from their speculative character as well as their basis in the 'sexual system' – that is, his hypothesis about plant sexuality. For his 'natural' classification he retained the term 'method', but he also used 'method' for more 'artificial' classifications.[75] Ambiguous though it was, Linnaeus's usage gave the word 'system' a previously unknown currency in botanical discourse. While the use of 'system' in taxonomy derived from the word's application to more clear-cut hypotheses, the two kinds of systems, general and taxonomical, were seen as different, if related, by the mid-eighteenth century. Taxonomists were accused of many things, but almost never of being blinded by *esprit de système*, the customary insult for hypothesisers. General treatments of 'system-building' tended to overlook taxonomical systems or mention them marginally as exceptional cases.[76] But if the two kinds of systems were not seen as identical, their

71. Ferrari, 'Kant, lecteur de Buffon', p.160-61; *HN*, i.399-400.

72. The quarrel is summarised in *HNS*, vii.125-31. On the theory behind Buffon's 'rule' see *HN*, ix.118-19; *HNS*, v.176-79. For a similar reversal of Diderot's rhetoric see Barruel, *Helviennes*, i.2n.

73. See for example *HN*, v.99-102, vii.19-25; *HNO*, ix.411-12.

74. Foucault, 'Situation de Cuvier', p.65-66. Notice the context of Tournefort's first use of 'system' in Tournefort, *Eléments de botanique*, i.40.

75. Stevens, *Development of biological systematics*, p.12.

76. Like many works of reference of the mid-eighteenth century, the *Encyclopédie* ignores taxonomy in its entries on systems. Condillac evokes taxonomy once in his *Traité des systèmes*, p.123-24.

proximity to each other was easy to highlight. Couched in a common language that pitted concrete reality against arbitrary abstraction, Buffon's attacks on systems, whether taxonomical or theoretical, were mutually reinforcing.[77]

Buffon's critique of taxonomy thus drew on and supported his critiques of mathematics and of hypotheses. It was also bound up with his appeals to nature's continuity. As noted in chapter 3, one of his preferred arguments against classifying was the inevitability of intermediates between any two taxa. Nature being continuous, taxonomy was doomed to be an artificial exercise. Likewise the principle of the plenitude of nature functioned as an element of his critique of taxonomy and abstraction in general. When natural phenomena turned out to be more varied than expected, Buffon could cite his principle 'tout ce qui peut être, est' and moralise about the futility of trying to capture nature with human abstractions. But when nature appeared less abundant than might be anticipated, he could blame the apparent shortfall on unrealistic abstractions and avoid the whole subject of nature's erstwhile plenitude. In a typical instance in the *Histoire des minéraux* Buffon calculates (incorrectly) that there are twenty-five possible combinations of his five primitive glasses, but notes that just eleven are discoverable in nature. Rather than question his 'belief' in natural plenitude, he writes off the difference as a failure of abstract thinking: 'Tant il est vrai que la nature ne s'est jamais soumise à nos abstractions!'[78] However nature behaved, Buffon managed to safeguard his rhetorical principles.

As we have seen, Buffon's attacks on hypotheses grew fewer with time, but his diatribes against mathematics and his evocations of natural plenitude were longstanding features of the *Histoire naturelle*. The same can be said of his interconnected complaints about abstraction and taxonomy. The critique of abstraction is especially visible in volume ii, where it serves to demolish previous accounts of reproduction (*HN*, ii.74-78, 155-57) and pre-empt 'abstract' objections to organic molecules (*HN*, ii.21-24). But all through the series Buffon denounces 'ideal', 'chimerical', and otherwise abstract entities, above all taxonomy, mathematics, and infinite quantity, but also bad analogies (*HN*, i.9-10), truth (*HN*, i.53), physical virginity in women (*HN*, ii.492-99), crystal forms (*HNM*, i.241-42, iii.433), perfectly hard atoms (*HN*, xiii.xvii-xviii; *HNM*, v.2n.), and *êtres de méthode* such as phlogiston and crystallising water (*HNM*, i.354). Whole theories are pronounced abstract and thus patently worthless, among others the theory that the brain is the site of feeling (*HN*, vii.13-19) and the theory that the Earth is made of different, concentric layers (*HN*, i.161-62).

77. Haller, conversely, argued that Buffon's tacit systematising was incompatible with his stand against taxonomical systems. See below, p.179. Another interested assimilation of taxonomical systems to ordinary hypotheses can be found in Adanson, *Familles des plantes*, i.xciv. For more on the connections between the two kinds of systems see Stevens, *Development of biological systematics*, p.12-13.

78. *HNM*, i.77-84. There are twenty-six possible combinations of Buffon's primitive glasses.

In the same way, the objections to taxonomy in the 'Premier discours' echo throughout the series, even in the sections written by Daubenton.[79] Taxonomy, we are told, is arbitrary, abstract, and conducive to error. Beyond these consistent grievances, Buffon's complaints about taxonomy typically emphasise one of two points. When he finds himself arguing for the identity of animals considered different by others, he rails against the 'excess' of nomenclature relative to reality. Already in the 'Premier discours' he writes: 'La langue est devenue plus difficile que la science!'[80] But when he disagrees with others' efforts to lump animals together he condemns nomenclature for being vague and for trying to force nature into unreal categories. Even here he sometimes grumbles about the 'excess' nomenclature in rivals' additive generic names, but he does so only marginally since he too is proposing to multiply names.[81] This second means of denouncing taxonomy builds on his 'nominalism' and fits with his sketch of the evolution of language in the article 'Rat'. Here, in agreement with the common eighteenth-century association of linguistic refinement with scientific and cultural progress, Buffon claims that primitive languages have only vague, general words ('tree', for example, to designate all trees), whereas modern languages also have specific, precise ones ('hemlock', 'fir', and others, in his example).[82] But the fact that he criticised nomenclature for being overly complex as well as overly simple shows that his complaints about it were less coherent and philosophical than he pretended. They were, within limits, a way of adding force to his specific disagreements with other taxonomists.

Visible here and there in attacks on taxonomy, nominalistic analysis persists to the end of the *Histoire naturelle*. But the pre-eminent nominalistic formula of the 'Premier discours' never returns: 'Il n'existe réellement dans la nature que

79. Roger, *Buffon*, p.408-409; Limoges, 'Daubenton', p.112; Louis Roule, *Daubenton et l'exploitation de la nature* (Paris 1925), p.150-51. See for example Daubenton, *HN*, iii.1-7, iv.113-18, 142-68, 268-69. Lacépède as well upheld Buffon's nominalism. See Burkhardt, *Spirit of system*, p.202-203. While maintaining a kind of nominalism, Daubenton later criticised the 'Premier discours' and Buffon's views on species. See Daubenton, 'Botanique', p.340-42; Sloan, 'Buffon–Linnaeus controversy', p.357; Daubenton, 'Histoire naturelle', in *Séances des écoles normales. Leçons*, i.292-93, 417-19, 437-38, iv.130-34.

80. *HN*, i.16. See also *HN*, i.9; Anne-Claire Déré, 'La réception de la nomenclature reformée par le corps médical français', in *Lavoisier in European context: negotiating a new language for chemistry*, ed. Bernadette Bensaude-Vincent and Ferdinando Abbri (Canton, Mass. 1995), p.216. For denunciations of taxonomy that support Buffon's arguments for consolidating species see *HN*, vii.108, x.284, 293, xi.370-74; *HNO*, i.71-72, 484, vii.286; *HNM*, ii.113, iii.6-7.

81. For denunciations of taxonomy that support Buffon's arguments for splitting others' taxa see *HN*, vii.43, ix.78; *HNO*, vii.409, 451; *HNM*, i.74-75, iv. 180. When Buffon favours splitting some taxa and consolidating others, he either denounces taxonomy vaguely (*HN*, ix.163) or switches his grievances from page to page (*HN*, ix.116-23).

82. *HN*, vii.278-80. See also *HN*, i.41-43; *HNS*, v.32. On eighteenth-century associations of language and knowledge see Foucault, *Les Mots et les choses*, p.98-103, 129-30; Daniel Brewer, *The Discourse of Enlightenment in eighteenth-century France: Diderot and the art of philosophizing* (Cambridge 1993), p.36-55; Auroux, *Sémiotique des encyclopédistes*, p.102. Unlike many theorists of primitive language, Buffon does not indicate that the first nouns were proper, but his account is neither detailed nor well-developed.

des individus' (*HN*, i.38). A similar formula ten volumes later is significantly modified: 'Dans la nature, il n'existe que des individus et des suites d'individus, c'est-à-dire des espèces' (*HN*, xi.369). This 'renunciation' of nominalism can be understood as an element of a larger evolution in Buffon's approach to taxonomy – an evolution defined by both practical exigencies and theoretical insights. Important among the latter were Buffon's 1748 microscopic observations, subsequent to at least the first version of the 'Premier discours'. By 'confirming' his theory of reproduction, these observations gave species an immutable real essence, the interior mould, and so justified taxonomy, albeit abstractly. Meanwhile Buffon was reflecting on the biological criterion for determining species. Already in the volumes of 1749 he briefly defined species by their capacity for interbreeding (*HN*, ii.10-11, 18). By 1751 he was doing experiments with animal cross-breeding (*HNS*, iii.3). In volume iv of the *Histoire naturelle* he pronounced the ability to produce fertile offspring the key to real species.[83]

Practical considerations also affected Buffon's attitude toward taxonomy. As one critic noted in a footnote to a German translation of the *Histoire naturelle*, only God could write a truly nominalistic natural history.[84] Buffon's 1748 plan to cover animals, plants, and minerals in just fifteen volumes suggests underestimation of the abundance of nature, an error that may have bolstered his scorn for taxonomy. In his prospectus Buffon proposed a 'système de végétation' and a 'méthode simple, naturelle et invariable pour connaître les pierres précieuses', but no system for animals.[85] Between 1749 and the early 1760s he allowed that taxonomy had a role to play where living beings were too numerous and indistinct to be treated in a 'real' way, but for 'higher' animals he refused taxa above the level of the species (*HN*, iv.383-84, ix.10). By the end of the *Histoire des quadrupèdes*, however, he had accepted animal genera on his own terms, making them groups of genetically linked animals.[86] This acceptance, which reduced all animals to just thirty-eight genera, came as he turned from mostly European animals to a cumbersome multitude of exotic species. Birds, even more numerous and plagued with 'varieties', pushed him toward a more overtly 'convenient' classification:

> Au lieu de traiter les oiseaux un à un, c'est-à-dire, par espèces distinctes et séparées, je les réunirai plusieurs ensemble sous un même genre [...]. Par ce moyen, j'ai beaucoup abrégé, et j'ai réduit à une assez petite étendue cette histoire des oiseaux qui serait devenue trop volumineuse, si [...] j'eusse traité de chaque espèce en particulier.[87]

83. *HN*, iv.384-86. On these developments see Farber, 'Buffon and the concept of species', p.263-68; Roger, *Buffon*, p.410-14.

84. *Herrn von Buffons Allgemeine Naturgeschichte*, i.73-74n. Predictably, even enthusiasts of Buffon's 'nominalism' resorted to familiar classificational distributions to organise their books on natural history. See Spary, '"Nature" of Enlightenment', p.279; Sloan, 'Buffon–Linnaeus controversy', p.357.

85. *Journal des savants* (October 1748), p.639-40.

86. See Roger, *Buffon*, p.410-41; Farber, 'Buffon and the concept of species', p.268-78.

87. *HNO*, i.xx. See also Roger, *Buffon*, p.438-39; Farber, 'Buffon and the concept of species', p.269-70, 278-79.

While hopeful that his genera consisted of species with a common ancestry, Buffon admitted that they would be defined by resemblance, not historical kinship or interbreeding (*HNO*, i.xx-xxi). Without mentioning it, in fact, he had long used morphology and ethology to determine species and genera which he nonetheless defended as being based on the 'reality' of reproduction. Such fudging, we have seen, was common among classifiers. Like other taxonomists' official criteria, Buffon's criterion for 'real' species and genera was valuable polemically and philosophically, but it was not very practical. As he discovered, experiments with interbreeding were not merely difficult but also inconclusive, since animals might refuse to mate for any number of reasons. Testing their reproductive compatibility on a worldwide scale was simply not feasible.[88] While Buffon's experiments with interbreeding shaped his ideas about change in the living world, his theoretical criterion for species and genera had little bearing on the way he actually classified. Henry Home, Lord Kames, even accused him of contradicting his criterion for species, since Buffon suggested that apes and humans could produce fertile offspring, yet he balked at considering them of the same kind.[89] The contemporary Spanish naturalist Félix de Azara charged him with using his criterion of interbreeding opportunistically, bringing it up or ignoring it as it suited his needs.[90] Certainly Buffon's use of the word 'real' was rhetorical. Limited to genera based on ancestry in the *Histoire des quadrupèdes*, the word is used to characterise genera based on habitus once Buffon acknowledges that his genera are so constituted (*HNO*, vi.355-56). Through his polemical appropriation of pertinent vocabulary, Buffon always managed to place himself on the side of reality.

Whatever the ambiguities of his taxonomies, the author of the 'Premier discours' had become a prolific classifier by the 1750s – 'un nomenclateur infatigable', as Condorcet put it.[91] This paradox has led some scholars to see a major shift in Buffon's thinking from early, naive nominalism to mature realism.[92] As we have seen, Buffon's nominalistic statements continue to the end of the

88. Stemerding, 'How to make oneself nature's spokesman?', p.204-10; Stemerding, *Plants, animals and formulae*, p.70-74, 95-98; Larson, *Interpreting nature*, p.62, 79; Roger, *Buffon*, p.414-16, 437-41; Michel Delsol and Janine Flatin, 'L'espèce existe-t-elle? question aux paléontologues et aux philosophes', in *Buffon 88*, p.532-35. For a dissenting view see Ernst Mayr, 'The species as category, taxon and population', in *Histoire du concept d'espèce*, p.306-307.

89. Phillip Sloan, 'The gaze of natural history', in *Inventing human science: eighteenth-century domains*, ed. Christopher Fox, Roy Porter and Robert Wokler (Berkeley, Calif. 1995), p.137-38; Larson, *Interpreting nature*, p.85.

90. Andrés Galera and Marcelo Frías, 'Félix de Azara y Georges Lucien Leclerc: dos formas de *iluminar* la naturaleza americana', *Asclepio* 48 (1996), p.32.

91. *HARS* 1788 (1791), p.61. Buffon also defined taxa by morphological *minutiae*, despite his earlier outrage at the use of such characters. See Martinez-Contreras, 'Des mœurs des singes', p.559-60.

92. Flourens, *Histoire*, p.15-20; Lovejoy, 'Buffon and the problem of species', p.87-95; Jean Chaline, 'Evolution des concepts de l'espèce et de la formation des espèces', in *Buffon 88*, p.516; Roger, *Sciences de la vie*, p.565-66. Roger's more recent book makes Buffon's development seem less dramatic. See Roger, *Buffon*, p.408-12.

Histoire naturelle. They are perhaps mostly rhetoric, but this suggests a continuity of sorts between the 'Premier discours' and the rest of the series. Significantly, the 'Premier discours' labels genera, orders, and classes 'imaginary' taxa but passes no judgement on species, a preferred target for nominalists from Locke to Adanson.[93] The 'Premier discours' does assert that only individuals exist, a contention that seems to rule out the possibility of species, but it also alludes to species without mentioning any doubts about their ontological status. The task of description is explicitly premised on species: 'L'histoire d'un animal doit être non pas l'histoire de l'individu, mais celle de l'espèce entière de ces animaux' (*HN*, i.30). The 'Premier discours' defines species only once, and then platitudinously alongside genera and classes as a taxa determined by overall similarity (*HN*, i.20-22). The fact that species are absent from Buffon's list of 'imaginary' taxa suggests that he was considering a different criterion.

In effect Phillip Sloan has argued that the rudiments of Buffon's subsequent understanding of species are present in the 'Premier discours'. In particular the concepts used to define and defend biological taxa are developed in the concluding discussion of truth. In opposition to the mathematical truth which he criticises as sterile and tautological, Buffon proposes that physical, 'real' truth must be founded on repeated phenomena (*HN*, i.53-57). In the *Histoire des animaux* he defines 'real' species in the same terms, stressing their existence as successions of individuals. Buffon's re-use of the concepts of the 'Premier discours' to define biological species shows that his evolution was not a simple about-face from nominalism to realism. The discussion of truth in the 'Premier discours' was probably added after 1744, but even so the 'Premier discours' offered as much realism as nominalism by 1749. Why the notion of physical, 'real' truth was not applied to species in the 'Premier discours' remains problematic. Sloan speculates that Buffon was waiting for additional proof, or did not think of 'real' truth in the context of taxonomy before encountering species as serial repetition in Harvey's interpretation of Aristotelian philosophy.[94]

Sloan's work brings to light a considerable consistency between the 'Premier discours' and Buffon's later, 'realistic' approach to taxonomy. It is a tribute to the obscurity of the 'Premier discours' that this consistency was overlooked for more than two hundred years. Many of Buffon's contemporaries read the 'Premier discours' as a misguided, contradictory manifesto of nominalism, but its potential for foreshadowing a 'real' concept of species slipped by unnoticed. To the extent that it was one, this misunderstanding of the 'Premier discours' was mostly Buffon's fault. Everything suggests that his concept of taxonomy was neither unified nor very solid at the time of the final draft of the 'Premier discours'. Not only does he propose incompatible

93. See Atran, *Cognitive foundations of natural history*, p.232n.; *HN*, i.38.

94. Sloan, 'Buffon–Linnaeus controversy', p.366-75; Sloan, 'From logical universals to historical individuals', p.118-26. For more on realism in the 'Premier discours' see Barsanti, 'Linné et Buffon', p.92-95; Barsanti, 'Buffon et l'image de la nature', p.259; Larson, 'Linné's French critics', p.73-74.

natural systems, he undercuts them all with his critique of universals. Much of the text's obscurity probably comes from his hesitation between different possibilities, perhaps over several years while the 'Premier discours' grew by accretion instead of rewriting. Sometimes Buffon's debate was 'less with Linnaeus than himself', as Lorin Anderson has put it.[95] It was not obvious to Buffon, as it may be retrospectively, that he would end up adopting the criterion for species vaguely adumbrated at the end of the 'Premier discours'.

Along with his ambivalence and apparently additive style of editing, Buffon's polemical ambitions made the 'Premier discours' contradictory and hard to interpret. More than anything else he wrote, it was designed to be aggressive and provocative. As Jacques Roger observed, Buffon 'semble avoir calculé son discours de manière à se faire le plus d'ennemis possible dans le monde savant'.[96] But Buffon wrote the 'Premier discours' above all to nullify Linnaean systematics. In 1744, when Buffon pronounced the first version of the 'Premier discours', Linnaeus was just beginning to interest Parisian botanists. Tournefort notwithstanding, taxonomy had not been emphasised in the French scientific community in the first third of the century.[97] In 1738 Linnaeus visited the Jardin du roi. In 1739 the Jardin's renowned botanist Bernard de Jussieu pronounced Linnaeus's system superior to that of Tournefort. But in his article in the *Mémoires* of the Académie des sciences Jussieu acknowledged the inconvenience of systems defined, like Linnaeus's, by microscopic floral organs. The use of microscopic *differentiae* had been rejected by Ray and Tournefort and would likewise be rejected in the *Histoire naturelle*. Buffon was perhaps thinking of Jussieu's article when he composed the first version of the 'Premier discours'. In 1744, the year of Buffon's discourse, Jussieu had Linnaeus's *Systema naturae* published in Paris, glossed with French equivalents for Linnaean names.[98]

Read before the Académie on 15 April 1744, the first version of the 'Premier discours' created little stir. Linnaeus was presumably infected with his friend

95. *Charles Bonnet*, p.38. On the almost universal misunderstanding of the 'Premier discours' see Sloan, 'From logical universals to historical individuals', p.118n.; Larson, *Interpreting nature*, p.15, 78; Farber, 'Buffon and the concept of species', p.261-63.

96. *Buffon*, p.128-29, 133-34, 419. For more on the polemical aspect of the 'Premier discours' see Flourens, *Histoire*, p.10; Sloan, 'From logical universals to historical individuals', p.119; Lyon, '"Initial discourse"', p.137n., 140; Farber, 'Buffon and the concept of species', p.260-61; Atran, 'Common-sense basis', p.233n.; Larson, 'Linné's French critics', p.68-69; Daudin, *De Linné à Jussieu*, p.125-32.

97. Roger, *Sciences de la vie*, p.749; Daudin, *De Linné à Jussieu*, p.49-50, 124-25. See also the discussion of Réaumur's views on classifying in Gerhard H. Müller, '"Distinguer les uns des autres". Le concept de l'espèce chez Réaumur: pragmatisme et utilité', in *Histoire du concept d'espèce*, p.61-77.

98. See Daudin, *De Linné à Jussieu*, p.118-19; Roger, *Buffon*, p.81-82; Larson, *Reason and experience*, p.59; Bernard de Jussieu, 'Histoire d'une plante, connue par les botanistes sous le nom de *Pilularia*', *MARS* 1739 (1741), p.240, 251. On Ray, Tournefort, and microscopic characters see Stevens, *Development of biological systematics*, p.231; Adanson, *Familles des plantes*, i.cliii; Tournefort, *Eléments de botanique*, i.14.

Bäck's indignation,[99] but few others appear to have had strong reactions. Réaumur, for example, did not mention Buffon's presentation in his letter to Trembley the following month.[100] As we have seen, the *Mercure de France*, which often expanded on newsworthy aspects of the Académie's public meetings, devoted just a sentence to Buffon's discourse. The discussion of taxonomy in Jean-Etienne Guettard's *Observations sur les plantes* (1747) ignores Buffon's claims about Linnaeus's system.[101] In contrast, we shall see, the printed 'Premier discours' of 1749 prompted Guettard to write a scathing rebuttal. The relatively cool reception of Buffon's 1744 discourse suggests that he had not yet developed his scandalous, 'Pyrrhonistic' attacks on taxonomy. The one argument we know he used in 1744 – his argument that natural groupings will be violated by systems relying on limited *differentiae* – had been familiar to botanists since the time of Ray and Tournefort. Nor was an attack on Linnaeus necessarily cause for scandal. Others had quibbled with Linnaeus on at least minor points, and the major Parisian botanists remained sceptical of his system, if not his nomenclature, throughout the century. Despite his 'endorsement', even Bernard de Jussieu kept his distance from Linnaeus's reforms.[102]

If Buffon's discourse of 1744 attracted little attention, his published 'Premier discours' had the opposite effect. As noted above, there was something to offend almost everyone somewhere in the 'Premier discours'. Among non-botanists, however, reactions to Buffon's 'nominalism' ranged from indifference to enthusiasm. D'Alembert applauded Buffon's attack on taxonomy and echoed his insistence that only particulars exist. Diderot extended the argument by evoking continuities that Buffon avoided, and hinting that even individuals were just convenient abstractions. Whether or not they approved of the 'Premier discours', sensationistic philosophers from Condillac to Bonnet agreed with Buffon's premise that taxa were nominal.[103] Less philosophical, the *Universal magazine of knowledge and pleasure* published an English translation of two-thirds of the 'Premier discours' without mentioning Buffon, recommending it to readers as 'A dissertation on the dignity and usefulness of

99. Apparently against Bäck's wishes, the 1744 French edition of Linnaeus's *Systema naturae* lacks an explicit refutation of Buffon or Lister, but this hardly means that Linnaeus was indifferent. See Bäck to Linnaeus, 1 May 1744, in Fries (ed.), *Bref till och från Abraham Bäck*, p.20-21.

100. Réaumur to Trembley, 21 May 1744, in Trembley (ed.), *Correspondance inédite*, p.185-87.

101. Guettard admits the inconvenience of microscopic observation, but nothing proves he is trying to neutralise an earlier argument. See Guettard, *Observations sur les plantes*, i.xxxi. Nor does Grandjean de Fouchy's review of the *Observations* clearly engage with Buffon's attack on Linnaeus. See *HARS* 1747 (1752), p.79-81.

102. See Sloan, 'Buffon–Linnaeus controversy', p.358; Stafleu, *Linnaeus and the Linnaeans*, p.281; Larson, *Reason and experience*, p.59-60.

103. On d'Alembert and Diderot see d'Alembert, 'Discours préliminaire', p.xv-xvi, xix; Diderot, 'Animal', p.468-74; Diderot, *Rêve de d'Alembert*, p.137-40. On Condillac and Bonnet see Condillac, *Traité des sensations*, p.307n.; Knight, *Geometric spirit*, p.67-70; Bonnet, *Palingénésie philosophique*, i.202-203; Bonnet, *Contemplation de la nature*, i.52-55; Daudin, *De Linné à Jussieu*, p.101-103. Condillac accused Buffon of reintroducing essence with his uncritical references to animals' 'natures'. See Thomas, 'Philosophy, science, and style', forthcoming.

natural history, with the method of studying it to advantage'.[104] In France the *Journal de Trévoux* praised Buffon's resistance to Linnaean taxonomy as favourable to genius and accessible natural history.[105] Buffon made the same associations in the *Histoire naturelle.* French naturalists had long assumed that taxonomy was repulsive to popular readers, and apologies for nomenclature were a platitude of vulgarisation. Exaggerated though it may have been by less taxonomical naturalists, the public's distaste for nomenclature and classification was clearly real in some measure.[106] While it seems unlikely that Buffon repudiated Linnaean taxonomy to please Parisian readers, assumptions about his audience probably reinforced his rhetoric. Like his critique of mathematics, his critique of taxonomy could count on a sympathetic audience. Ironically, Buffon's 'populist' complaints about Linnaean natural history – specifically, his complaints about nomenclature and microscopic *differentiae* – were adopted by Linnaeans against their own enemies once their taxonomy was reduced to 'popular science' around 1800.[107] This reversal points again to the rhetorical aspect of scientific polemic.

In contrast with the broader public, many botanists found the 'Premier discours' offensive and baffling. As mentioned above, Adanson later claimed it as an inspiration for his natural method, and Antoine-Laurent de Jussieu borrowed its image of method as a 'pierre philosophale'.[108] Parisian botanists in general remained sceptical of Linnaeus, but their scepticism was more limited than Buffon's in the 'Premier discours'. Working as they did in the Jardin du roi, the Jussieus may not have wanted to criticise Buffon too overtly,[109] but none of them defended the 'Premier discours' either. An amateur botanist and a pupil of Bernard de Jussieu, Malesherbes devoted more than a hundred pages of his *Observations* to refuting Buffon's ideas on

104. *Universal magazine of knowledge and pleasure* (August 1751), p.49-57 (July 1752), p.3-9. On the first half of this translation see Lyon and Sloan, *From natural history to the history of nature*, p.93-94. William Christmas announced his discovery of the second part of the translation at a meeting of the American Society for Eighteenth-Century Studies in a paper entitled '"Beautifully coloured from nature": color illustration, the periodical press and the popularization of natural history' (25 March 1999). The translation is quite reliable but omits the last third of the 'Premier discours' – specifically, the sections dealing with ancient natural history and the concept of truth.

105. September 1749, p.1854-55, 1861.

106. For caveats on this 'distaste' see Drouin and Bensaude-Vincent, 'Nature for the people', p.414-17. But see also *HARS* 1734 (1736), p.22; Torlais, *Réaumur*, p.120-21, 131; Gilles-Auguste Bazin, *Abrégé de l'histoire des insectes pour servir de suite à l'histoire naturelle des abeilles*, 4 vols (Paris, Guerin 1747-1751), i.vi-ix; Stephen Werner, 'Abraham Rees's eighteenth-century *Cyclopaedia*', *SVEC* 315 (1994), p.192-93; Duris, *Linné et la France*, p.104-105; Pluche, *Spectacle de la nature*, ii.xxii-xxiii; *Mercure de France* (March 1771), p.148-49; *HN*, x.156. On taxonomy as a barrier to genius see *HN*, i.7, xiii.7-9, 14-15; *HNM*, ii.158-59; Gómez Torres, *Ideas de Buffon sobre retórica*, p.67n.

107. Duris, *Linné et la France*, p.209. At roughly the same time Cuvier rejected the notion of popularly accessible scientific taxa. See Larson, *Interpreting nature*, p.53.

108. On Jussieu's and others' appropriation of Buffon's image see Stevens, *Development of biological systematics*, p.28-29. Compare *HN*, i.14.

109. But see Bernard de Jussieu's suggestions in Bäck to Linnaeus, 1 May 1744, in Fries (ed.), *Bref till och från Abraham Bäck*, p.20.

classification.[110] The naturalist Guettard included his own refutation as a lengthy digression in an article on caterpillars in the *Mémoires* of the Académie des sciences. His tone was caustic: 'Il aurait sans doute, avant ce temps, paru ridicule à toute personne, même la moins instruite, d'entrer dans une semblable discussion.'[111] Haller malignly defended taxonomy as a hypothesis no more objectionable than Buffon's hypothesis about the Earth's origin.[112] Unhappy with Buffon's treatment of Linnaeus, the amateur botanist William Smellie omitted the 'Premier discours' from his 1785 English translation of the *Histoire naturelle*.[113] All in all the 'Premier discours' came in for considerable criticism for its arguments against taxonomy, unlike Buffon's presentation of 1744. This change in reception was undoubtedly due in large part to the publicity surrounding Buffon and the *Histoire naturelle*, but it seems likely that Buffon's arguments had grown more provocative as well. If it took courage to attack Linnaeus in 1749, as Jacques Roger has asserted, it took a great deal of courage to attack Linnaeus as Buffon did in 1749 – philosophically, relentlessly, and in all generality.[114]

Several factors probably contributed to the vehemence of Buffon's attack on Linnaeus in the 'Premier discours'. Scorn for Linnaean classifying was certainly important. By the 1760s Buffon would be citing Linnaeus as a trustworthy naturalist even while grumbling about the ills of nomenclature, but in the 1740s his attitudes appear to have been consistently negative.[115] By linking his attack on Linnaeus with denunciations of abstraction, minute details, and mathematics, Buffon infused it with indignation drawn from a whole range of 'causes' he believed he represented. The provocations of the 'Premier discours' were also meant to attract attention. To believe Lesley Hanks, Buffon's presentation had been growing more aggressive since the 1740s.[116] Having attained European celebrity by challenging one rationalistic scientific giant in 1747 – his burning mirrors discredited one of Descartes's speculations – Buffon may have felt eager to vanquish another one. In all likelihood none of this was consciously decided, but he rose to the occasion with his well-known ability to be carried away by his rhetoric. As his critics

110. Malesherbes, *Observations sur l'histoire naturelle*, i.5-148.

111. 'Description de deux espèces de nids', p.198-205.

112. Sloan, 'Buffon, German biology', p.122.

113. Likewise, Daubenton's article 'Botanique' was left out of the Yverdon *Encyclopédie* for repeating the anti-Linnaean arguments of the 'Premier discours'. See Sloan, 'History of the concept of biological species', p.388n.; Kathleen Hardesty Doig, 'The Yverdon *Encyclopédie*', *SVEC* 315 (1994), p.95. For other botanists' reactions to the 'Premier discours' see Sloan, 'Buffon–Linnaeus controversy', p.360-61; Daudin, *De Linné à Jussieu*, p.119. Smellie's reaction should be compared with that of Oliver Goldsmith, a less botanically trained adaptor of the *Histoire naturelle*. See Ritvo, *Platypus and the mermaid*, p.23.

114. Roger, *Buffon*, p.408. See also Sloan, 'Buffon–Linnaeus controversy', p.358-59.

115. Early praise for Linnaeus can be found in *HN*, x.155, xii.102. In the 1740s, in contrast, Buffon considered Linnaeus unworthy of trust. See Buffon to Jalabert, 2 August 1745, in Ritter (ed.), 'Lettres de Buffon', p.653-54; Buffon to Cramer, 14 December 1748, in Weil (ed.), 'Correspondance Buffon–Cramer', p.130.

116. Hanks, *Buffon avant l'"Histoire naturelle"*, p.182-87.

observed, Buffon's provocative over-confidence would continue through the series, but it was nowhere so striking as in the 'Premier discours', the best site strategically for making an impression. In sum it seems highly probable, as Giulio Barsanti has argued, that Buffon's 1744 discourse was broadened and sharpened to create a revolutionary philosophical statement. No doubt it was broadened in directions reflecting Buffon's developing ideas. As it turned out, one late addition to the 'Premier discours' – the section on truth – anticipates the criterion for species Buffon would subscribe to for much of his life. But taken as a whole, with its many contradictions and categorical principles, the 'Premier discours' only makes sense in light of its polemical aspirations.

In the end, if Buffon exploited his differences with Linnaeus to make him the villain in an elaborate, biased philosophical drama, he was not the only naturalist putting rhetoric to use in disputes over classifying. When Buffon charged Linnaeus with creating an unnatural system, Malesherbes could reply that he was attacking a straw man, since the *Systema naturae* had been designed to be artificial.[117] But as we have seen, Linnaeus and others subtly waffled on the precise status of their systems and regularly took advantage of charged words such as 'natural'. Malesherbes himself suggested that artificial systems were somehow more than artificial.[118] In this sense the 'Premier discours' merely turned the tables on taxonomists. Likewise Buffon's proposal to consider every part of an organism in determinations of taxa can be derided as impossible and thus as a deceptive rhetorical ploy. But as noted above, the proposed bases for others' systems were no less ideal and no less duplicitous. No doubt the 'Premier discours' is more intensely rhetorical than taxonomists' usual discourse. Buffon's arguments against Linnaeus are so absolute and contradictory that scholars have long dismissed them as incoherent. But if Buffon misconstrued taxonomy's goals and terminology for strategic advantage, others, including Linnaeus, had done the same thing.

117. Malesherbes, *Observations sur l'histoire naturelle*, i.38-39.
118. See especially Malesherbes, *Observations sur l'histoire naturelle*, i.48-49.

Conclusion

As we saw in the first half of this study, eighteenth-century readers read natural histories not just for knowledge, but also for pleasure, distraction, and entertainment. What they sought and found in the *Histoire naturelle* in particular was not always what is emphasised in treatments of the series in our own time. In the last forty years scholars have focused on the series' overall philosophical and scientific content, especially as represented in Buffon's prominent general discourses. Contemporaries also took notice of Buffon's ideas and the celebrated discourses in which they usually appeared. To judge by reviews, early anthologies, and personal testimony, this interest was all the more marked when the ideas in question promised sublime views of nature. But other emphases were different. Buffon's evolving views on species and degeneration, tantalising to scholars in the wake of Darwin's 'revolution', were considerably less attended to by contemporaries of the *Histoire des quadrupèdes* and the *Histoire des oiseaux*. In contrast, several now neglected aspects of the *Histoire naturelle* attracted preponderant attention from eighteenth-century readers – above all the series' style, sublimity, moralising, and curious facts. After 1753, the approximate date of Buffon's identification as a superlative stylist, the truth of his theories was regularly judged secondary to his presentation and anecdotes, as in this judgement from the *Journal de Trévoux*: 'Toutes les critiques qu'on a faites [...] de ce système ne raviront jamais à son auteur la gloire de s'être montré [...] un écrivain du mérite le plus rare.'[1] For many of Buffon's readers what counted most was his genius, his way of writing, and the piquant subjects he wrote on.

Important as they must have been in attracting a broad readership, such 'popular' aspects of the *Histoire naturelle* are discounted by modern scholars as incidental to Buffon's project. But the series' popularity, far from being a felicitous after-effect of Buffon's decision to communicate his ideas on natural history, was something he actively cultivated from the 1740s onward. Much of the distinctiveness of the *Histoire naturelle* comes from 'superficial' appeals to readers, whether via style, sexual titillation, anthropomorphism, or grandiose moralising about nature and animals. As we have seen, Buffon was not averse to assuming positions that contradicted his deeper beliefs in order to be more poetic or understandable. Furthermore, the audience-pleasing 'ornaments' of the *Histoire naturelle* cannot always be separated from his philosophy and epistemology. Consider for example his *grandes vues* of nature, a product of sublime style among other factors. *Grandes vues* allowed Buffon to entertain readers, but just as importantly they reflected his commitment to speculative, intelligible natural history. Likewise his evocations of final causes and

1. January 1754, p.165.

continuities which he disavowed elsewhere were not always just vacuous literary embellishments but also, in some cases, efforts to 'experiment' with ideas he saw as unjustified but potentially fecund. At the same time, like other vulgarisers, he drew on these 'principles' for their aesthetic and pedagogical value even when they lacked any heuristic function. Whether reinforcing or compromising his theoretical ideals, popularity was a central goal of the *Histoire naturelle*.

As noted repeatedly in chapters 1, 2, and 3, Buffon was not alone among popular naturalists in reaching out to readers with rhetorical techniques. Naturally, not everyone employed the same means to make natural history appealing. Many 'vulgar' natural histories were written as dialogues on the model of the *Entretiens sur la pluralité des mondes*, although Fontenelle's flirtations were typically replaced by more chaste, homey banter. Some naturalists relied on erudition and literary citations to make their works attractive to humanistically schooled readers. Buffon, for his part, did not use explicit dialogue and only sparingly brought in poetry and literary precedent, but elements of his rhetoric were standard for the genre. Moralising and teleology were ubiquitous in popular natural history, even in the works of authors not noted for their piety. Both tendencies varied from fervently Christian to vaguely deistic, with active nature sometimes taking on God's usual roles. Following in the footsteps of many lesser-known naturalists, Buffon engaged in a great deal of variously Christian and more generic moralising, sometimes about final causes or the plenitude of nature, sometimes about the 'lessons' of physiology or ethology. At the same time he was expansive in discussing the sexuality of humans and animals. No more than other, similarly provocative naturalists did he write his famous racy passages in ignorance of the reactions they would elicit. Again like other vulgarisers, Buffon made a conscious effort to balance details and nomenclature with intelligible overviews and other 'easy' material. Above all, his style and its mythic reputation were principal contributors to his popularity. With his sporadic forays into grandiose style Buffon departed from the conversational, playful styles of Fontenelle and Pluche, but he was not the only writing naturalist to sense the advantages of sober, dramatic eloquence.

Reactions to the series show that Buffon's rhetoric of seduction affected people differently. Undoubtedly representative of many contemporary readers, Stéphanie-Félicité de Genlis was fascinated above all with Buffon's style.[2] Honoré-Gabriel Riqueti de Mirabeau judged it more critically: 'La pompe continuelle de Buffon dégénère peut-être quelquefois en monotonie.'[3] No less an admirer than Genlis of the *Histoire naturelle*, Mirabeau seems to have appreciated it as a source of practical counsel, notably on paediatrics.[4]

2. See *Mémoires inédits de Mme de Genlis*, i.363-65, ii.277-79, 374-76, viii.167-68, 171.

3. To Paul-Pierre de La Fage, 7 December 1780, in Dauphin Meunier and Georges Leloir (eds), *Lettres à Julie écrites du donjon de Vincennes, publiées et commentées d'après les manuscrits originaux et inédits* (Paris 1903), p.205.

4. See for example P. Manuel (ed.), *Lettres originales de Mirabeau écrites du donjon de Vincennes, pendant les années 1777, 78, 79, et 80*, 4 vols (Paris, J. B. Garnery 1792), ii.256-59, 309-11, 353-55.

Henri-Joseph Dulaurens singled out Buffon's anecdotes and moralising about animals as especially interesting.[5] Marie Du Deffand, a friend of d'Alembert's, pronounced the series unpalatable for roughly the same reason: 'Buffon est d'une monotonie insupportable; il sait bien ce qu'il sait, mais il ne s'occupe que des bêtes.'[6]

With regard to aspects of the series that were potentially immoral, reactions were also split. Buffon's accounts of reproduction and different animals' 'loves' delighted the relatively free-thinking *Journal encyclopédique*, and provided Diderot and Ferdinando Galiani with matter for bawdy humour.[7] As we have seen, this eroticism was mostly ignored by the *Journal de Trévoux* as well as by prudes such as the youthful Roland. In contrast Buffon's final causes and moralistic conclusions were picked out and amplified in support of religion by the *Journal de Trévoux*, while they were welcomed as fine images and expressions of genius by the *Journal encyclopédique*. In these matters as in others Buffon managed to balance orthodoxy and heterodoxy so as to maintain relations with both the moderately free-thinking and the moderately devout. Erotic it might be, but the *Histoire naturelle* avoids the pointed lewdness of Diderot or La Mettrie. In the eyes of many pious critics, though certainly not all of them, Buffon's arguments against biblical chronology, final causes, and other quasi-Christian dogmas were redeemed by the 'good intentions' manifest in his church-going, his professions of faith, and his literary invocation of final causes. The *philosophes*, for their part, could reverse the interpretation with no less plausibility. As the *Nouvelles ecclésiastiques* argued with regard to Buffon's 'irreligious' *Histoire et théorie de la terre*: 'Ceux qui pensent comme lui savent le concilier avec lui-même, et regardent comme un hors-d'œuvre des traits de catholicité que tout le système dément. Ceux qui pensent différemment, mais qui n'approfondissent rien, se contentent de ces marques équivoques de catholicité et en demeurent là.'[8]

Buffon's success with the *Histoire naturelle* surely owed something to this ambiguous and delicate balance, but the *Nouvelles ecclésiastiques* exaggerates his readers' passiveness in making them the helpless victims of a two-tongued rhetorician. In the case of the Jesuits, active good will, not to mention self-interest, motivated their efforts to include France's greatest naturalist in their Catholic Enlightenment. Many people were clearly too preoccupied with Buffon's writing to succumb to his subversions at the level of ideas. As this

On Mirabeau's admiration for Buffon see Manuel (ed.), *Lettres originales*, ii.413-14; Antonina Vallentin, *Mirabeau*, translated by E. W. Dickes (New York 1948), p.479.

5. [Henri-Joseph Dulaurens], *Imirce, ou la fille de la nature* (Berlin, Philosophe de Sans-souci 1765), p.175.

6. Marie de Vichy-Chamrond Du Deffand to Horace Walpole, 9 March 1777, in Paget Toynbee (ed.), *Lettres de la marquise Du Deffand à Horace Walpole (1766-1780)*, 3 vols (London 1912), iii.308. See also Deffand to Walpole, 6 June 1774, ii.617-18.

7. See for example *Journal encyclopédique* 4.3 (June 1772), p.343-46, 350-52, 5.1 (July 1772), p.46-48, 49-50; Diderot, *Rêve de d'Alembert*, p.129, 195, 205; Galiani to Epinay, 13 June 1772, in Asse (ed.), *Lettres de l'abbé Galiani*, i.351.

8. 6 February 1750, p.21.

study has emphasised, natural history was a literary genre as well as a science, and so it was frequently both read and written, even in the case of works as 'serious' as the *Histoire naturelle*. To much of Buffon's public the matter of his alleged heresies must have seemed as academic as debates over the use of pagan imagery in Christian poetry. Significantly, early anthologies of the *Histoire naturelle* made little effort to respect the coherence of Buffon's major theories, whether they were unorthodox or not at all so – a sign that his ideas were considered peripheral. It is also worth remembering that those issuing warnings about his dangerous, 'bewitching' rhetoric seem to have had no trouble finding the errors it supposedly camouflaged. Such warnings may have been directed toward gullible, lesser readers, of whom a number no doubt existed, but as we have seen, a great many readers were apparently less naive than engaged in a pleasurable 'suspension of disbelief'.

The same point can be made with regard to Buffon's epistemological and polemical rhetoric, the subject of chapters 4, 5, and 6. As noted in chapter 5, the abundance of methodological and epistemological moralising in vulgarisations such as the *Histoire* of the Académie des sciences suggests that it was seen as important to the public image of the sciences. Reviews of the *Histoire naturelle* in eighteenth-century periodicals often duly cited its far-reaching claims about the dangers of systems, mathematics, classification, and abstraction, rarely bothering to analyse, let alone criticise them. Probably such claims about method and knowledge had some persuasive power, especially when premised on commonplaces of Newtonian and sensationistic philosophy, as Buffon's mostly were. Perhaps Buffon even convinced some that he avoided all the epistemological ills he denounced in the series. But as his failure to elude a reputation for systematising shows, he did not always 'bewitch' the public as much as some suggested. Even his admirers regularly rejected his theories.[9] No doubt the *Histoire naturelle* had 'frivolous' readers, 'ces lecteurs peu faits pour méditer, qui, n'aimant que la variété, se rebutent dès que l'ouvrage exige une attention trop suivie', as Ferri described them in *Génie de Buffon*.[10] But frivolity and gullibility are not the same thing. Readers in general were clearly more interested in the series' style, anecdotes, and particular arguments than in the epistemology that supposedly underlay them. Even if susceptible in some sense to scientific rhetoric, the French literary public seems to have read natural histories on its own terms, picking through texts' offerings with noncommittal pragmatism.

If Buffon's stated epistemology appeared plausible though little interesting to many in the public, it was treated with scepticism by critics of the series, especially by natural history's scientific elite. Not only were his claims about knowledge and methodology rejected directly, they were all used against him at one time or another. The enemy of system-building was accused of building

9. See for example Roger, *Buffon*, p.442-45; Buffon to Charles de Brosses, 16 February 1750, Buffon to Suzanne Curchod Necker, 12 July 1782, in Nadault (ed.), *Correspondance générale [de] Buffon*, i.64, ii.131; Manuel (ed.), *Lettres originales*, iii.327-28.

10. [Ferri], intro. to *Génie de Buffon*, p.xxiii.

systems, the enemy of mathematics was accused of rushing inappropriately into mathematical analysis, the enemy of abstraction was accused of being abstract, and the enemy of nomenclature was accused of discussing little else. Buffon's failure to persuade colleagues with his epistemological 'principles' was due in part to the weakness of scientific rhetoric in general. Widespread though they were in the Enlightenment, self-valorising claims about scientific programmes were rarely taken seriously unless uttered by allies or potentially helpful authorities. In part Buffon's failure was personal, a consequence of his ambitious aims, his inconsistency, and his often fragmented presentation. For all his virtuosity as a writer, he never managed to communicate a detailed rationale for his numerous judgements about good and bad forms of knowledge. Above all his criterion for physical truth – the key to his proposed epistemology in Sloan's reconstruction – was never developed beyond a sketchy possibility, probably because Buffon saw the difficulty of going farther. What was visible to contemporaries reading his scattered views on knowledge was a series of charged dichotomies, all of them staples of scientific rhetoric – good systems versus bad systems, reality versus abstraction, mathematics versus physics, etc. It is easy to understand why veterans of scientific quarrelling paid little heed.

This is not to say that Buffon had no epistemology or that the *Histoire naturelle* lacks general coherence. In a sense he had two epistemologies, both of them fairly stable from the 1740s onward. First was the one he professed. Minus some caveats it rejected mathematics, abstraction, taxonomy, finalism, and systems in the name of such virtues as reality, naturalness, and epistemological humility. The elements of this epistemology seem to have developed in the 1730s and 1740s as Buffon struggled with scientific problems and assimilated ideas from a variety of authors and philosophical traditions. The goals of fortifying the 'Premier discours', denouncing Linnaeus, and justifying historical geology probably pushed him toward a synthesis in the 1740s. In its outlines his proposed epistemology was unoriginal. The contrast between abstractions and concrete reality had been drawn for millennia. His arguments against final causes, systems, and taxonomical universals followed lines familiar since the seventeenth century. What was new in this epistemology was his proposal for justifying certain 'good' forms of taxonomy, abstraction, and hypothesis as probabilistically determinant facts of induction, but this was never developed beyond verbal generalities and simple examples. Thanks to its vagueness and Manichean oppositions, Buffon's official epistemology was a perfect weapon for promoting his views and denouncing those of rivals.

Underneath this stylised and highly rhetorical epistemology there were regularities in the way Buffon approached knowledge in the *Histoire naturelle*, some of them in rough harmony with his conscious principles. He avoided all calculus but exploited and trusted in arithmetic, algebra, and probability theory. A thoroughgoing Newtonian in some sense of the word, he saw Newton's law of gravity not as a convenient formalism but as a term-by-term representation and mirror of reality. Not surprisingly attraction became one of his principal models for understanding the universe. He was drawn at the same time to the concrete and the real, to the abstract and the hypothetical. Rationalising his efforts to balance these tendencies may be impossible. Certainly his own claim that his abstractions and hypotheses depended on probabilistically calibrated repeated events was unsubstantiated. His most famous systems were associated with areas beyond feasible observation – sexual reproduction and the Earth's prehistory – while he often refused abstractions when concrete alternatives could be found close at hand, but the haziness of such notions as abstraction and concreteness makes it unwise to pursue the correlation too far. Though his explicit epistemology postulated absolute boundaries between right and wrong ways of knowing, Buffon was implicitly less doctrinaire, perhaps because of his commonsensical, 'realistic' perspective. Notwithstanding his denunciation of teleology, for example, he countenanced certain final causes as worthwhile heuristics if nothing else. Likewise his evocations of laws of plenitude and continuity which he did not literally believe in can sometimes be understood as heuristic gestures. In this sense, like many naturalists of the next generation, he vaguely anticipated Kant's reduction of finality and natural continuity to regulative principles.[11] Such pragmatism makes it hard to isolate principled features of his approach to natural history.

More could be said about the underlying epistemology of the *Histoire naturelle,* but this summary is adequate for illustrating the contrast with Buffon's conscious stances. Clearly these stances expanded on tendencies he sometimes exemplified, but they were more rigid and idealistic than the assumptions going into his work as a naturalist. This was a typical situation in the history of the sciences. All through the Enlightenment thinkers condemned systems, 'unnatural' classification, the glut of nomenclature, and arid abstractions, while commending experience, simplicity, and anything 'natural'. Probably few of these exhortations were consciously hypocritical, but many who pronounced them were accused with good reason of practising what they preached against and not what they preached. Adepts of scientific rhetoric could no doubt reconcile their principles with their practical work, but the exercise required a favourable eye and an appropriate interpretation of words such as 'systematic', 'abstract', 'natural', and 'real'. Authors' claims about epistemology and methodology regularly offered warped views of their approaches. Buffon's claims were more obviously rhetorical than most, perhaps because of his ambitions for the *Histoire naturelle*, but they grew out of a long history of analogous rhetoric.

'Pourquoi mettre de la poésie et faire des suppositions métaphysiques où il ne faut qu'un simple exposé des choses?' asked Galiani midway through his reading of the *Histoire naturelle.*[12] Large parts of the answer to Galiani's

11. See Larson, *Interpreting nature*, p.185-86.
12. Galiani to Epinay, 13 June 1772, in Asse (ed.), *Lettres de l'abbé Galiani*, i.171.

question – actually an accusation – have been sketched out in the preceding chapters. In short Buffon sought to attract and persuade readers. These goals, both rhetorical, are not always easy to disentangle, but his major techniques can be roughly divided. Literariness, intelligibility, and the promise of entertainment won Buffon readers who might otherwise have hesitated to read natural history, while his moralistic pronouncements about good and bad forms of knowledge aimed to bolster his arguments and undermine those of rivals. Whether as persuasion or as marketing to the public, rhetoric went well beyond the *Histoire naturelle*. As we have seen, interested manifestos of method and epistemology were a constant of the sciences in Buffon's time. Tactics for drawing in the public may not have been important in certain publications, but they were critical to authors who sought a broader audience. Buffon, like Réaumur, has for too long been studied as an academic naturalist unconnected with vulgarisers such as Pluche and Beaurieu. This has led to inattention to the audience-pleasing surface of the *Histoire naturelle*. Conversely, 'vulgar' natural history has been too quickly dismissed as naive and insubstantial on the basis of its concessions to genre and audience. Buffon no less than Pluche offered his readers 'silly' material as well as more serious natural history.

Scientific rhetoric, studied as it is here, is not meant as a tool for evoking hypocrisy, for diminishing reputations, or for proving that science is no more than words. It is meant as a complement to intellectual history, where minor authors especially are prone to being pigeon-holed on the basis of a small number of self-conscious pronouncements. When we acknowledge the importance of textual strategies and contrast famous discourses with less focalised texts, it becomes clear that much nominalism, finalism, empiricism, and so on is as much rhetorical as representative of thinkers' positions. In addition studying rhetoric constitutes a means of transcending the historiographical boundary between 'amateur' and 'professional' natural history, between natural history as knowledge and natural history as pedagogical and literary entertainment. These oppositions crudely polarise what should be seen in the Enlightenment as a more gradual range. The age-old allegation that literature and science were incompatible grew shriller and in some ways truer in the eighteenth century, but natural history remained an area where academic and more general interests could coexist under tension.

Works cited

Aaron, R. I., *The Theory of universals* (Oxford 1952).

Adanson, Michel, *Familles des plantes*, 2 vols (Paris, Vincent 1763).

Alembert, Jean Le Rond d', *Encyclopédie, ou dictionnaire raisonné des sciences, des arts et des métiers*, ed. Denis Diderot and Jean Le Rond d'Alembert, 17 vols of text ([Paris] 1751-1765). 'Discours préliminaire', i.i-xxxiv; and articles: 'Astronomie', i.783-93; 'Attraction', i.846-55; 'Attraction des montagnes', i.855-56; 'Fleuve', vi.868-74; 'Fluide', vi. 881-90; 'Géomètre', vii.627-29; 'Mars', x.149-50; 'Newtonianisme', xi.122-25; 'Physico-mathématiques', xii.536-37.

– 'Introduction aux recherches sur la précession des équinoxes, et sur la nutation de l'axe de la terre dans le système newtonien' (1749), in *Œuvres*, 5 vols (Paris 1821-1822), i.437-50.

– *Traité de dynamique* (Paris, David 1758).

Allen, David Elliston, *The Naturalist in Britain: a social history* (New York 1976).

Amoreux, Pierre-Joseph, *Notice des insectes de la France, réputés venimeux* (Paris, Hôtel Serpente 1789).

Anderson, Elisabeth, 'La collaboration de Sonnini de Manoncourt à l'*Histoire naturelle* de Buffon', *SVEC* 120 (1974), p.329-58.

Anderson, Lorin, *Charles Bonnet and the order of the known* (Dordrecht 1982).

Anderson, Wilda C., *Between the library and the laboratory* (Baltimore, Md. 1984).

– 'Error in Buffon', *MLN* 114 (1999), p.691-701.

Argenson, René-Louis de Voyer de Paulmy, marquis d', *Mémoires et journal inédit du marquis d'Argenson*, 5 vols (Paris 1857-1858).

Arnauld, Antoine, and Pierre Nicole, *La Logique, ou l'art de penser* (1662), ed. Emile Charles (Paris 1878).

Ashworth, Jr, William B., 'Natural history and the emblematic world view', in *Reappraisals of the scientific revolution*, ed. David C. Lindberg and Robert S. Westman (Cambridge 1990), p.303-32.

Asse, Eugène (ed.), *Lettres de l'abbé Galiani à Mme d'Epinay, Voltaire, Diderot, le baron d'Holbach, Morellet, Suard, d'Alembert, Marmontel, la vicomtesse de Belsunce, etc.*, 2 vols (Paris 1903).

Atran, Scott, *Cognitive foundations of natural history: towards an anthropology of science* (Cambridge 1990).

– 'The common-sense basis of Buffon's *méthode naturelle*', in *Buffon 88*, p.223-40.

Auroux, Sylvain, *La Sémiotique des encyclopédistes. Essai d'épistémologie historique des sciences du langage* (Paris 1979).

Ayers, Michael, *Locke: epistemology and ontology*, 2 vols (London 1991).

Bachelard, Gaston, *La Formation de l'esprit scientifique. Contribution à une psychanalyse de la connaissance* (Paris 1993).

Badinter, Elisabeth (ed.), *Correspondance inédite de Condorcet et Mme Suard, M. Suard et Garat (1771-1791)* (Paris 1988).

– and Robert Badinter, *Condorcet (1743-1794). Un intellectuel en politique* (Paris 1988).

Baker, Keith Michael, *Condorcet: from natural philosophy to social mathematics* (Chicago 1975).

– 'An unpublished essay of Condorcet on technical methods of classification', *Annals of science* 18 (1962), p.99-123.

Barberet, Denis, 'Dissertation sur la nature et la formation de la grêle', *Mémoires de l'Académie de Dijon* 1 (1769), p.1-32.

Barruel, Augustin, *Les Helviennes, ou lettres provinciales philosophiques* (1781), 4 vols (Paris 1823).

Barsanti, Giulio, 'Buffon et l'image de la nature. De l'échelle des êtres à la carte géographique et à l'arbre généalogique', in *Buffon 88*, p.255-96.
– 'Linné et Buffon. Deux visions différentes de la nature et de l'histoire naturelle', *Revue de synthèse* 113-14 (1984), p.83-111.
– *La Scala, la mappa, l'albero: immagini e classificazioni della natura fra Sei e Ottocento* (Florence 1992).
Bassy, Alain-Marie, 'A l'heure des grandes synthèses. L'œuvre de Buffon à l'Imprimerie royale, 1749-1789', in *L'Art du livre à l'Imprimerie nationale* (Paris 1973), p.170-89.
– 'Le texte et l'image', in *Histoire de l'édition française*, ed. Henri-Jean Martin, Roger Chartier, and Jean-Pierre Vivet, 4 vols (Paris 1984), vol. ii, *Le Livre triomphant, 1660-1830*, p.140-61.
Bazin, Gilles-Auguste, *Abrégé de l'histoire des insectes pour servir de suite à l'histoire naturelle des abeilles*, 4 vols (Paris, Guerin 1747-1751).
[Beaurieu, Gaspard Guillard de], *Abrégé de l'histoire des insectes dédié aux jeunes personnes*, 2 vols (Paris, Panckoucke 1764).
Beer, Gillian, '"The face of nature": anthropomorphic elements in the language of *The Origin of species*', in *Languages of nature: critical essays on science and literature*, ed. L. J. Jordanova (New Brunswick, N.J. 1986), p.207-43.
Beeson, David, *Maupertuis: an intellectual biography*, *SVEC* 299 (1992).
Belaval, Yvon, 'La crise de la géométrisation de l'univers dans la philosophie des Lumières', *Revue internationale de philosophie* 6 (1952), p.337-55.
– 'Leibniz et la chaîne des êtres', *Analecta Husserliana* 11 (1981), p.59-68.
Bender, John, 'Enlightenment fiction and the scientific hypothesis', *Representations* 61 (1998), p.6-28.
Bensaude-Vincent, Bernadette, *Lavoisier. Mémoires d'une révolution* (Paris 1993).
Beretta, Marco, *The Enlightenment of matter* (Canton, Mass. 1993).
Bernardin de Saint-Pierre, Jacques-Henri, *Etudes de la nature* (1784-1788), in *Œuvres complètes*, ed. L. Aimé-Martin, 12 vols (Paris 1818), vol. iii-vii.
– *Paul et Virginie* (1788), in *Œuvres complètes*, vol. vi.
– *Voyage à l'Ile-de-France* (1773), in *Œuvres complètes*, vol. i-ii.
Bernoulli, Daniel, 'Quelle est la cause physique de l'inclinaison des plans des orbites des planètes' (1735), in *Werke*, ed. D. Speiser *et al.* (Basel 1987), iii.303-26.
Bexon, Gabriel-Léopold, 'Discours sur les phénomènes de l'univers', BMHN, Ms. 804.
Binet, Jacques-Louis and Jacques Roger (eds), *Un autre Buffon* (Paris 1977).
Blanckaert, Claude, 'La valeur de l'homme. L'idée de nature humaine chez Buffon', in *Buffon 88*, p.583-600.
Blunt, Wilfrid, *The Compleat naturalist: a life of Linnaeus* (New York 1971).
Bongie, Laurence L., 'A new Condillac letter and the genesis of the *Traité des sensations*', *Journal of the history of ideas* 16 (1978), p.83-94.
Bonnet, Charles, *Considérations sur les corps organisés*, 2 vols (Amsterdam, Marc-Michel Rey 1762).
– *Contemplation de la nature* (1764), in *Œuvres d'histoire naturelle et de philosophie* (Neuchâtel, Samuel Fauche 1779-1783), vol. vii-ix.
– *La Palingénésie philosophique, ou idées sur l'état passé et sur l'état futur des êtres vivants*, 2 vols (Geneva, Philibert and Chirol 1769).
– *Traité d'insectologie* (1745), in *Œuvres d'histoire naturelle et de philosophie*, vol. i.
Bos, Henk J. M., *Lectures in the history of mathematics* (Providence, R.I. 1993).
– 'Mathematics and rational mechanics', in *The Ferment of knowledge*, ed. G. S. Rousseau and Roy Porter (Cambridge 1980), p.327-55.
Bouchard, Marcel, *De l'humanisme à l'Encyclopédie. L'esprit public en Bourgogne sous l'Ancien Régime* (Paris 1930).
Bowler, Peter, 'Bonnet and Buffon: theories of generation and the problem of species', *JHB* 6 (1973), p.259-81.

Boyle, Robert, *A Disquisition about the final causes of natural things* (London, John Taylor 1688).

Bradley, Richard, *A Philosophical account of the works of nature* (London, W. Mears 1721).

Brahimi, Denise, 'La sexualité dans l'anthropologie humaniste de Buffon', *Dix-huitième siècle* 12 (1980), p.113-26.

Bremner, Geoffrey, 'Buffon and the casting out of fear', *SVEC* 205 (1982), p.75-88.

Brewer, Daniel, *The Discourse of Enlightenment in eighteenth-century France: Diderot and the art of philosophizing* (Cambridge 1993).

Brian, Eric, 'L'Académie royale des sciences de l'absolutisme à la Révolution', in *Histoire et mémoire de l'Académie des sciences. Guide de recherches*, ed. Eric Brian and Christiane Demeulenaere-Douyère (Paris 1996), p.15-32.

Broberg, Gunnar, 'The broken circle', in *The Quantifying spirit in the eighteenth century*, ed. Tore Frängsmyr, J. L. Heilbron and Robin E. Rider (Berkeley, Calif. 1990), p.45-71.

– '*Homo sapiens*: Linnaeus's classification of man', in *Linnaeus: the man and his work*, ed. Tore Frängsmyr, revised edn (Canton, Mass. 1994), p.156-94.

Brockliss, Lawrence, and Colin Jones, *The Medical world of early modern France* (Oxford 1997).

Bru, Bernard, 'Laplace et la critique probabiliste des mesures géodésiques', in *La Figure de la terre du XVIIIe siècle à l'ère spatiale*, ed. Henri Lacombe and Pierre Costabel (Paris 1988), p.223-44.

– 'Statistique et bonheur des hommes', *Revue de synthèse* 109 (1988), p.69-95.

Bruneau, Charles, 'Buffon et le problème de la forme', in *Œuvres philosophiques de Buffon*, ed. Jean Piveteau (Paris 1954), p.491-99.

Brunet, Pierre, *Les Physiciens hollandais et la méthode expérimentale en France au XVIIIe siècle* (Paris 1926).

– *La Vie et l'œuvre de Clairaut (1713-1765)* (Paris 1952).

Buffier, Claude, *Eléments de métaphysique à la portée de tout le monde* (1725), in *Cours de sciences sur des principes nouveaux et simples* (Paris, Cavelier and Giffart 1732), p.893-936.

– *Traité des premières vérités et de la source de nos jugements* (1724), in *Cours de sciences*, p.553-744.

– *Traité philosophique et pratique d'éloquence* (1728), in *Cours de sciences*, p.293-419.

Buffon, Georges-Louis Leclerc de, 'Addition au mémoire qui a pour titre: Réflexions sur la loi de l'attraction', *MARS* 1745 (1749), p.551-52.

– *Correspondance générale [de] Buffon*, ed. H. Nadault de Buffon, intro. by J.-L. de Lanessan, 2 vols (Geneva 1971), i.i-xx.

– *Herrn von Buffons Allgemeine Naturgeschichte*, 7 vols (Troppau, Joseph Georg Trassler 1785).

– *Histoire naturelle des minéraux*, 5 vols (Paris 1783-1788).

– 'Invention de miroirs ardents pour brûler à une grande distance', *MARS* 1747 (1752), p.82-101.

– 'Lettres de Buffon et de Maupertuis adressées à Jalabert', ed. E. Ritter, *Revue d'histoire littéraire de la France* 8 (1901), p.650-56.

– *Natural history, general and particular, by the Count de Buffon*, trans. and ed. William Smellie (1780-1785), 4th edn, 10 vols (London 1812).

– preface to Isaac Newton, *La Méthode des fluxions et des suites infinies*, translated by Buffon (Paris, Debure 1740), p.iii-xxx.

– preface to Stephen Hales, *Statique des végétaux*, translated by Buffon (Paris, Debure 1735), p.iii-viii.

– 'Réflexions sur la loi de l'attraction', *MARS* 1745 (1749), p.493-500.

– 'Seconde addition au mémoire qui a pour titre: Réflexions sur la loi de l'attraction', *MARS* 1745 (1749), p.580-83.

– *Supplément à l'histoire naturelle*, 7 vols (Paris, Imprimerie royale 1774-1789).

– and Louis-Jean-Marie Daubenton, *Histoire naturelle, générale et particulière, avec*

la description du Cabinet du roi, 15 vols (Paris, Imprimerie royale 1749-1767).
– and Philibert Guéneau de Montbeillard, *Histoire naturelle des oiseaux*, 9 vols (Paris, Imprimerie royale 1770-1783).
Buffon 88. Actes du colloque international pour le bicentenaire de la mort de Buffon (Paris, Montbard, Dijon, 14-22 juin 1988), ed. Jean Gayon *et al.* (Paris 1992).
Bulliard, Pierre, 'METHODE botanique', in *Dictionnaire élémentaire de botanique*, 3 vols (Paris, Didot *et al.* 1783), ii.108-20.
Burke, Kenneth, *A Rhetoric of motives* (New York 1950).
Burkhardt, Jr, Richard W., 'Le comportement animal et l'idéologie de domestication chez Buffon et chez les éthologues modernes', in *Buffon 88*, p.569-82.
– *The Spirit of system: Lamarck and evolutionary biology*, revised edn (Cambridge, Mass. 1995).
Burrell, Paul, 'Pierre Bayle's *Dictionnaire historique et critique*', *SVEC* 194 (1981), p.83-103.

Cain, A. J., 'Logic and memory in Linnaeus's system of taxonomy', *Proceedings of the Linnaean Society of London* 169 (1958), p.144-63.
Callot, Emile, 'Système et méthode dans l'histoire de la botanique', *RHS* 18 (1965), p.45-53.
Carr, J., 'The biological work of Martin Lister (1639-1712)', Doctoral dissertation, University of Leeds, 1974.
Carr, Jr, Thomas M., *Descartes and the resilience of rhetoric: varieties of Cartesian rhetorical theory* (Carbondale, Ill. 1990).
Carramolino del Valle, David, 'El espejo ustorio de Buffon: entre la leyenda de Siracusa y la tradición del análisis químico', *Asclepio* 45 (1993), p.269-89.
Casini, Paolo, 'Buffon et Newton', in *Buffon 88*, p.299-308.
– 'D'Alembert, l'économie des principes et la "métaphysique des sciences"', in *Jean d'Alembert, savant et philosophe. Portrait à plusieurs voix*, ed. Monique Emery and Pierre Monzani (Paris 1989), p.135-52.
Cassirer, Ernst, *The Philosophy of the Enlightenment*, translated by Fritz C. A. Koelln and James P. Pettegrove (Princeton, N.J. 1979).
Castiglioni, Arturo, *A History of medicine*, translated by E. B. Krumbhaar, 2nd edn (New York 1947).
Chaline, Jean, 'Evolution des concepts de l'espèce et de la formation des espèces', in *Buffon 88*, p.515-25.
Chandler, Philip, 'Clairaut's critique of Newtonian attraction: some insights into his philosophy of science', *Annals of science* 32 (1975), p.369-78.
Chateaubriand, François-René de, *Mémoires d'outre-tombe*, ed. Maurice Levaillant, 4 vols (Paris 1948).
Chatelain, N., *Pastiches, ou imitations libres du style de quelques écrivains du XVIIe et XVIIIe siècles* (Paris 1855).
Cherwitz, Richard A. and Thomas J. Darwin, 'Beyond reductionism in rhetorical theories of meaning', *Philosophy and rhetoric* 27 (1994), p.313-29.
Christmas, William, '"Beautifully coloured from nature": color illustration, the periodical press and the popularization of natural history', presentation at the 30th annual meeting of the American Society for Eighteenth-Century Studies, 25 March 1999.
Clairaut, Alexis-Claude, 'Du système du monde dans les principes de la gravitation universelle', *MARS* 1745 (1749), p.329-64.
– 'Réponse à la réplique de M. de Buffon', *MARS* 1745 (1749), p.578-80.
– 'Réponse au nouveau mémoire de M. de Buffon', *MARS* 1745 (1749), p.583-86.
– 'Réponse aux réflexions de M. de Buffon sur la loi de l'attraction et sur le mouvement des apsides', *MARS* 1745 (1749), p.529-48.
Clavillart, Narcisse, 'Description botanico-mathématique de la Clavi-lakea', BMHN, Ms. 1063.
Cohen, I. Bernard, 'Hypotheses in Newton's philosophy', *Physis* 8 (1966), p.163-84.

Cohen, Jonathan, 'Teleological explanation', in *Purposive behaviour and teleological explanations*, ed. Frank George and Les Johnson (New York 1985), p.40-68.

Condillac, Etienne Bonnot de, *La Langue des calculs* (1798), in *Œuvres philosophiques de Condillac*, ed. Georges Le Roy, 3 vols (Paris 1947-1951), ii.417-529.

– *Traité des animaux* (1755), in *Œuvres philosophiques*, i.337-79.

– *Traité des sensations* (1754), in *Œuvres philosophiques*, i.219-319.

– *Traité des systèmes* (1749), in *Œuvres philosophiques*, i.119-217.

Conley, Thomas M., *Rhetoric in the European tradition* (New York 1990).

Connell, Richard J., *Nature's causes* (New York 1995).

Cooter, Roger and Stephen Pumfrey, 'Separate spheres and public places: reflections on the history of science popularization and science in popular culture', *History of science* 32 (1994), p.237-67.

Correspondance littéraire, philosophique et critique de Grimm et de Diderot, 16 vols (Paris 1829).

Corsi, Pietro, *The Age of Lamarck*, translated by Jonathan Mandelbaum (Berkeley, Calif. 1988).

– 'Buffon sous la Révolution et l'Empire', in *Buffon 88*, p.639-48.

Crombie, A. C., *Styles of scientific thinking in the European tradition*, 3 vols (London 1994).

Curran, Andrew, and Patrick Graille, 'The faces of eighteenth-century monstrosity', *Eighteenth-century life* 21 (1997), p.1-15.

Dacier, Anne Lefèvre, *Des causes de la corruption du goût* (Paris, Rigaud 1714).

Darnton, Robert, *The Literary underground of the Old Regime* (Cambridge, Mass. 1982).

Daston, Lorraine, *Classical probability in the Enlightenment* (Princeton, N.J. 1988).

– 'Marvelous facts and miraculous evidence in early modern Europe', *Critical inquiry* 18 (1991), p.92-124.

Daubenton, Louis-Jean-Marie, 'Botanique', in *Encyclopédie*, ii.340-45.

– 'Coquille', in *Encyclopédie*, iv.184-92.

– 'Cours de zoologie (mammifères), par Daubenton, professé après 1786, soit à l'Ecole normale, soit au Collège de France, soit à l'Ecole d'Alfort', BMHN, Ms. 809.

– 'Histoire naturelle', in *Séances des écoles normales, recueillies par des sténographes et revues par les professeurs. Leçons*, new edn, 8 vols (Paris, Imprimerie du Cercle-social 1796-1798).

– 'Histoire naturelle', in *Séances des écoles normales recueillies par des sténographes et revues par les professeurs. Débats*, new edn, 3 vols (Paris, Imprimerie du Cercle-social 1800-1801).

Daudin, Henri, *De Linné à Jussieu. Méthodes de la classification et idée de série en botanique et en zoologie (1740-1790)* (Paris [1926]).

Dawson, Virginia P., *Nature's enigma: the problem of the polyp in the letters of Bonnet, Trembley and Réaumur* (Philadelphia, Penn. 1987).

Dear, Peter, *Discipline and experience: the mathematical way in the scientific revolution* (Chicago 1995).

– (ed.), intro. to *The Literary structure of scientific argument* (Philadelphia, Penn. 1991), p.1-9.

– 'Miracles, experiments, and the ordinary course of nature', *Isis* 81 (1990), p.663-83.

Delaporte, François, *Le Second règne de la nature* (Paris 1979).

Delisle de Sales, Jean-Claude-Izoard, *De la philosophie de la nature*, 3 vols (Amsterdam, Arkstée & Merkus 1770).

Delsol, Michel, and Janine Flatin, 'L'espèce existe-t-elle? question aux paléontologues et aux philosophes', in *Buffon 88*, p.529-39.

Déré, Anne-Claire, 'La réception de la nomenclature reformée par le corps médical français', in *Lavoisier in European context: negotiating a new language for chemistry*, ed. Bernadette Bensaude-Vincent and Ferdinando Abbri (Canton, Mass. 1995), p.207-24.

Descartes, René, *Discours de la méthode* (1637), in *Œuvres*, ed. Charles Adam and Paul Tannery, 9 vols in 11 pts (Paris, 1964-1974), vi.1-78.

– *Meditationes de prima philosophia* (1640), in *Œuvres*, vol. vii.

Dewhurst, Kenneth, *John Locke (1632-1704), physician and philosopher: a medical biography* (London 1963).

Dhombres, Jean, 'Quelques rencontres de Diderot avec les mathématiques', in *Colloque international Diderot (1713-1784), Paris-Sèvres-Reims-Langres, 4-11 juillet 1984*, ed. Anne-Marie Chouillet (Paris 1985), p.269-80.

Diderot, Denis, 'Animal', in *Encyclopédie*, i.469-74.

– *Lettre sur les aveugles* (1749), in *DPV*, iv.1-107.

– *Pensées sur l'interprétation de la nature* (1753), in *DPV*, ix.1-111.

– *Le Rêve de d'Alembert* (1831), in *DPV*, xvii.23-209.

Di Meo, Antonio, 'Macquer e Buffon: una controversia scientifica del Settecento', *Cultura e scuola* 29 (1990), p.240-60.

Doig, Kathleen Hardesty, 'The Yverdon *Encyclopédie*', *SVEC* 315 (1994), p.85-116.

Douglas, Aileen, 'Popular science and the representation of women: Fontenelle and after', *Eighteenth-century life* 18 (1994), p.1-14.

Drouin, Jean-Marc and Bernadette Bensaude-Vincent, 'Nature for the people', in *Cultures of natural history*, ed. N. Jardine, J. A. Secord and E. C. Spary (Cambridge 1996), p.408-25.

Du Châtelet, Gabrielle-Emilie Le Tonnelier de Breteuil, marquise, *Institutions physiques. Nouvelle édition (1742)*, ed. Sonia Carboncini (Hildesheim 1988).

Duchesneau, François, 'Buffon et la physiologie', in *Buffon 88*, p.451-62.

– *Philosophie de la biologie* (Paris 1997).

– *La Physiologie des Lumières. Empirisme, modèles et théories* (The Hague 1982).

Duchez, Marie-Elisabeth, 'D'Alembert diffuseur de la théorie harmonique de Rameau. Déduction scientifique et simplification musicale', in *Jean d'Alembert, savant et philosophe. Portrait à plusieurs voix*, ed. Monique Emery and Pierre Monzani (Paris 1989), p.475-96.

Du Deffand, Marie de Vichy-Chamrond, marquise, *Lettres de la marquise Du Deffand à Horace Walpole (1766-1780)*, ed. Paget Toynbee, 3 vols (London 1912).

[Dulaurens, Henri-Joseph], *Imirce, ou la fille de la nature* (Berlin, Philosophe de Sans-souci 1765).

Du Marsais, César Chesneau, *Des tropes, ou des différents sens dans lesquels on peut prendre un même mot dans une même langue* (1730), in *DES TROPES ou des différents sens, FIGURE et vingt autres articles de l'Encyclopédie, suivis de L'ABREGE DES TROPES de l'abbé Ducros*, ed. Françoise Douay-Soublin (Paris 1988).

Duris, Pascal, *Linné et la France (1780-1850)* (Geneva 1993).

Dutka, Jacques, 'On the St Petersburg paradox', *Archive for history of exact sciences* 39 (1988), p.13-39.

Eddy, Jr, John H., 'Buffon's *Histoire naturelle*: history? A critique of recent interpretations', *Isis* 85 (1994), p.644-61.

Ehrard, Jean, 'Diderot, l'*Encyclopédie*, et l'*Histoire et théorie de la terre*', in *Buffon 88*, p.135-42.

– *L'Idée de nature en France dans la première moitié du XVIIIe siècle*, revised edn (Paris 1994).

Eisenstein, Elizabeth L., *The Printing press as an agent of change*, 2 vols (Cambridge 1979).

Ellenberger, François, 'Les sciences de la terre avant Buffon', in *Buffon 88*, p.327-42.

Encyclopédie, ou dictionnaire raisonné des sciences, des arts et des métiers, ed. Denis Diderot and Jean Le Rond d'Alembert, 17 vols of text ([Paris] 1751-1765).

Eriksson, Gunnar, 'Linnaeus the botanist', in *Linnaeus: the man and his work*, ed. Tore Frängsmyr, revised edn (Canton, Mass. 1994), p.63-109.

'Essais sur l'étude des montagnes', *Observations et mémoires sur la physique, sur l'his-*

toire naturelle, et sur les arts et métiers 2 (November 1773), p.416-32.

Falconet, Camille, intro. to Bernard Le Bovier de Fontenelle, *Théorie des tourbillons cartésiens; avec des réflexions sur l'attraction* (Paris, Hippolyte-Louis Guérin 1752), p.iii-xxxi.

Farber, Paul Lawrence, 'Buffon and the concept of species', *JHB* 5 (1972), p.259-84.

– *Discovering birds: the emergence of ornithology as a scientific discipline, 1768-1850* (Baltimore, Md. 1997).

Félice, Fortuné-Barthélemy de, 'Phénomène', in *Supplément à l'Encyclopédie*, ed. Jean-Baptiste Robinet, 5 vols (Amsterdam, Rey 1777), iv.319-22.

Fellows, Otis E., and Stephen F. Milliken, *Buffon* (New York 1972).

Fénelon, François de Salignac de La Mothe-, *Traité de l'existence et des attributs de Dieu* (1713), in *Œuvres de Fénelon* (Versailles 1820), i.1-284.

Ferrari, Jean, 'Kant, lecteur de Buffon', in *Buffon 88*, p.155-63.

[Ferri de Saint-Constant, Jean-L.], intro. to *Génie de M. de Buffon* (Paris, Panckoucke 1778), p.v-xxiii.

Feyerabend, Paul K., 'Classical empiricism', in *The Methodological heritage of Newton*, ed. Robert E. Butts and John W. Davis (Toronto 1970), p.150-70.

Finch, Robert, and John Elder (eds), intro. to *The Norton book of nature writing* (New York 1990), p.19-28.

Findlen, Paula, 'Francis Bacon and the reform of natural history in the seventeenth century', in *History and the disciplines: the reclassification of knowledge in early modern Europe*, ed. Donald R. Kelley (Rochester, N.Y. 1999), p.239-60.

– *Possessing nature: museums, collecting, and scientific culture in early modern Italy* (Berkeley, Calif. 1994).

– review of Scott Atran, *Cognitive foundations of natural history*, *JHB* 24 (1991), p.537-40.

Flourens, Pierre, *Histoire des travaux et des idées de Buffon* (Paris 1844).

– *Des manuscrits de Buffon avec des fac-simile de Buffon et de ses collaborateurs* (Paris 1860).

Fontenelle, Bernard Le Bovier de, *Eléments de la géométrie de l'infini* (1727), ed. Michel Blay and Alain Niderst (Paris 1995).

– *Théorie des tourbillons cartésiens; avec des réflexions sur l'attraction* (Paris, Hippolyte-Louis Guérin 1752).

Formey, Jean-Henri-Samuel, *La Belle Wolfienne*, 6 vols (The Hague, Jean Neaulme 1741-1753).

Foucault, Michel, *Les Mots et les choses* (Paris 1966).

– 'La situation de Cuvier dans l'histoire de la biologie', *RHS* 23 (1970), p.63-92.

France, Peter, *Politeness and its discontents* (Cambridge 1990).

– *Rhetoric and truth in France: Descartes to Diderot* (Oxford 1972).

Fréchet, Maurice, 'Buffon comme philosophe des mathématiques', in *Œuvres philosophiques de Buffon*, ed. Jean Piveteau (Paris 1954), p.435-46.

French, Roger, *Ancient natural history* (London 1994).

– *William Harvey's natural philosophy* (Cambridge 1994).

Fréron, Elie-Catherine, *Lettres sur quelques écrits de ce temps* (1749-1754).

Fries, Th. M. (ed.), *Bref till och från Abraham Bäck*, in *Bref och Skrifvelser af och till Carol von Linné* (Stockholm 1910), ser. 1, vol. iv.

Fritzell, Peter A., *Nature writing in America* (Ames, Ia. 1990).

Fuchs, Max, 'La langue des sciences', in *Histoire de la langue française des origines à nos jours*, ed. Ferdinand Brunot (Paris 1966), vol. vi, pt 1, fasc. 2.

Fuller, Steve, '"Rhetoric of science": a doubly vexed expression', *Southern communication journal* 58 (1993), p.306-11.

– 'The strong program in the rhetoric of science', in *Science, reason, and rhetoric*, ed. Henry Krips, J. E. McGuire and Trevor Melia (Pittsburgh, Penn. 1995), p.95-117.

Furetière, Antoine, 'Système', in *Dictionnaire universel*, 4th edn, ed. Henri Basnage de Beauval and Jean-Baptiste Brutel de La Rivière, 4 vols (The Hague, Husson *et al.* 1727), iv.Ggg2*v*-Ggg3*r*.

Galera, Andrés, and Marcelo Frías, 'Félix de Azara y Georges Lucien Leclerc: dos formas de *iluminar* la naturaleza americana', *Asclepio* 48 (1996), p.27-36.

Gaonkar, Dilip Parameshwar, 'The idea of rhetoric in the rhetoric of science', *Southern communication journal* 58 (1993), p.258-95.

Garrison, Fielding H., *An Introduction to the history of medicine*, 4th edn (Philadelphia 1929).

Garrone, Alessandro Galante, *Gilbert Romme. Histoire d'un révolutionnaire (1750-1795)*, translated by Anne and Claude Manceron (Paris 1971).

Gascar, Pierre, *Buffon* (Paris 1983).

Gaukroger, Stephen, *Descartes: an intellectual biography* (Oxford 1995).

Gay, Peter, *The Enlightenment: an interpretation*, 2 vols, vol. i, *The Rise of modern paganism* (New York 1966); vol. ii, *The Science of freedom* (New York 1969).

Genet-Varcin, E. and Jacques Roger, 'Bibliographie de Buffon', in *Œuvres philosophiques de Buffon*, ed. Jean Piveteau (Paris 1954), p.513-70.

Genette, Gérard, 'La rhétorique restreinte', in *Figures III* (Paris 1972), p.21-40.

Genlis, Caroline-Stéphanie-Félicité Du Crest, comtesse de, *Mémoires inédits de Mme la comtesse de Genlis, sur le dix-huitième siècle et la Révolution française*, 10 vols (Paris 1825).

'Geology', in *Encyclopaedia Britannica*, 4th edn (Edinburgh 1810), ix.550-628.

Gérardin, Sébastien, 'Cours élémentaire de l'histoire naturelle des animaux, ou leçons extraites des plus célèbres naturalistes', 2 vols, BMHN, Ms. 1537-38.

Gillispie, Charles Coulston, *The Edge of objectivity: an essay in the history of scientific ideas* (Princeton, N.J. 1960).

Gohau, Gabriel, 'La *Théorie de la terre* de 1749', in *Buffon 88*, p.343-52.

Golinski, J. V., 'Language, discourse, and science', in *Companion to the history of modern science*, ed. R. C. Colby *et al.* (London 1990), p.110-23.

Gómez Torres, Ana María, *Las Ideas de Buffon sobre retórica y poética en los inicios de la teoría literaria moderna* (Málaga 1996).

Goodman, David, *Buffon's natural history* (Milton Keynes 1980).

Gough, J. B., 'Réaumur, René-Antoine Ferchault de', in *Dictionary of scientific biography*, ed. Charles Coulston Gillispie, 16 vols (New York 1970-1980), xi.327-35.

Greene, Edward Lee, *Landmarks of botanical history*, ed. Frank N. Egerton, 2 vols (Stanford, Calif. 1983).

Grimsley, Ronald, *Jean d'Alembert, 1717-83* (Oxford 1963).

Grinevald, Paul-Marie, 'Les éditions de l'*Histoire naturelle*', in *Buffon 88*, p.631-37.

Gross, Alan G., *The Rhetoric of science* (Cambridge, Mass. 1990).

Guénon, Anne-Sylvie, 'Le XVIIIe siècle', in *Histoire et mémoire de l'Académie des sciences. Guide de recherches*, ed. Eric Brian and Christiane Demeulenaere-Douyère (Paris 1996), p.113-27.

Guettard, Jean-Etienne, 'Description de deux espèces de nids singuliers faits par des chenilles', *MARS* 1749 (1753), p.163-205.

– 'Mémoire sur les causes qui peuvent rendre les chevaux poussifs', *MARS* 1745 (1749), p.80-92.

– 'Mémoire sur quelques montagnes de la France qui ont été des volcans', *MARS* 1752 (1756), p.27-59.

– *Observations sur les plantes*, 2 vols (Paris, Durand 1747).

Guillaume, James, 'Le berger Daubenton. Encore une légende contre-révolutionnaire', in *Etudes révolutionnaires* (Paris 1908-1909), i.178-97.

Hagner, Michael, 'Enlightened monsters', in *The Sciences in Enlightened*

Europe, ed. William Clark, Jan Golinski, and Simon Schaffer (Chicago 1999), p.175-217.

Hahn, Roger, *The Anatomy of a scientific institution: the Paris Academy of Sciences, 1666-1803* (Berkeley, Calif. 1971).

Haller, Albrecht von, *Tagebuch seiner Beobachtungen*, 2 vols (Bern, in der Hallerschen Buchhandlung 1787).

Hampson, Norman, *The Enlightenment* (New York 1968).

Hankins, Thomas L., *Jean d'Alembert: science and the Enlightenment* (Oxford 1970).

– *Science and the Enlightenment* (Cambridge 1985).

Hanks, Lesley, *Buffon avant l'"Histoire naturelle'* (Paris 1966).

– 'Buffon et les fusées volantes', *RHS* 14 (1961), p.137-54.

Hanson, N. R., 'Hypotheses fingo', in *The Methodological heritage of Newton*, ed. Robert E. Butts and John W. Davis (Toronto 1970), p.14-33.

Harré, Rom, 'Knowledge', in *The Ferment of knowledge*, ed. G. S. Rousseau and Roy Porter (Cambridge 1980), p.1-54.

Harris, Randy Allen (ed.), intro. to *Landmark essays on rhetoric of science: case studies* (Mahwah, N.J. 1997), p.xi-xlv.

Harrison, Peter, *The Bible, protestantism, and the rise of natural science* (Cambridge 1998).

Harth, Erica, *Cartesian women: versions and subversions of rational discourse in the Old Regime* (Ithaca, N.Y. 1992).

– *Ideology and culture in seventeenth-century France* (Ithaca, N.Y. 1983).

Hartshorne, Charles and Paul Weiss (eds), *Collected papers of Charles Sanders Peirce*, 6 vols (Cambridge, Mass. 1931-1935).

Harvey, Joy, 'Buffon and nineteenth-century French anthropologists', in *Buffon 88*, p.649-65.

Hatfield, Gary, 'Metaphysics and the new science', in *Reappraisals of the scientific revolution*, ed. David C. Lindberg and Robert S. Westman (Cambridge 1990), p.93-166.

Hayes, Julie C., *Reading the French Enlightenment: system and subversion* (Cambridge 1999).

– 'Sequence and simultaneity in Diderot's *Promenade Vernet* and *Leçons de clavecin*', *Eighteenth-century studies* 29 (1996), p.291-305.

Hazard, Paul, *European thought in the eighteenth century*, translated by J. Lewis May (Cleveland, Ohio 1963).

Heilbron, J. L., 'Experimental natural philosophy', in *The Ferment of knowledge*, ed. G. S. Rousseau and Roy Porter (Cambridge 1980), p.357-87.

– intro. to *The Quantifying spirit in the eighteenth century*, ed. Tore Frängsmyr, J. L. Heilbron and Robin E. Rider (Berkeley, Calif. 1990), p.1-23.

Helvétius, Claude-Adrien, *De l'homme* (London, Société typographique 1773).

– *Correspondance générale d'Helvétius*, ed. Alan Dainard, *et al.*, 3 vols (Toronto 1981-1991).

Hine, Ellen McNiven, 'Dortous de Mairan and eighteenth century "systems theory"', *Gesnerus* 52 (1995), p.54-65.

Histoire de l'Académie royale des sciences (Paris 1702-1797).

'Histoire naturelle', in *Encyclopédie*, viii.225-30.

Hobbes, Thomas, *Leviathan, or the matter, form, and power of a commonwealth ecclesiastical and civil* (1651), ed. J. C. A. Gaskin (Oxford 1996).

Hodge, M. J. S., 'Two cosmogonies (theory of the Earth and theory of generation) and the unity of Buffon's thought', in *Buffon 88*, p.241-54.

Höhne, Ernst, *Der Stil Buffons: eine stilistische Untersuchung der 'Histoire naturelle, générale et particulière'* (Marburg 1914).

Holgate, Philip, 'Studies in the history of probability and statistics XXXIX: Buffon's cycloid', *Biometrika* 68 (1981), p.712-16.

Hooke, Robert, *Micrographia, or some physiological descriptions of minute bodies made with magnifying glasses* (London, J. Martyn and J. Allestry 1665).

Houston, John Porter, *The Traditions of French prose style* (Baton Rouge, La. 1981).

'Hypothèse', in *Encyclopédie*, viii.417-18.

Ibrahim, Annie, 'La pensée de Buffon. Système ou anti-système?', in *Buffon 88*, p.175-90.

Jardine, Nicholas, 'Epistemology of the sciences', in *The Cambridge history of Renaissance philosophy*, ed. Charles B. Schmitt *et al.* (Cambridge 1988), p.685-711.

–, J. A. Secord and E. C. Spary (eds), *Cultures of natural history* (Cambridge 1996).

Jefferson, Thomas, *Notes on the state of Virginia* (1785), ed. William Peden (Chapel Hill, N.C. 1954).

Joly, André, intro. to James Harris, *Hermès, ou recherches philosophiques sur la grammaire universelle*, translated by François Thurot (Geneva 1972), p.1-144.

Jorland, Gérard, 'The Saint Petersburg paradox, 1713-1937', in *The Probabilistic revolution*, ed. Lorenz Krüger, Lorraine J. Daston and Michael Heidelberger, 2 vols (Cambridge, Mass. 1987), i.157-90.

Journal des savants (Paris 1665-1792).

Journal encyclopédique (Liège and Bouillon 1756-1794).

Jurin, James, *The Correspondence of James Jurin (1684-1750): physician and secretary to the Royal Society*, ed. Andrea A. Rusnock (Amsterdam 1996).

Jussieu, Bernard de, 'Histoire d'une plante, connue par les botanistes sous le nom de *Pilularia*', *MARS* 1739 (1741), p.240-56.

Kafker, Frank A., and Serena L. Kafker, *The Encyclopedists as individuals: a biographical dictionary of the authors of the 'Encyclopédie'*, *SVEC* 257 (1988).

Kitcher, Philip, 'The cognitive functions of scientific rhetoric', in *Science, reason, and rhetoric*, ed. Henry Krips, J. E. McGuire and Trevor Melia (Pittsburgh, Penn. 1995), p.47-66.

– 'The naturalists return', *The Philosophical review* 101 (1992), p.53-114.

– 'Persuasion', in *Persuading science: the art of scientific rhetoric*, ed. Marcello Pera and William R. Shea (Canton, Mass. 1991), p.3-27.

Kline, Morris, *Mathematical thought from ancient to modern times* (New York 1972).

Knight, Isabel F., *The Geometric spirit: the abbé de Condillac and the French Enlightenment* (New Haven, Conn. 1968).

Koerner, Lisbet, 'Carl Linnaeus in his time and place', in *Cultures of natural history*, ed. N. Jardine, J. A. Secord and E. C. Spary (Cambridge 1996), p.145-62.

Krantz, Emile, *Essai sur l'esthétique de Descartes. Rapports de la doctrine cartésienne avec la littérature classique française au XVIIe siècle*, 2nd edn (New York 1974).

Kuhn, Thomas S., *The Essential tension* (Chicago 1977).

Lacépède, Bernard-Germain-Etienne de La Ville-sur-Illon de, *Histoire naturelle des quadrupèdes ovipares et des serpents*, 2 vols (Paris, Hôtel de Thou 1788-1789).

Laclos, Pierre-Ambroise-François Choderlos de, *Des femmes et de leur éducation* (1903), in *Œuvres complètes* (Paris 1979), p.387-443.

La Condamine, Charles-Marie de, 'Mémoire sur l'inoculation de la petite vérole', *MARS* 1754 (1759), p.615-70.

Laissus, Yves, 'Les cabinets d'histoire naturelle', in *Enseignement et diffusion des sciences en France au XVIIIe siècle*, ed. René Taton (Paris 1964), p.659-712.

– 'Le Jardin du roi', in *Enseignement et diffusion des sciences en France*, p.287-341.

Lalande, Joseph-Jérôme Le François de, 'Hypothèse', in *Supplément à l'Encyclopédie*, ed. Jean-Baptiste Robinet, 5 vols (Amsterdam, Rey 1777), iii.487.

La Motte, Antoine Houdar de, *Discours sur Homère* (1714), in *Œuvres complètes*, 10 vols in 11 pts (Paris, Prault 1754), ii.1-137.

Larson, James L., *Interpreting nature: the science of living form from Linnaeus to Kant* (Baltimore, Md. 1994).

– 'Linné's French critics', *Svenska Linné-Sällskapets Årsskrift* (1978), p.67-79.

– *Reason and experience: the representation of natural order in the work of Carl von Linné* (Berkeley, Calif. 1971).

La Salle, Antoine de, *Le Désordre régulier, ou avis au public sur les prestiges de ses précepteurs et sur ses propres illusions* (Bern [Auxerre] 1786).

Laudan, Larry, 'The nature and sources of Locke's views on hypotheses', *Journal of the history of ideas* 28 (1967), p.211-23.

– *Science and hypothesis* (Dordrecht 1981).

Leclaire, Lucien, 'L'*Histoire naturelle des minéraux* ou Buffon géologue universaliste', in *Buffon 88*, p.353-69.

Leibniz, Gottfried Wilhelm, *Discours de métaphysique* (1846), in *Opera philosophica*, ed. Johann Eduard Erdmann and Renate Vollbrecht (Meisenheim an Glan 1959), p.816-33.

– *Discourse on metaphysics, correspondence with Arnauld, monadology*, translated by George Montgomery (La Salle, Ill. 1902).

– *Essais de théodicée* (1710), in *Opera philosophica*, p.468-629.

– *La Monadologie* (1720), in *Opera philosophica*, p.705-12.

– *Nouveaux essais sur l'entendement humain* (1765), in *Opera philosophica*, p.194-418.

Lémery, Louis, 'Remarques sur un nouveau monstre', *MARS* 1740 (1742), p.607-12.

Lenay, Charles, 'Le hasard chez Buffon. Une probabilité "anthropologique"', in *Buffon 88*, p.613-27.

Lenglet Du Fresnoy, Nicolas, 'Pont militaire', in *Encylopédie*, xiii.49-71.

Lenoir, Timothy, 'The Göttingen school and the development of transcendental Naturphilosophie in the Romantic era', *Studies in history of biology* 5 (1981), p.111-205.

Lepenies, Wolf, *Das Ende der Naturgeschichte: Wandel kultureller Selbstverständlichkeiten in den Wissenschaften des 18. und 19. Jahrhunderts* (Munich 1976).

Le Roy, Charles, 'Mémoire sur l'électricité résineuse', *MARS* 1755 (1761), p.264-83.

Lesch, John E., 'Systematics and the geometrical spirit', in *The Quantifying spirit in the eighteenth century*, ed. Tore Frängsmyr, J. L. Heilbron and Robin E. Rider (Berkeley, Calif. 1990), p.73-111.

Licoppe, Christian, *La Formation de la pratique scientifique. Le discours de l'expérience en France et en Angleterre (1630-1820)* (Paris 1996).

[Lignac, Joseph-Adrien Lelarge de], *Lettres à un Américain sur l'histoire naturelle, générale et particulière de M. de Buffon*, 3 pts and 9 letters in 1 vol. (Hamburg 1751).

Limoges, Camille, 'Daubenton, Louis-Jean-Marie', in *Dictionary of scientific biography*, ed. Charles Coulston Gillispie, 16 vols (New York 1970-1980), xv.111-14.

Lindroth, Sten, 'The two faces of Linnaeus', in *Linnaeus: the man and his work*, ed. Tore Frängsmyr, revised edn (Canton, Mass. 1994), p.1-62.

Linnaeus, Carl, *Philosophia botanica* (Stockholm, Kiesewetter 1751).

Locke, David, *Science as writing* (New Haven, Conn. 1992).

Locke, John, *An Essay concerning human understanding* (1690), ed. Alexander Campbell Fraser, 2 vols (New York 1959).

Loux, Michael J., 'Nominalism', in *Routledge encyclopedia of philosophy*, ed. Edward Craig, 10 vols (London 1998), vii.17-28.

Lovejoy, Arthur O., 'Buffon and the problem of species', in *Forerunners of Darwin, 1745-1859*, ed. Bentley Glass, Owsei Temkin and William L. Straus, Jr (Baltimore, Md. 1968), p.84-113.

– *The Great chain of being* (Cambridge, Mass. 1936).

Loveland, Jeff, 'French thought in William Smellie's natural history: a Scottish reception of Buffon and Condillac', in *Scotland and France in the Enlightenment*, ed. D. Dawson, forthcoming.

– 'Rhetoric and science in Buffon's natural history', Doctoral dissertation, Duke University, 1994.

Lowe, E. J., *Locke on human understanding* (New York 1995).

Lyon, John, 'The "Initial discourse" to Buffon's *Histoire naturelle*: the first complete English translation', *JHB* 9 (1976), p.133-81.

Lyons, Martyn, 'Les best-sellers', in *Histoire de l'édition française*, ed. Henri-Jean Martin, Roger Chartier and Jean-Pierre Vivet, 4 vols (Paris 1982-1986), vol. iii, *Le Temps des éditeurs. Du romantisme à la Belle Epoque*, p.368-97.

McClellan III, James E., *Science reorganized: scientific societies in the eighteenth century* (New York 1985).

McGrath, Alister, *The Intellectual origins of the European Reformation* (Oxford 1987).

McGuire, J. E., and Trevor Melia, 'The rhetoric of the radical rhetoric of science', *Rhetorica* 9 (1991), p.301-16.

– 'Some cautionary strictures on the writing of the rhetoric of science', *Rhetorica* 7 (1989), p.87-99.

Mack, Maynard, *Alexander Pope: a life* (New Haven, Conn. 1985).

McMullen, Emerson Thomas, *William Harvey and the use of purpose in the scientific revolution: cosmos by chance or universe by design* (Lanham, Md. 1998).

McMullin, Ernan, 'Rhetoric and theory choice in science', in *Persuading science: the art of scientific rhetoric*, ed. Marcello Pera and William R. Shea (Canton, Mass. 1991), p.55-76.

Mairan, Jean-Jacques Dortous de, 'Discours sur la propagation du son dans les différents tons qui le modifient', *MARS* 1737 (1740), p.1-59.

– *Dissertation sur la glace* (Paris, Imprimerie royale 1749).

– 'Lettre de Mr de Mairan', in Gabrielle-Emilie Le Tonnelier de Breteuil Du Châtelet, *Institutions physiques. Nouvelle édition (1742)*, ed. Sonia Carboncini (Hildesheim 1988), p.476-504.

Malesherbes, Chrétien-Guillaume de Lamoignon de, *Observations sur l'histoire naturelle générale et particulière de Buffon et Daubenton*, 2 vols (Paris, Pougens 1798).

Mamiani, Maurizio, 'The rhetoric of certainty: Newton's method in science and in the interpretation of the apocalypse', in *Persuading science: the art of scientific rhetoric*, ed. Marcello Pera and William R. Shea (Canton, Mass. 1991), p.157-72.

Manlove, Colin, 'Parts and wholes: Pope and poetic structure', in *Alexander Pope: essays for the tercentenary*, ed. Colin Nicholson (Aberdeen 1988), p.139-50.

Marmontel, Jean-François, *Mémoires d'un père pour servir à l'instruction de ses enfants*, in *Œuvres de Marmontel* (Paris 1819), vol. i-ii.

Marsak, Leonard M., *Bernard de Fontenelle: the idea of science in the French Enlightenment* (Philadelphia, Penn. 1959).

Martinez-Contreras, Jorge, 'Des mœurs des singes. Buffon et ses contemporains', in *Buffon 88*, p.557-68.

'Masculin/féminin. Discours sur le sexe et sexe du discours', in *Aimer en France 1760-1860*, ed. Paul Viallaneix and Jean Ehrard, 2 vols (Clermont-Ferrand 1980), ii.295-306.

Maupertuis, Pierre-Louis Moreau de, *Essai de cosmologie* (1750), in *Œuvres de Maupertuis*, 4 vols (Lyon, Bruyset 1768), i.1-78.

– *Lettres*, in *Œuvres de Maupertuis*, ii.217-372.

– 'Sur les lois de l'attraction', *MARS* 1732 (1735), p.343-62.

– *Système de la nature* (1756), in *Œuvres de Maupertuis*, ii.135-84.

May, Gita, *Madame Roland and the age of revolution* (New York 1970).

Mayer, Jean, *Diderot, homme de science* (Rennes 1959).

Mayr, Ernst, 'The species as category, taxon and population', in *Histoire du concept d'espèce dans les sciences de la vie* (Paris 1987), p.303-20.

Mémoires de mathématiques et de physique ... tirés des registres de cette Académie (Paris 1702-1797; bound together with *HARS*).

Mémoires pour l'histoire des sciences et des beaux arts (Trévoux and Paris 1701-1767).

Mengal, Paul, 'La psychologie de Buffon à travers le traité *De l'homme*', in *Buffon 88*, p.601-12.

Mercier, Louis-Sébastien, *Tableau de Paris, critiqué par un solitaire du pied des Alpes*, 6 vols (Nyon, Natthey 1783).

Mercure de France (Paris 1724-1791).

Milliken, Stephen, 'Buffon and the British', Doctoral dissertation, Columbia University, 1965.

– 'The lyricism of the intellect: Buffon's *Epoques de la nature*', *Diderot studies* 6 (1964), p.293-303.

Mirabeau, Honoré-Gabriel Riqueti, comte de, *Lettres à Julie écrites du donjon de Vincennes, publiées et commentées d'après les manuscrits originaux et inédits*, ed. Dauphin Meunier and Georges Leloir (Paris 1903).

– *Lettres originales de Mirabeau écrites du donjon de Vincennes, pendant les années 1777, 78, 79, et 80*, ed. P. Manuel, 4 vols (Paris, J. B. Garnery 1792).

Montesquieu, Charles-Louis de Secondat, baron de, *Correspondance de Montesquieu*, ed. François Gébelin and André Morize, 2 vols (Paris 1914).

Montgomery, Scott L., *The Scientific voice* (New York 1996).

Morellet, André, *Mémoires de l'abbé Morellet de l'Académie française sur le dix-huitième siècle et sur la Révolution*, ed. Jean-Pierre Guicciardi (Paris 1988).

Mornet, Daniel, 'Les enseignements des bibliothèques privées (1750-1780)', *Revue d'histoire littéraire de la France* 17 (1910), p.449-96.

– *Les Sciences de la nature en France, au XVIIIe siècle* (New York 1911).

Morton, A. G., *History of botanical science: an account of the development of botany from ancient times to the present day* (London 1981).

Mortureux, Marie-Françoise, *La Formation et le fonctionnement d'un discours de la vulgarisation scientifique au XVIIIe siècle à travers l'œuvre de Fontenelle* (Paris 1983).

Müller, Gerhard H., '"Distinguer les uns des autres". Le concept de l'espèce chez Réaumur: pragmatisme et utilité', in *Histoire du concept d'espèce dans les sciences de la vie* (Paris 1987), p.61-77.

– 'L'histoire d'un manuscrit allemand du Muséum. "La Diss[ertatio] contra Buff[onem]" de Georg Forster', in *Le Muséum au premier siècle de son histoire*, ed. Claude Blanckaert *et al.* (Paris 1997), p.581-89.

Murphy, Terence, 'Jean-Baptiste-René Robinet: the career of a man of letters', *SVEC* 150 (1976), p.183-250.

Niderst, Alain, *Fontenelle* (Paris 1991).

– *Fontenelle à la recherche de lui-même (1657-1702)* (Paris 1972).

Nollet, Jean-Antoine, 'Conjectures sur les causes de l'électricité des corps', *MARS* 1745 (1749), p.107-51.

Nouvelles ecclésiastiques, ou mémoires pour servir à l'histoire de la constitution Unigenitus ([Paris] 1728-1803).

Nyhart, Lynn K., 'Natural history and the "new" biology', in *Cultures of natural history*, ed. N. Jardine, J. A. Secord and E. C. Spary (Cambridge 1996), p.426-43.

Observations et mémoires sur la physique, sur l'histoire naturelle, et sur les arts et métiers (Paris 1771-1793).

O'Neal, John C., *The Authority of experience: sensationist theory in the French Enlightenment* (University Park, Penn. 1996).

Osler, Margaret J., 'John Locke and the changing ideal of scientific knowledge', *Journal of the history of ideas* 31 (1970), p.3-16.

Padley, G. A., *Grammatical theory in western Europe, 1500-1700*, 2 vols (Cambridge 1985).

Palmer, R. R., *Catholics and unbelievers in eighteenth-century France* (Princeton, N.J. 1939).

Pappas, John, 'Buffon matérialiste? Les critiques de Berthier, Feller et les *Nouvelles ecclésiastiques*', in *Etre matéria-*

liste à l'âge des Lumières. Hommage offert à Roland Desné (Paris 1999), p.233-49.
– 'Buffon vu par Berthier, Feller et les *Nouvelles ecclésiastiques*', *SVEC* 216 (1983), p.26-28.
Pascal, Blaise, *Pensées* (1670), ed. Dominique Descotes (Paris 1976).
Pascal, Jean-Noël (ed.), intro. to *La Fable au siècle des Lumières, 1715-1815. Anthologie des successeurs de La Fontaine, de La Motte à Jauffret* (St Etienne 1991), p.1-28.
Paty, Michel, 'D'Alembert et la théorie physique', in *Jean d'Alembert, savant et philosophe. Portrait à plusieurs voix*, ed. Monique Emery and Pierre Monzani (Paris 1989), p.233-60.
Peiffer, Jeanne, 'L'engouement des femmes pour les sciences au XVIIIe siècle', in *Femmes et pouvoirs sous l'Ancien Régime*, ed. Danielle Haase-Dubosc and Eliane Viennot (Paris 1991), p.196-222.
Pera, Marcello, *The Discourses of science*, translated by Clarissa Botsford (Chicago 1994).
Perelman, Chaim, *Traité de l'argumentation* (Paris 1958).
Persell, Stuart M., 'The revival of Buffon in the early Third Republic', *Biography: an interdisciplinary quarterly* 14 (1991), p.12-24.
'Philosophie', in *Encyclopédie*, xii.511-15.
Pliny, *Naturalis historia*, ed. Carl Mayhoff, 6 vols (Stuttgart 1967-1970).
Pluche, Antoine-Noël, *Le Spectacle de la nature, ou entretiens sur les particularités de l'histoire naturelle, qui ont paru les plus propres à rendre les jeunes gens curieux, et à leur former l'esprit* (1732-1750), 9 pts in 8 vols (Paris, Etienne 1763).
Pratt, Vernon, 'System-building in the eighteenth century', in *The Light of nature*, ed. J. D. North and J. J. Roche (Dordrecht 1985), p.421-31.
Prelli, Lawrence J., *A Rhetoric of science* (Columbia, S.C. 1989).
Priestley, F. E. L., 'Pope and the great chain of being', in *Essays in English literature from the Renaissance to the Victorian age*, ed. Millar Maclure and F. W. Watt (Toronto 1964), p.213-28.

Quemada, Bernard, *Les Dictionnaires du français moderne, 1539-1863. Etude sur leur histoire, leurs types et leurs méthodes* (Paris 1967).
Quintilian, *Institutionis oratoriae libri XII*, ed. Ludwig Radermacher, 2 vols (Leipzig 1965).

Rappaport, Rhoda, 'Borrowed words: problems of vocabulary in eighteenth-century geology', *British journal for the history of science* 15 (1982), p.27-44.
Ray, John, *Methodus plantarum nova* (London, Faitborne & Kersey 1682).
Raynal, Guillaume, 'Nouvelles littéraires', in *Correspondance littéraire, philosophique et critique par Grimm, Diderot, Raynal, Meister, etc.* (Paris 1877), vol. i-ii.
Réaumur, René-Antoine Ferchault de, *Réaumur's memoirs on steel and iron*, ed. Cyril Stanley Smith, translated by Anneliese Grünhaldt Sisco (Chicago 1956).
– 'Sur la manière de conserver les œufs', *MARS* 1735 (1738), p.465-72.
Redondi, Pietro, 'D'Alembert et la technologie. L'affaire du canal de Picardie', in *Jean d'Alembert, savant et philosophe. Portrait à plusieurs voix*, ed. Monique Emery and Pierre Monzani (Paris 1989), p.433-60.
Rey, Roselyne, 'Buffon et le vitalisme', in *Buffon 88*, p.399-413.
– 'Les paradoxes du matérialisme de Robinet', *Dix-huitième siècle* 24 (1992), p.137-51.
Ricken, Ulrich, *'Gelehrter' und 'Wissenschaft' im Französischen: Beiträge zu ihrer Bezeichnungsgeschichte vom 12.-17. Jahrhundert* (Berlin 1961).
Ritvo, Harriet, *The Platypus and the mermaid and other figments of the classifying imagination* (Cambridge, Mass. 1997).
Roche, Daniel, 'Natural history in the academies', in *Cultures of natural history*, ed. N. Jardine, J. A. Secord and E. C. Spary (Cambridge 1996), p.127-44.

Roe, Shirley, 'Buffon and Needham: diverging views on life and matter', in *Buffon 88*, p.439-50.

Roger, Jacques, 'Buffon, Georges-Louis Leclerc, comte de', in *Dictionary of scientific biography*, ed. Charles Coulston Gillispie, 16 vols (New York 1970-1980), ii.576-82.

– *Buffon. Un philosophe au Jardin du roi* (Paris 1989).

– 'Diderot et Buffon en 1749', *Diderot studies* 4 (1963), p.221-36.

– intro. to Georges-Louis Leclerc de Buffon, *Les Epoques de la nature* (Paris 1962), p.ix-cxlix.

– 'Leibniz et les sciences de la vie', in *Akten des internationalen Leibniz-Kongresses, Hannover, 14.-19. November 1966* (Wiesbaden 1969), ii.209-19.

– *Les Sciences de la vie dans la pensée française au XVIIIe siècle*, 2nd edn (Paris 1993).

Roland, Marie-Jeanne Philipon, *The Private memoirs of Madame Roland*, ed. Edward Gilpin Johnson (New York 1976).

Rollin, Charles, *Traité des études. De la manière d'enseigner et d'étudier les belles-lettres par rapport à l'esprit et au cœur*, 2 vols (Paris, Veuve Estienne 1740).

Rouelle, Guillaume-François, 'Sur le sel marin (première partie)', *MARS* 1745 (1749), p.57-79.

Roule, Louis, *Daubenton et l'exploitation de la nature* (Paris 1925).

Rousseau, G. S., 'Science books and their readers in the eighteenth century', in *Books and their readers in eighteenth-century England*, ed. Isabel Rivers (New York 1982), p.197-255.

Rousseau, Jean-Jacques, *Correspondance complète de Jean-Jacques Rousseau*, ed. R. A. Leigh, 52 vols (Geneva, etc. 1965-1998).

Royou, Thomas-Marie, *Le Monde de verre réduit en poudre, ou analyse et réfutation des Epoques de la nature de M. le comte de Buffon* (Paris, Mérigot 1780).

Ruestow, Edward G., *Physics at seventeenth- and eighteenth-century Leiden* (The Hague 1973).

Saban, Roger, 'Le testament de Buffon', in *Buffon 88*, p.97-115.

Sabra, A. I., *Theories of light from Descartes to Newton* (London 1967).

Schiebinger, Londa, 'Gender and natural history', in *Cultures of natural history*, ed. N. Jardine, J. A. Secord and E. C. Spary (Cambridge 1996), p.163-77.

– *The Mind has no sex? Women in the origins of modern science* (Cambridge, Mass. 1989).

Schuster, John A., 'Methodologies as mythic structures: a preface to the future historiography of method', *Metascience* 1/2 (1984), p.15-36.

– and Richard R. Yeo (eds.), intro. to *The Politics and rhetoric of scientific method* (Dordrecht 1986), p.ix-xxxvii.

Scott, Robert L., 'On viewing rhetoric as epistemic', *Central states speech journal* 18 (1967), p.9-17.

Séchelles, Marie-Jean Hérault de, 'Journey to Montbard' (1801), translated by John Lyon, in *From natural history to the history of nature*, ed. John Lyon and Phillip R. Sloan (Notre Dame, Ind. 1981), p.355-77.

Secord, Anne, 'Artisan botany', in *Cultures of natural history*, ed. N. Jardine, J. A. Secord and E. C. Spary (Cambridge 1996), p.378-93.

Secord, James A., 'The crisis of nature', in *Cultures of natural history*, ed. N. Jardine, J. A. Secord and E. C. Spary (Cambridge 1996), p.447-59.

Seguin, Jean-Pierre, *La Langue française au XVIIIe siècle* (Paris 1972).

Seidengart, Jean, 'Le traitement du problème cosmologique dans l'œuvre de Buffon', in *Buffon 88*, p.309-25.

Shapin, Steven, 'Talking history: reflections on discourse analysis', *Isis* 75 (1984), p.125-30.

Singer, Charles, 'Biology', in *Encyclopaedia Britannica*, 14th edn (Chicago 1948), iii.598-618.

Sloan, Phillip R., 'Buffon, German biology, and the historical interpretation of biological species', *British journal for the history of science* 12 (1979), p.109-53.

– 'The Buffon–Linnaeus controversy', *Isis* 67 (1976), p.356-75.
– 'Buffon on Newton's law of attraction (1749)', in *From natural history to the history of nature*, ed. John Lyon and Phillip R. Sloan (Notre Dame, Ind. 1981), p.77-78.
– 'Buffon studies today', *History of science* 32 (1994), p.469-77.
– 'From logical universals to historical individuals: Buffon's idea of biological species', in *Histoire du concept d'espèce dans les sciences de la vie* (Paris 1987), p.101-40.
– 'The gaze of natural history', in *Inventing human science: eighteenth-century domains*, ed. Christopher Fox, Roy Porter and Robert Wokler (Berkeley, Calif. 1995), p.112-51.
– 'The history of the concept of the biological species in the seventeenth and eighteenth centuries, and the origin of the species problem', Doctoral dissertation, University of California, San Diego, 1970.
– 'L'hypothétisme de Buffon. Sa place dans la philosophie des sciences du dix-huitième siècle', in *Buffon 88*, p.207-22.
– 'John Locke, John Ray, and the problem of the natural system', *JHB* 5 (1972), p.1-53.
– 'Natural history', in *Cambridge history of eighteenth-century philosophy*, ed. Knud Haakonssen (Cambridge, forthcoming).
– 'Natural history, 1670-1802', in *Companion to the history of modern science*, ed. R. C. Colby *et al.* (London 1990), p.295-313.
– 'Organic molecules revisited', in *Buffon 88*, p.415-38.
– 'The question of natural purpose', in *Evolution and creation*, ed. Ernan McMullin (Notre Dame, Ind. 1985), p.121-50.
– review of Wolf Lepenies, *Das Ende der Naturgeschichte*, *Isis* 72 (1981), p.123-24.
– and John Lyon (eds.), intro. to *From natural history to the history of nature: readings from Buffon and his critics* (Notre Dame, Ind. 1981), p.1-32.
Smellie, William, *The Philosophy of natural history*, 2 vols (Edinburgh 1790-1799).
Sneath, Peter H. A., 'Mathematics and classification, from Adanson to the present', in *Adanson: the bicentennial of Michel Adanson's 'Familles des plantes'*, ed. George H. M. Lawrence, 2 vols (Pittsburgh, Penn. 1963), ii.471-98.
Sonntag, Otto (ed.), *The Correspondence between Albrecht von Haller and Charles Bonnet* (Bern 1983).
Spary, Emma, 'Codes of passion: natural history specimens as a polite language in late eighteenth-century France', in *Wissenschaft als kulturelle Praxis, 1750-1900*, ed. Hans Erich Bödeker, Peter Hanns Reill, and Jürgen Schlumbohm (Göttingen 1999), p.105-35.
– 'The "nature" of Enlightenment', in *The Sciences in Enlightened Europe*, ed. William Clark, Jan Golinski, and Simon Schaffer (Chicago 1999), p.272-304.
– 'Political, natural and bodily economies', in *Cultures of natural history*, ed. N. Jardine, J. A. Secord and E. C. Spary (Cambridge 1996), p.178-96.
– 'Le spectacle de la nature. Contrôle du public et vision républicaine dans le Muséum jacobin', in *Le Muséum au premier siècle de son histoire*, ed. Claude Blanckaert *et al.* (Paris 1997), p.457-79.
Stace, Clive A., *Plant taxonomy and biosystematics*, 2nd edn (London 1989).
Stafleu, Frans, 'Adanson and his *Familles des plantes*', in *Adanson: the bicentennial of Michel Adanson's 'Familles des plantes'*, ed. George H. M. Lawrence, 2 vols (Pittsburgh, Penn. 1963), i.123-264.
– *Linnaeus and the Linnaeans: the spreading of their ideas in systematic botany, 1735-1789* (Utrecht 1971).
Stamos, David N., 'Was Darwin really a species nominalist?', *JHB* 29 (1996), p.127-44.
Stemerding, Dirk, 'How to make oneself nature's spokesman? A Latourian account of classification in eighteenth- and early nineteenth-century natural history', *Biology and philosophy* 8 (1993), p.193-223.

– *Plants, animals, and formulae: natural history in the light of Latour's 'Science in action' and Foucault's 'The Order of things'* (Entschede 1991).
Stevens, Peter F., *The Development of biological systematics: Antoine-Laurent de Jussieu, nature, and the natural system* (New York 1994).
– 'Haüy and A.-P. Candolle: crystallography, botanical systematics, and comparative morphology, 1780-1840', *JHB* 17 (1984), p.49-82.
– 'Metaphors and typology in the development of botanical systematics 1690-1960, or the art of putting new wine in old bottles', *Taxon* 33 (1984), p.169-211.
Stigler, Stephen M., *The History of statistics: the measurement of uncertainty before 1900* (Cambridge, Mass. 1986).
Stroup, Alice, *A Company of scientists* (Berkeley, Calif. 1990).
Sturdy, David J., *Science and social status: the members of the Académie des sciences, 1666-1750* (Woodbridge 1995).
Sutton, Geoffrey V., *Science for a polite society: gender, culture, and the demonstration of Enlightenment* (Boulder, Col. 1995).
Svalgelski, Jean, 'Buffon et les "intermittences de la nature"', in *Buffon 88*, p.165-74.
'Système', in *Dictionnaire historique de la langue française*, ed. Alain Rey, 2 vols (Paris 1993), ii.2069-70.

Taylor, Kenneth L., 'The *Epoques de la nature* and geology during Buffon's later years', in *Buffon 88*, p.371-85.
Terrall, Mary, 'Emilie Du Châtelet and the gendering of science', *History of science* 33 (1995), p.283-310.
– 'Generation and desire in Maupertuis's science of life', *Isis* 87 (1996), p.217-29.
– 'Metaphysics, mathematics, and the gendering of science in eighteenth-century France', in *The Sciences in Enlightened Europe*, ed. William Clark, Jan Golinski, and Simon Schaffer (Chicago 1999), p.246-71.
'Théorie', in *Encyclopédie*, xvi.253.
Thomas, Downing, 'Philosophy, science, and style: Condillac and Buffon on figurative language', *Common knowledge*, forthcoming.
Thomas, Jean-Jacques, *La Langue volée. Histoire intellectuelle de la formation de la langue française* (Bern 1989).
Thomas, Keith, *Man and the natural world: changing attitudes in England, 1500-1800* (Oxford 1983).
Thomson, Ann, 'L'homme-machine, mythe ou métaphore?', *Dix-huitième siècle* 20 (1988), p.367-75.
Todhunter, I., *A History of the mathematical theory of probability from the time of Pascal to that of Laplace* (New York 1949).
Torlais, Jean, *Réaumur. Un esprit encyclopédique en dehors de l'*'*Encylopédie*' (Paris 1961).
Tort, Patrick, *L'Ordre et les monstres. Le débat sur l'origine des déviations anatomiques au XVIII*^{e} *siècle*, 2nd edn (Paris 1998).
Toulmin, Stephen, *The Uses of argument* (Cambridge 1958).
Tournefort, Joseph Pitton de, *Eléments de botanique, ou méthode pour connaître les plantes*, 2 vols (Paris, Imprimerie royale 1694).
Trembley, Maurice (ed.), *Correspondance inédite entre Réaumur et Abraham Trembley* (Geneva 1943).
Trinkle, Dennis, 'Noël-Antoine Pluche's *Le Spectacle de la nature*: an encyclopedic best-seller', *SVEC* 358 (1997), p.93-134.

Universal magazine of knowledge and pleasure (London 1747-1803).

Vallentin, Antonina, *Mirabeau*, translated by E. W. Dickes (New York 1948).
Valmont de Bomare, Jacques-Christophe, *Dictionnaire raisonné universel d'histoire naturelle*, 5 vols (Paris, Didot *et al.* 1764).
Van der Pas, Peter W., 'Goedaert, Johannes', in *Dictionary of scientific biography*, ed. Charles Coulston Gillispie, 16 vols (New York 1970-1980), v.439-40.
Varloot, Jean, 'Buffon et Diderot', in *Présence de Diderot*, ed. Siegfried Jüttner (Frankfurt am Main 1990), p.298-315.

Vartanian, Aram, 'Buffon and Diderot', in *Buffon 88*, p.119-33.
– *Diderot and Descartes: a study of scientific naturalism in the Enlightenment* (Princeton, N.J. 1953).
Vickers, Brian, *In defence of rhetoric* (Oxford 1988).
– 'The Royal Society and English prose style: a reassessment', in *Rhetoric and the pursuit of truth: language change in the seventeenth and eighteenth centuries*, ed. Brian Vickers and Nancy S. Struever (Los Angeles 1985), p.1-76.
Vicq d'Azyr, Félix, 'Eloge de Buffon', in *Œuvres complètes de Buffon augmentées par M. F. Cuvier* (Paris 1829), i.xlvii-lxxviii.
Voltaire, François-Marie Arouet de, *Dictionnaire philosophique* (1764), in M.xvii-xx.
– *Dictionnaire philosophique* (1764), ed. Christiane Mervaud *et al.*, Voltaire 35-36 (1994).
– *Eléments de la philosophie de Newton* (1738), ed. Robert L. Walters and W. H. Barber, Voltaire 15 (1992).
– *Lettres philosophiques*, ed. René Pomeau (Paris 1964).
– *Correspondence and related documents*, ed. Th. Besterman, Voltaire 85-135 (1968-1977).

Walters, Alice N., 'Conversation pieces: science and politeness in eighteenth-century England', *History of science* 35 (1997), p.121-54.
Wartburg, W. von, *Evolution et structure de la langue française*, 10th edn (Bern 1971).
Wattles, Gurdon, 'Buffon, d'Alembert and materialist atheism', *SVEC* 266 (1989), p.285-341.
Weil, Françoise, 'La correspondance Buffon–Cramer', *RHS* 14 (1961), p.97-136.
Werner, Stephen, 'Abraham Rees's eighteenth-century *Cyclopaedia*', *SVEC* 315 (1994), p.183-99.
Wheeler, William Morton (ed. and translator), intro. to René-Antoine Ferchault de Réaumur, *The Natural history of ants* (New York 1926), p.1-40.
Widmalm, Sven, 'Accuracy, rhetoric, and technology: the Paris–Greenwich triangulation, 1784-88', in *The Quantifying spirit in the eighteenth century*, ed. Tore Frängsmyr, J. L. Heilbron and Robin E. Rider (Berkeley, Calif. 1990), p.179-206.
Wild, John, *Introduction to realistic philosophy* (New York 1948).
Winslow, Jacques-Bénigne, 'Observations anatomiques sur un enfant', *MARS* 1740 (1742), p.586-606.
– 'Remarques sur deux dissertations touchant les monstres', *MARS* 1742 (1745), p.91-120.
Wohl, Robert, 'Buffon and his project for a new science', *Isis* 51 (1960), p.186-99.
Wokler, Robert, 'Tyson and Buffon on the orang-utan', *SVEC* 155 (1976), p.2301-19.
Wood, P. B., 'Buffon's reception in Scotland: the Aberdeen connection', *Annals of science* 44 (1987), p.169-90.
– 'Methodology and apologetics: Thomas Sprat's history of the Royal Society', *British journal for the history of science* 13 (1980), p.3-26.
Wright, Larry, 'Functions', in *Purposive behaviour and teleological explanations*, ed. Frank George and Les Johnson (New York 1985), p.141-64.

Yeo, Richard R., 'The principle of plenitude and natural theology in nineteenth-century Britain', *British journal for the history of science* 19 (1986), p.263-82.
– 'Reading encyclopedias: science and the organization of knowledge in British dictionaries of arts and sciences, 1730-1850', *Isis* 82 (1991), p.24-49.

Zabell, Sandy L., 'Buffon, Price, and Laplace: scientific attribution in the 18th century', *Archive for history of exact sciences* 39 (1988), p.173-81.
– 'The continuum of inductive methods revisited', in *The Cosmos of science: essays of exploration*, ed. John Earman and John D. Norton (Pittsburgh, Penn. 1997), p.351-85.
– 'The rule of succession', *Erkenntnis* 31 (1989), p.283-321.

Index

www.ingramcontent.com/pod-product-compliance
Ingram Content Group UK Ltd.
Pitfield, Milton Keynes, MK11 3LW, UK
UKHW021538120326
468918UK00003B/60

9 780729 407472